Models of Information Processing in the Basal Ganglia

Computational Neuroscience
Terrence J. Sejnowski and Tomaso A. Poggio, editors

Methods in Neuronal Modeling: From Synapses to Networks, edited by Christof Koch and Idan Segev, 1989

Neural Nets in Electric Fish, Walter Heiligenberg, 1991

The Computational Brain, Patricia S. Churchland and Terrence J. Sejnowski, 1992

Dynamic Biological Networks: The Stomatogastric Nervous System, edited by Ronald M. Harris-Warrick, Eve Marder, Allen I. Selverston, and Maurice Moulins, 1992

The Neurobiology of Neural Networks, edited by Daniel Gardner, 1993

Large-Scale Neuronal Theories of the Brain, edited by Christof Koch and Joel L. Davis, 1994

The Theoretical Foundation of Dendritic Function: Selected Papers of Wilfrid Rall with Commentaries, edited by Idan Segev, John Rinzel, and Gordon M. Shepherd, 1995

Models of Information Processing in the Basal Ganglia, edited by James C. Houk, Joel L. Davis, and David G. Beiser, 1995

Models of Information Processing in the Basal Ganglia

edited by James C. Houk, Joel L. Davis, and David G. Beiser

A Bradford Book
The MIT Press
Cambridge, Massachusetts
London, England

This book was set in Palatino by DEKR Corporation and was printed and bound in the United States of America.

Library of Congress Cataloging-in-Publication Data

Models of information processing in the basal ganglia / edited by
James C. Houk, Joel L. Davis, and David G. Beiser.
p. cm. — (Computational neuroscience)
"A Bradford book."
Includes bibliographical references and index.
ISBN 0-262-08234-9
1. Basal ganglia—Physiology. 2. Basal ganglia—Computer simulation. 3. Neural networks (Neurobiology) I. Houk, James C. II. Davis, Joel L., 1942– . III. Beiser, David G. IV. Series.
[DNLM: 1. Basal Ganglia—physiology. 2. Models, Neurological. 3. Mental Processes—physiology. WL 307 M6885 1994]
QP383.3.M63 1994
612.8'25—dc20
DNLM/DLC
for Library of Congress 94-14822
CIP

Contents

Series Foreword

Computational neuroscience is an approach to understanding the information content of neural signals by modeling the nervous system at many different structural scales, including the biophysical, the circuit, and the systems levels. Computer simulations of neurons and neural networks are complementary to traditional techniques in neuroscience. This book series welcomes contributions that link theoretical studies with experimental approaches to understanding information processing in the nervous system. Areas and topics of particular interest include biophysical mechanisms for computation in neurons, computer simulations of neural circuits, models of learning, representation of sensory information in neural networks, systems models of sensory-motor integration, and computational analysis of problems in biological sensing, motor control, and perception.

Terrence J. Sejnowski
Tomaso A. Poggio

Preface

The chapters of this book were written by a group of individuals representing a diverse combination of disciplines to stimulate fresh conceptual and analytical syntheses of new knowledge about the basal ganglia. Some chapters summarize current understanding, others describe exciting new findings, and yet others attempt to synthesize these data into models of the information processing that may occur in basal ganglia networks. In all cases, the authors were encouraged to present their ideas in a manner that would be accessible to readers without special expertise. We assumed only that our readers would be interested in learning about the exciting new concepts that are rapidly emerging in this previously neglected area of neuroscience.

The book is organized into four parts, each of which deals with a separate theme area. Part I reviews *fundamentals* and serves as an introduction to the remainder of the book. Part II explores neuronal architecture, signals, and models as they relate to the *motor functions and working memories* that are represented in sensorimotor regions of the basal ganglia. Part III addresses the emerging realization that *reward mechanisms* are strongly represented in the basal ganglia and may be key to achieving a functional understanding. Part IV considers several *cognitive and memory operations* that are implicated as products of information processing in basal ganglia networks. At the conclusion of each part, the editors provide brief *commentaries* that highlight key issues contained in the individual chapters and illustrate how different issues may relate to each other.

One of our goals in organizing the book was to assemble a foundation of knowledge that might promote the construction of models of information processing in this brain region. Computational analysis of the basal ganglia and its linkages with the cerebral cortex can be expected to provide critical insights as to the adaptive mechanisms the brain uses to control complex interactions between an organism and its environment. These ideas could also have considerable relevance to recent research that attempts to apply connectionist approaches to classes of engineering control problems that have been very difficult to solve using conventional methods. By encouraging cross-communication between neurobiologists studying the basal ganglia and theorists who are specialists in network approaches to engineering control, this book may help to chart new directions for fruitful interdisciplinary collaborative research.

The editors wish to acknowledge the many individuals who have contributed in various ways to this book. We acknowledge those scientists who have toiled in the brain and have broken new intellectual ground and crossed traditional disciplinary boundaries. We acknowledge our publisher, who early on saw the wisdom of producing a book in this area. Last, we acknowledge the Office of Naval Research, which has served as a catalyst in helping to stimulate the early phases of many computational neuroscience endeavors.

I Fundamentals

1 Information Processing in Modular Circuits Linking Basal Ganglia and Cerebral Cortex

James C. Houk

In recent years, there has been a remarkable expansion of knowledge about the anatomical organization of the basal ganglia, the signal processing that occurs in these structures, and their many relations both to molecular mechanisms and to cognitive functions. The various findings are pointing toward an unexpected role for the basal ganglia in the contextual analysis of the environment and in the adaptive use of this information for the planning and execution of intelligent behaviors. This book is about these emerging concepts. As an introductory chapter, I provide here an overview that highlights some of the more salient findings, and I outline a simple conceptual model for their interpretation. Let me begin with a brief review of the basic anatomy, since it provides a useful foundation for the interpretation of the other findings.

ANATOMICAL PLAN OF INFORMATION FLOW

Virtually the whole cerebral cortex projects to the basal ganglia, and the outputs then funnel back to the frontal areas of cortex, or in some cases directly to motor systems in the midbrain and hindbrain. In this overview, I focus on the pathways that return to the frontal cortex. The fundamental plan of information flow is shown in figure 1.1. This diagram is based on a pivotal review in which several parallel circuits, each having the basic design shown in figure 1.1, were identified (Alexander et al., 1986). Diverse areas of cerebral cortex are shown to converge upon regions of striatum that, via pallidum and thalamus, project back to the region of frontal cortex that contributed to the striatal input. There is also a less direct pathway from striatum to pallidum via the external pallidum and subthalamus, and there is a shorter route from thalamus to striatum that bypasses the cerebral cortex. The thin arrowheads in figure 1.1 represent predominantly excitatory connections whereas the solid arrowheads signify inhibition.

The impressive topographic specificity within these pathways, as described in chapters 4 through 7 of this book and reviewed elsewhere (Goldman-Rakic, 1984; Graybiel, 1991), led Houk and Wise (1993) to go

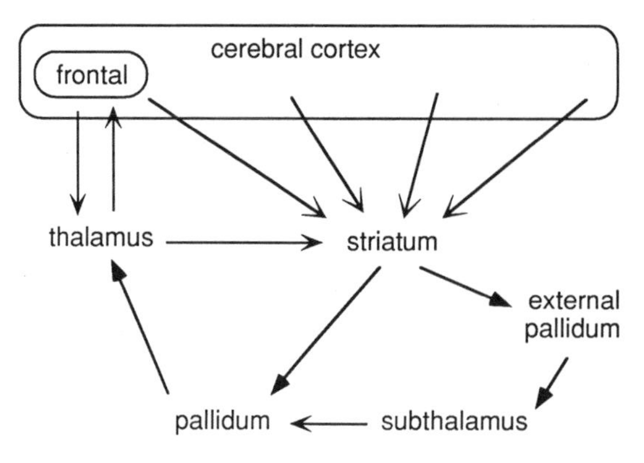

Figure 1.1 Schema of information flow between the cerebral cortex and several divisions of the basal ganglia.

one step further. We postulated that each of the segregated circuits identified by Alexander et al. (1986) is in turn comprised of a large array of modules, each module being organized according to the same plan as shown in figure 1.1 (Wise and Houk, 1994). This hypothesis fits well with the general observation that each area of frontal cortex contains a population of units with response properties that are similar to one another in some generic sense, whereas one or several parametric features of the responses differ among units (see Goldman-Rakic, 1991, and chapter 7). The parametric differences might result from slight variations in the inputs to modules that have the same structure and essentially the same information-processing capability. An hypothesis concerning the generic processing function of individual cortical-basal ganglionic modules was advanced by Houk and Wise (1993), based on the cellular properties of striatal spiny neurons discussed in the next section.

POTENTIAL FOR PATTERN RECOGNITION BY STRIATAL SPINY NEURONS

The input stage of the basal ganglia is the striatum, and the principal neurons of the striatum are called spiny neurons because of the great density of synaptic spines on their long dendrites, as discussed in chapters 3 and 4 (see also Wilson, 1990). Each spiny neuron receives input from about 10,000 different afferent fibers, a remarkable degree of convergence that is second only to that for Purkinje cells in the cerebellar cortex. This type of cellular architecture is analogous to the network architecture used in perceptrons, a pattern-recognizing network that was extensively studied by theorists a few decades ago (Rosenblatt 1962; Albus, 1971; Minsky and Papert, 1969 but see 1987). A perceptron requires a substantial convergence of different kinds of information onto individual units, analogous to the convergence of cortical inputs onto striatal spiny neurons. In addition, a perceptron requires a special input that adjusts the synaptic weights of the convergent inputs along lines discussed in chapters 4 and 10 through 13 of this book. This adjustment mechanism trains individual units to recognize particular patterns of input. There is growing evidence, summarized in chapters 4, 10, and 12, that dopamine fibers provide a reinforcement input to striatal spiny neurons that trains them to recognize patterns in their cerebral cortical input (also see Ljungberg et al., 1989; Wickens, 1990). Finally, a perceptron utilizes an abrupt activation threshold in the response properties of the unit, which forms a sharp decision line for pattern recognition. Spiny neurons also have abrupt thresholds between "up" and "down" states owing to the highly nonlinear ionic properties of their membranes (see Wilson, 1990, and chapter 3). These three features—convergence of diverse inputs, specialized training signals, and dual-state behavior—suggest that spiny neurons may be particularly well suited for pattern recognition tasks.

The upper part of figure 1.2 incorporates these features into a schematic diagram of a cortical-basal ganglionic information-processing module as defined by Houk and Wise (1993). The diagram shows convergent excitatory input onto a spiny neuron (*SP*) from five (out of 10,000) cortical neurons (the *C*'s and the *F*) originating from diverse regions of the cerebral cortex; *F* is a neuron from the frontal cortex. The more diffuse dopamine input is assumed to function as a training signal that reinforces the synaptic weights of C and F inputs to guide the pattern recognition process. Since reinforcement guides learning, the pattern that is eventually learned should reflect a context that is behav-

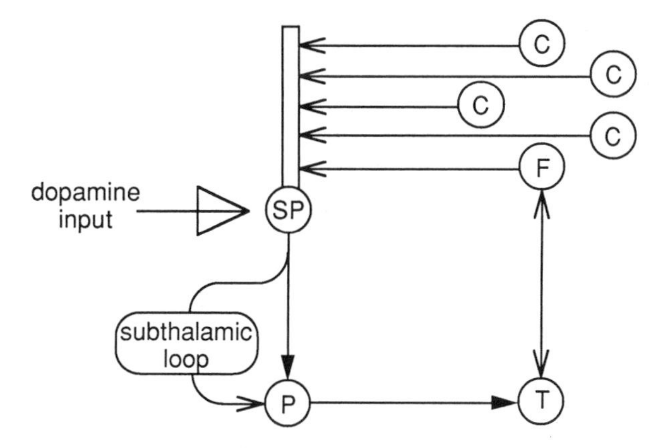

Figure 1.2 Schematic diagram of a cortical-basal ganglionic module. Thin arrowheads signify net excitation; the solid black arrowheads in the direct SP to P projection and in the P to T projection signify net inhibition; the triangular arrowhead for the dopamine input signifies neuromodulation. C, cerebral cortical neuron; F, neuron in frontal cortex; SP, spiny neuron of the striatum; P, pallidal neuron; T, thalamic neuron.

iorally significant, as discussed in chapters 10 through 13. Later, when the spiny neuron is exposed to the same or a similar contextual pattern, the resultant excitation would cause a transition to the up state of the neuron, whereupon it would fire a burst of action potentials and then return to its down state. In chapter 2 Schultz and colleagues describe how the burst discharges of spiny neurons relate to a variety of contextual situations that the animal confronts in performing behavioral tasks. Assuming that a burst generated by a spiny neuron signifies the detection of a behaviorally significant context, the remainder of the circuit in figure 1.2 could then serve to refine this computation and deposit it in working memory for use in planning subsequent behavioral actions, as discussed in the next section.

CORTICOTHALAMIC LOOPS AND WORKING MEMORY

Single-unit studies have shown patterns of persistent discharge in frontal cortical neurons that appear to represent transitory, working memories of behaviorally significant stimuli or events (Goldman-Rakic, 1991). In chapter 7 Goldman-Rakic describes how these signals provide distributed representations of contextual information consisting of stimulus features or internal states that need to be saved for a short duration so that they can be used in controlling an ongoing behavioral action. After the action is completed, this information is no longer needed, and the sustained discharge in frontal neurons correspondingly ceases.

Houk and Wise (1993) suggested a mechanism whereby cortical-basal ganglionic modules might serve to detect these contexts and register them in working memory. Bursts of spiny neurons are known to produce a pause in the sustained inhibitory output from pallidal neurons (Chevalier and Deniau, 1990), as illustrated by the solid traces in figure 1.3. This pause in inhibitory input to the thalamic neurons would be expected to initiate a brief burst of thalamic discharge, due to the inhibitory rebound properties of these neurons (Wang et al., 1991). We postulated that the rebound burst would then initiate positive feedback in the reciprocal corticothalamic loop shown in figure 1.2. This could mediate the sustained discharge of working memory. The sustained activity is a property of both thalamic and cortical neurons (Fuster and Alexander, 1973; Goldman-Rakic and Friedman, 1991).

The solid traces in figure 1.3 spell out the proposed relationships between these neural signals. The burst in spiny neuron discharge (trace *SP*), signifying the detection of a behaviorally significant context, produces a pause in pallidal neuron discharge (trace *P*), which through disinhibition initiates sustained positive feedback in reciprocally connected thalamic and frontal cortical neurons (trace *T-F*). The suggested mechanism can be thought of as a registration of the context detected by a spiny neuron into working memory, so that this information can

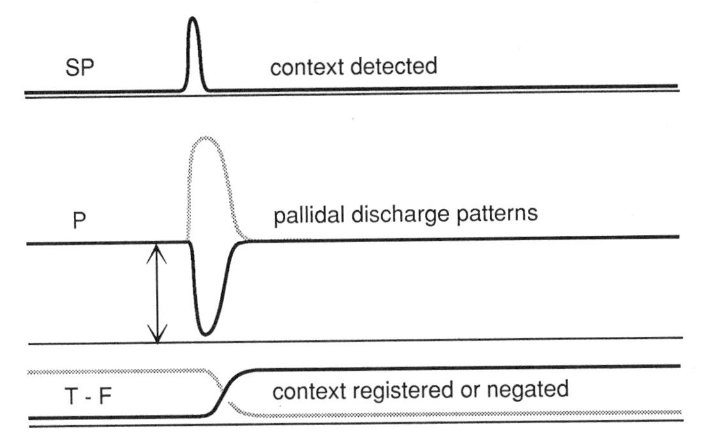

Figure 1.3 Postulated relationship between basal ganglia signals and the sustained discharge of thalamic and frontal cortical neurons (T-F). Solid traces show responses that would be mediated by the direct pathway in figure 1.2 and stippled traces show those mediated via the subthalamic sideloop. SP, spiny neuron discharge; P, pallidal neuron discharge.

be used later in the control of behavior. The stippled traces in figure 1.3 indicate how transmission that was specifically directed through the subthalamic sideloop (see figure 1.2) might serve to cancel a previously registered context. Concurrent activity in both the direct inhibitory and the excitatory sideloop pathways could control the competitive selection processes that are discussed in chapter 17 of this book.

USE OF REGISTERED CONTEXTS IN THE CONTROL OF BEHAVIOR

How might the working memories in the frontal cortex be used to control behavior? Three anatomical pathways for the use of this information were considered by Houk and Wise (1993, 1994). One is through the extensive corticocortical pathways that lead, by many routes, to the primary motor cortex, as discussed in chapter 7. These connections might be an effective way of initiating activity in the recurrent network that interconnects cerebellum, red nucleus, and motor cortex; positive feedback in this network has been suggested as the main driving force for generating the motor commands that control limb movement (Houk et al., 1993). A second major output pathway is via the pons to the cerebellum. This pathway might help to program the cerebellum to control an action (Berthier et al., 1993). A third pathway would be through corticostriatal inputs to different basal ganglionic modules, and this might permit the detection of one context, in the first module, to contribute to the detection of a subsequent context, in another module. This recursive process could be a very powerful mechanism for generating complex properties that might be useful in the high-level planning of actions (Houk and Wise, 1994).

CONCLUSION

In this introductory chapter, I have given an overview of the salient anatomy and physiology of basal ganglia circuits, and I have attempted to integrate these findings into a simple conceptual model of the essential information-processing operations performed in cortical-basal ganglionic modules. At the input stage of the basal ganglia, spiny neurons in the striatum receive a diversity of convergent signals from widespread areas of the cerebral cortex. The spiny neurons are postulated to function as pattern classifiers that learn, under the training influence of their dopamine fiber input, to recognize patterns of activity in their cortical inputs that signify occurrences of behaviorally significant contexts. When such a context is detected, the resultant burst of spiny neuron discharge is thought to disinhibit thalamocortical loops to initiate sustained activity in clusters of frontal cortical neurons. In this manner, detected contexts would be registered as working memories, whereupon this information would become available for subsequent use in the planning and control of motor behavior.

REFERENCES

Albus, J. S. (1971) A theory of cerebellar function. *Math. Biosci.* 10:25–61.

Alexander, G. E., DeLong, M. R., and Strick, P. L. (1986) Parallel organization of functionally segregated circuits linking basal ganglia and cortex. *Annu. Rev. Neurosci.* 9:357–381.

Berthier, N. E., Singh, S. P., Barto, A. G., and Houk, J. C. (1993) Distributed representation of limb motor programs in arrays of adjustable pattern generators. *J. Cogni. Neurosci.* 5:56–78.

Chevalier, G., and Deniau, J. M. (1990) Disinhibition as a basic process in the expression of striatal functions. *Trends Neurosci.* 13:277–280.

Fuster, J. M., and Alexander, G. E. (1973) Firing changes in cells of the nucleus medialis dorsalis associated with delayed response behavior. *Brain Res.* 61:79–91.

Goldman-Rakic, P. S. (1984) Modular organization of prefrontal cortex. *Trends Neurosci.* 7:419–429.

Goldman-Rakic, P. S. (1991) Prefrontal cortical dysfunction in schizophrenia: The relevance of working memory. In B. J. Carroll, and J. E. Barrett (eds.), *Psychopathology and the Brain.* New York: Raven Press, pp. 1–23.

Goldman-Rakic, P. S., and Friedman, H. R. (1991) The circuitry of working memory revealed by anatomy and metabolic imaging. In H. S. Levin, H. M. Eisenberg, and A. L. Benton (eds.), *Frontal Lobe Function and Dysfunction.* New York: Oxford University Press, pp. 72–91.

Graybiel, A. M. (1991) Basal ganglia—input, neural activity, and relation to the cortex. *Cur. Opin. Neurobiol.* 1:644–651.

Houk, J. C., and Wise, S. P. (1993) Outline for a theory of motor behavior: Involving cooperative actions of the cerebellum, basal ganglia, and cerebral cortex. In P. Rudomin,

M. A. Arbib, and F. Cervantes-Perez (eds.), *From Neural Networks to Artificial Intelligence* Heidelberg: Springer-Verlag, pp. 452–470.

Houk, J. C., and Wise, S. P. (1994) Distributed modular architectures linking basal ganglia, cerebellum and cerebral cortex: Their role in planning and controlling action. *Cerebral Cortex,* in press.

Houk, J. C., Keifer, J., and Barto, A. G. (1993) Distributed motor commands in the limb premotor network. *Trends Neurosci.* 16:27–33.

Ljungberg, T., Apicella, P., and Schultz, W. (1989) Responses of monkey dopamine neurons to external stimuli: Changes with learning. In G. Bernardi, M. B. Carpenter, G. Di Chiara, M. Morelli, and P. Stanzione (eds.), *The Basal Ganglia III.* New York: Plenum Press, pp. 487–494.

Minsky, M. L., and Papert, S. A. (1987) *Perceptrons.* Cambridge, Mass.: MIT Press.

Rosenblatt, F. (1962) *Principles of Neurodynamics.* New York: Spartan Books.

Wang, X.-J., Rinzel, J., and Rogawski, M. A. (1991) A model of the T-type calcium current and the low-threshold spike in thalamic neurons. *J. Neurophysiol* 66:839–850.

Wickens, J. (1990) Striatal dopamine in motor activation and reward-mediated learning: Steps towards a unifying model. *J. Neural Transm.* 80:9–31.

Wilson, C. J. (1990) Basal ganglia. In G. M. Shepherd (ed.), *The Synaptic Organization of the Brain.* New York: Oxford University Press, pp. 279–316.

Wise, S. P., and Houk, J. C. (1994) Modular neuronal architecture for planning and controlling behavior. *Biol. Commun. Dan. R. Acad. Sci. Lett.* (in press)

2 Context-dependent Activity in Primate Striatum Reflecting Past and Future Behavioral Events

Wolfram Schultz, Paul Apicella, Ranulfo Romo, and Eugenio Scarnati

The basal ganglia have classically been considered as primarily motor structures. This view is supported by the prominent motor deficits associated with lesions of their component structures, both in human patients and in experimental animals, as well as by the extensive basal ganglia output to the primary motor cortex and the superior colliculus. Furthermore, neurons in several parts of the basal ganglia, including the striatum, are activated during the execution of skeletal and ocular movements and can reflect certain movement parameters, such as direction and load (Crutcher and DeLong, 1984; Liles, 1985; Hikosaka et al., 1989a; Crutcher and Alexander, 1990). However, more detailed investigations of the anatomical connections have revealed that a major part of basal ganglia output is directed to most parts of the frontal lobe (Schell and Strick, 1984; Ilinsky et al., 1985), thus allowing the basal ganglia to operate in close functional association with this large cortical area involved in cognition, planning, execution, and control of behavioral output. A concomitant reassessment of behavioral alterations in parkinsonian patients and in experimental animals also suggests an involvement of the basal ganglia in cognitive and motivational components of the organization of behavior (Cools et al., 1984; Robbins and Everitt, 1992). One of the first neurophysiological correlates found for these higher functions was a form of memory-related activity in the pars reticulata of substantia nigra, a major output station of the basal ganglia (Hikosaka and Wurtz, 1983). Thus, the view of a primarily motor role of the basal ganglia may have to be enlarged to include nonmotor functions. This chapter provides an overview of the current state of knowledge regarding the involvement of a major input stage of the basal ganglia, the striatum, in these higher functions. We focus on the activity of the slowly discharging medium spiny output neurons which constitute the large majority of striatal neurons rather than the small population of tonically firing interneurons (Kimura et al., 1990) which show a much narrower spectrum of task-related activity.

One aspect of basal ganglia function that has become particularly apparent during the last decade is the context dependency of striatal cell activity. In most cases, the activity of striatal neurons is not sufficiently explained by the physical characteristics of the stimuli presented or the movements performed but depends on certain behavioral situations, certain conditions, or particular kinds of trials in a given task, thus showing relationships to the context in which the particular events occurred. Note that this definition of context is somewhat broader than that used in animal learning theory, where it denotes the surroundings in which the specifics of a task, such as lights, levers, and positions and modalities of reward or punishment are embedded.

RESPONSES TO PAST EVENTS

Signals for Movement Initiation

Investigations of movement-related activity revealed that some neurons in both caudate and putamen respond to external auditory, somatosensory, and visual stimuli only when they elicit arm or eye movements, and that responses to the same stimuli are absent or considerably reduced when no movement is elicited (figure 2.1) (Aldridge et al., 1980; Rolls et al., 1983; Hikosaka et al., 1989b; Kimura, 1990; Romo et al., 1992). These responses occur prior to the movement but are time-locked to the stimulus rather than to movement onset. These characteristics suggest that the responses are not purely sensory, and may be involved in a neuronal process in which the reception of a specific signal is used for initiating a rapid behavioral reaction. Similar responses dependent on mouth licking or specific arm movements occur in tonically discharging neurons (Kimura et al., 1984; Kimura, 1986).

Instruction Cues

In contrast to directly movement-triggering stimuli, instruction cues contain specific information that is used by the subject for a later reaction to a trigger stimulus which by itself does not contain this information and only serves to determine the time of reaction (what vs. when). In the various forms of delayed-response tasks, the instruction cue disappears before the trigger appears in a given trial, such that the information contained in the instruction needs to be kept in short-term memory until it can be used for the reaction.

Neurons in both caudate and putamen show phasic responses to visual instruction cues which last between 100 ms and several seconds and subside before the trigger stimulus. They occur selectively in movement or in no-movement trials of a delayed go/no-go task in which the animal needs to remember the color of the instruction light in order to react correctly to the trigger stimulus (Schultz and Romo, 1992). Spa-

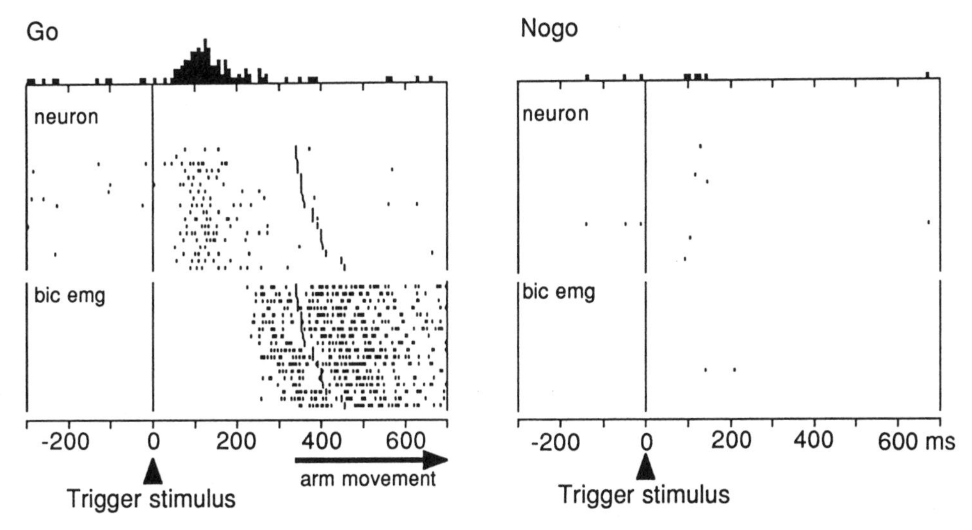

Figure 2.1 Response of a putamen neuron to a movement-triggering stimulus (*left*) and lack of response when the monkey withholds responding (*right*). The animal performs in a go/no-go task in which the rapid vertical opening of a small food box triggers a reaching movement from a resting key toward the box (go trials) or requires withholding arm movement (nogo). Movement and no-movement trials were alternated randomly during the experiment and are separated for display. Individual parts show perievent time histograms of neuronal impulses above raster displays of the same neuronal data and rasters from activity of the biceps brachii muscle (bic) recorded simultaneously. Each raster dot denotes the time of a neuronal impulse or electromyographic (emg) activity exceeding a preset level whose distance to stimulus onset corresponds to the real-time interval. Each raster line shows one trial. Go trials are rank-ordered according to reaction time, whereas no-go trials are shown in original order downward. Onset of arm movement (key release) is indicated by short vertical lines in raster displays. (From Romo et al. 1992.)

tially selective visual responses occur in an oculomotor delay task in which the animal remembers the position of the stimulus for a saccadic reaction a few seconds later (Hikosaka et al., 1989b). In the absence of further control tasks, it is difficult to say whether these responses to instruction cues in delay tasks are related to the preparation of forthcoming behavioral reactions or may reflect the transient storage of information in short-term memory. Tonically discharging neurons in the rostral putamen respond mostly with a short depression of activity to instruction cues, usually without differentiating between different types of trials (Apicella et al., 1991b).

Reward

Some caudate neurons are activated when monkeys see food objects and press a bar in order to obtain them (Nishino et al., 1984). These responses are not purely visual but depend on the motivational state of the animal: a neuron responding both to a piece of orange and to a bean loses its response to the orange but not to the beam when the monkey is satiated with oranges. This suggests that the neuronal response may be related to the appetitive value of the stimulus but does not exclude a relationship to the instructed preparation of bar press movements.

More direct evidence for the detection of task reinforcement is obtained when striatal neurons respond to a drop of liquid reward delivered to the animal's mouth in a learned behavioral task (Apicella et al., 1991a). These responses clearly differ from activity related to the simultaneously occurring mouth movements. Other control experiments revealed that responses are related to the occurrence of the liquid drop and not to the associated noise of the liquid delivery valve. Tonically discharging striatal neurons also respond to reward, although usually with a depression of activity (Apicella et al., 1991b). These data suggest that striatal neurons are informed about the outcome of each behavioral trial and would be able to use the information about the reception of the reinforcer for goal-directed behavior. In view of the role of the ventral striatum in learning, it is particularly interesting that reward responses are twice as frequent in this structure (nucleus accumbens and surrounding ventral striatum rostral to the anterior commissure), as compared to more dorsal parts of caudate and putamen.

ACTIVITY PRECEDING FUTURE EVENTS

Besides being involved in detecting and registering behaviorally important events, striatal neurons appear to be heavily engaged in the processing of information about particular events which are anticipated but have not yet occurred. A phasic response to an instruction cue indicates the reception of this signal (figure 2.2A). In a memory task, this may reflect the neuronal encoding of information, whereas a more sustained activity may be a neuronal correlate for maintaining this information in the memory store and bridging the gap between the entry into memory and its recall for use (figure 2.2B). Like short-term memory, that sustained activity should initially be high and degrade over time. Whereas these kinds of activity still refer back to the past event to be remembered, another type of activity reflects the anticipated occurrence of a future event, such as a movement or a stimulus (figure 2.2C). This is entirely different from the passive response to a past event (see figure 2.2A) in being related to an event that has not yet occurred when the neuronal activity begins. The occurrence of future events, in particular stimuli and movements, is predicted by learned external cues that evoke an expectation of the event. The neuronal activity anticipates the predicted event and is only related to the past predictor as far as it evokes

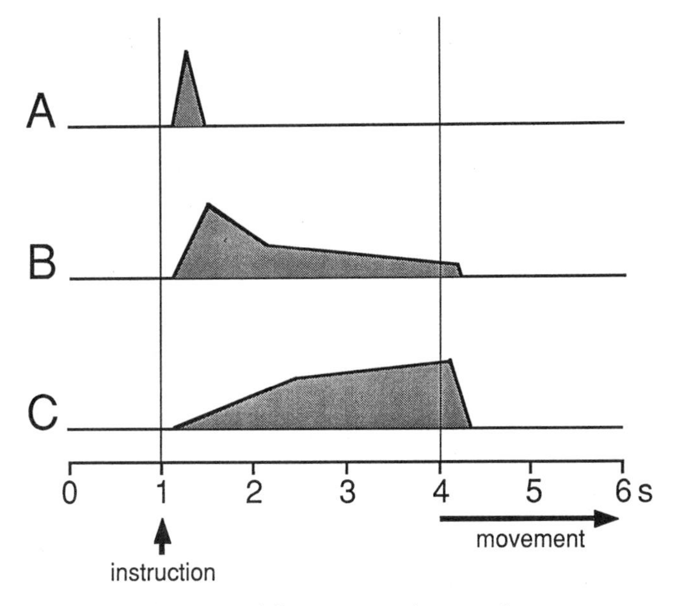

Figure 2.2 Schematic differentiation of neuronal response to a past event (*A*) from activity preceding a future event (*C*). *B* shows an intermediate form occurring in a memory task which would span the time between the storage of information into memory (instruction) until the information is recalled and discarded after use (e.g., a movement).

the expectation. When the future event concerns an action emitted by the subject itself, it can be prepared following an external cue, but it may also be generated by the subject itself without requiring predictive cues apart from the surrounding context.

Several studies have shown that striatal neurons display sustained activity in delays between events when instruction-specific information needs to be remembered for a few seconds. Although this might suggest the presence of short-term memory mechanisms in the striatum, decisive control tests for differentiating it against preparatory activity have not yet been done. These tests would need to show that the sustained activity reflects the past (remembered) event and not the future (expected or prepared) event. Examples for such tests are found in studies on prefrontal and premotor cortex. During learning and reversal errors in go/no-go task performance, some neurons show sustained activity reflecting the information contained in the preceding instruction cue, suggesting a memory mechanism, whereas other neurons reflect the future behavioral reaction, suggesting a preparatory mechanism (Niki et al., 1990). In a saccade-antisaccade task, some prefrontal neurons show cue-dependent delay activity unrelated to the direction of the impending movement, whereas others show movement direction-related activity, differentiating again between memory and movement preparatory processes (Funahashi et al., 1993).

Instruction-induced Preparation of Movements

In the absence of more specific memory tests, sustained striatal activity preceding instructed behavioral reactions is generally interpreted as being preparatory. This kind of activity is found during instructed go/no-go tasks where it occurs predominantly during movement trials, lasting up to several tens of seconds (Schultz and Romo, 1992). In more parametric motor tasks, striatal preparatory activity is related to the direction of the impending arm or eye movement (Hikosaka et al., 1989a; Alexander and Crutcher, 1990a; Apicella et al., 1992). This activity typically begins at some time after the instruction cue, increases during the trial, and ends abruptly when the movement begins, even if movement onset is delayed by a few seconds. Interestingly, there are a few striatal neurons that are activated only in no-go trials during the preparation for a rewarded or sometimes even an unrewarded withholding of movement, thus potentially reflecting the preparation of movement inhibition (Schultz and Romo, 1992; Apicella et al., 1992)

Self-initiated Movements

An intuitive example of activity related to future as opposed to past events is found in monkeys performing arm-reaching movements in the absence of external imperative or instruction cues (Schultz and Romo, 1992). Whereas the animal's liberty to move is restricted by the use of a single known target and by the appetitive nature of the food reward, the moment of moving is chosen by the animal itself (but rhythmic or automated movements are not allowed). Some neurons in the striatum are activated 0.5 to 5.0 seconds before the onset of such self-initiated reaching movements. The activity usually begins slowly with a few irregularly spaced impulses, builds up over hundreds of milliseconds toward movement onset, and abruptly ends either with movement onset or goal attainment (figure 2.3). Two thirds of these neurons are not activated when the movement is prepared in response to an external cue, suggesting a certain selectivity for an internal preparatory process. These data indicate that the striatum is engaged when an internal urge to move toward a known target is transformed into an overt behavioral act. The observed activity could serve to internally set preparatory states for individual behavioral acts on the basis of information about the environmental context (task contingencies, position of movement target, reward).

Expectation of External Signals

Whereas the described preparatory activities are related to actions emitted by the subject itself, other striatal neurons are activated during the

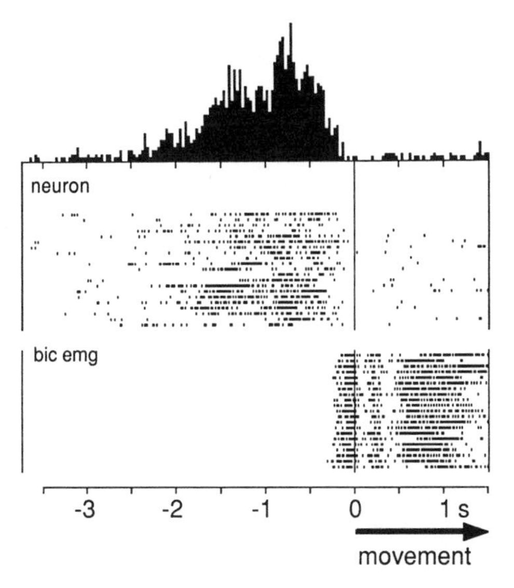

Figure 2.3 Activity in a putamen neuron preceding self-initiated arm movements. In the absence of any phasic external cue, neuronal activity starts about 2 seconds before the monkey releases a resting key and reaches into a food box to collect a hidden morsel of apple. The original sequence of trials is preserved downward. All data are referenced to movement onset (key release). The histogram and raster of neuronal activity (neuron) is shown above the raster of simultaneously recorded biceps muscle activity (bic emg). Note that the offset of neuronal activity roughly coincides with onset of muscle activity, suggesting a relationship to the internal preparation of movement. (From Schultz et al., 1992.)

expectation period before an externally imposed stimulus occurs. In a task in which several distinct events occur sequentially, different populations of striatal neurons are rather selectively activated prior to each task event (figure 2.4). Thus, some neurons increase their discharge rate before the first task-related signal in a given trial, usually an instruction or target cue, whose occurrence is predicted by the end of the previous trial in discrete trial schedules. Other striatal neurons are activated in delay tasks before the signal terminating the delay, such as an arm– or eye movement–triggering signal, a movement-withholding signal, or a visual matching signal, without reflecting the behavioral reaction itself (Hikosaka et al., 1989c; Johnstone and Rolls, 1990; Apicella et al., 1992). Activities of these types often begin without strict temporal relation to a preceding event and continue until the respective future event occurs. They are prolonged when the future event is delayed and terminate abruptly immediately afterward. Besides predicting the occurrence of a specific task-related signal, some striatal activity is related to the predictable position of the target (figure 2.5) (Hikosaka et al., 1989c) or to the direction of an impending target movement (figure 2.6) (Alexander and Crutcher, 1990b), thus adding further specificity to expectation-related activities.

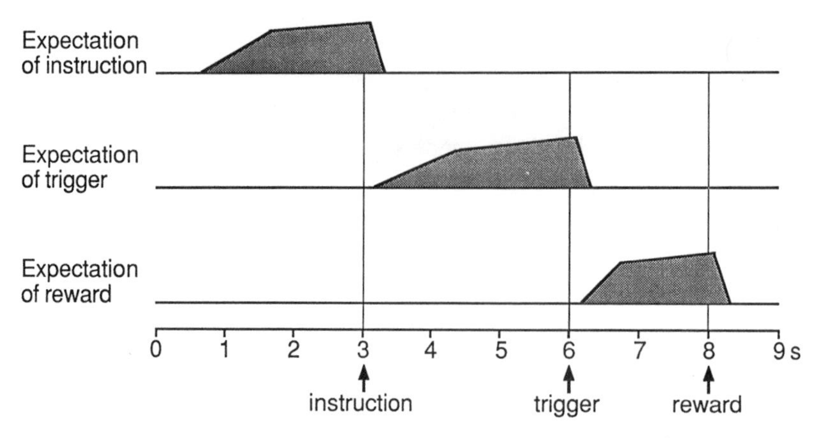

Figure 2.4 Schematics of expectation-related neuronal activity in the primate striatum preceding distinctive, sequentially occurring events in a behavioral task, such as a delayed-response paradigm. Each of the three rows indicates one class of striatal neurons being selectively active prior to the event indicated. Although activities occur between individual task events, they are better related in time to the future, not the past, event, suggesting that striatal neurons have access to stored knowledge about the upcoming events.

Expectation of Reward

The liquid reward given at the end of each trial of a task is a particularly effective event for activating striatal neurons. Activity during the expectation of the future reward occurs in close to 50% of dorsal striatal neurons showing any expectation- or preparation-related activity, and in about 75% of such neurons in the ventral striatum (Apicella et al., 1992; Schultz et al., 1992). As with all expectation-related activity, the increased discharge rate continues until reward is delivered and stops abruptly thereafter. If reward delivery is delayed for a series of trials, the onset of reward expectation activity will be similarly delayed, further illustrating the relationship of this activity to the temporal occurrence of reward (figure 2.7A). A specificity to the reward becomes apparent when reward liquids are changed, e.g., from apple juice to water, which quantitatively changes the expectation-related activity (figure 2.7B). Together with activities related to the expectation of other task events, these data show that expectation-related activities in striatal neurons concern a full sequence of task events (see figure 2.4).

CONCLUSIONS AND SPECULATIONS

The reported data suggest that the striatum as the major input structure of the basal ganglia is involved in a number of processes that go beyond those typically associated with a motor control structure. Whereas the activities related to movement-initiation signals and the preparation of

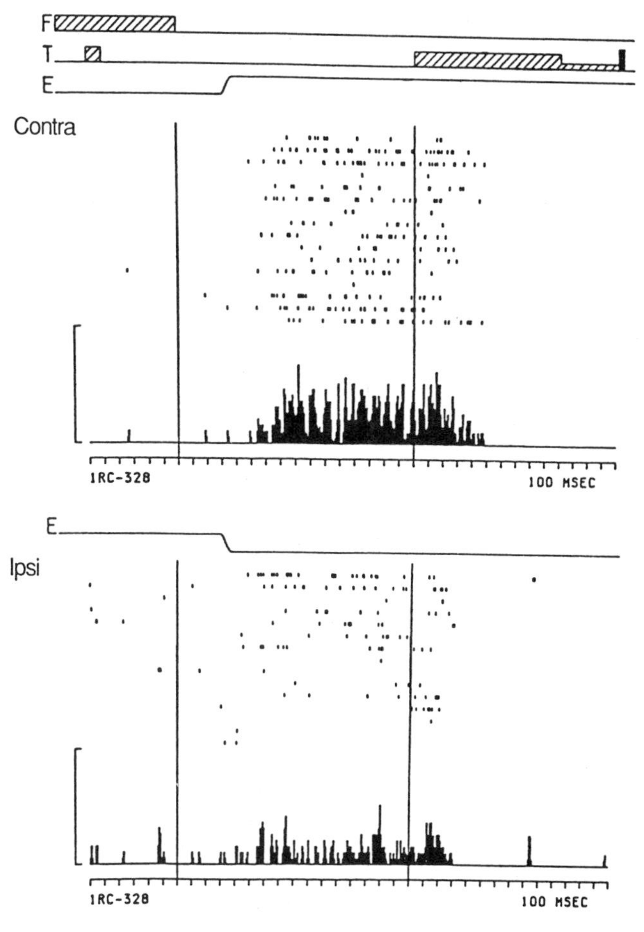

Figure 2.5 Neuronal activity in the monkey caudate related to the expectation of a target for eye movement. In the task, the monkey fixates a central spot of light (F) and a target light at 20 degrees lateral comes up briefly (T). Following fixation spot offset, the animal makes a saccadic eye movement (E) to the position of the previously seen target. Shortly thereafter, the target light reappears and the monkey releases a holding key for obtaining a drop of liquid reward (*vertical black bar in top scheme*). The top raster shows a block of consecutive trials in which the target location was contralateral to the neuronal recording position (contra). Without further indication to the animal, this was followed by a block of trials with ipsilateral target position (Ipsi). The neuronal activity shows directional specificity is occurring only when the contralateral target (*top*) as opposed to the ipsilateral one is expected (*bottom*). Activity in the first trials of the bottom raster suggests that the animal still expected the previously used contralateral target, and expectation-related activity disappeared gradually with repeated use of the ipsilateral target. (From Hikosaka et al., 1989c.)

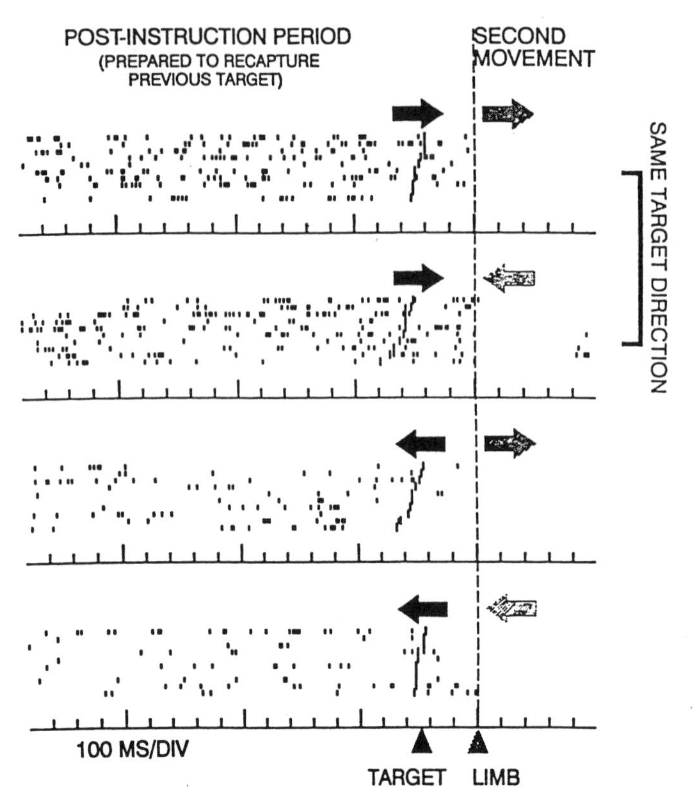

Figure 2.6 Neuronal activity in the monkey putamen related to the expected direction of visual cursor movement. In a delayed response task with arm movement about the elbow joint, the monkey receives a brief target cue on whether to flex or extend the forearm. Movement of the arm is accompanied by movement of a light cursor in the same or the opposite direction, thus allowing separation of the cursor from arm movement direction. Reappearance of both targets following a delay after cue offset requires that the monkey recapture the remembered target with the light cursor by moving its arm appropriately. The top two rasters show how this neuron is activated before cursor movement in the direction indicated by the left-hand arrows pointing to the right, independent of arm movement direction (indicated by the right-hand arrows). The bottom two rasters show that such activity is absent prior to cursor movements in the opposite direction, again independent of arm movement direction. (From Alexander and Crutcher 1990b.)

movements could be reconciled with motor functions, some of the described activities, particularly those reflecting the expectation of predictable events, suggest that the basal ganglia are involved in a much wider spectrum of behavioral functions.

Large Repertoire of Heterogeneous Activities

An overview of the described activities reveals a high number of functional classes (figure 2.8). Owing to the segregation of task relationships,

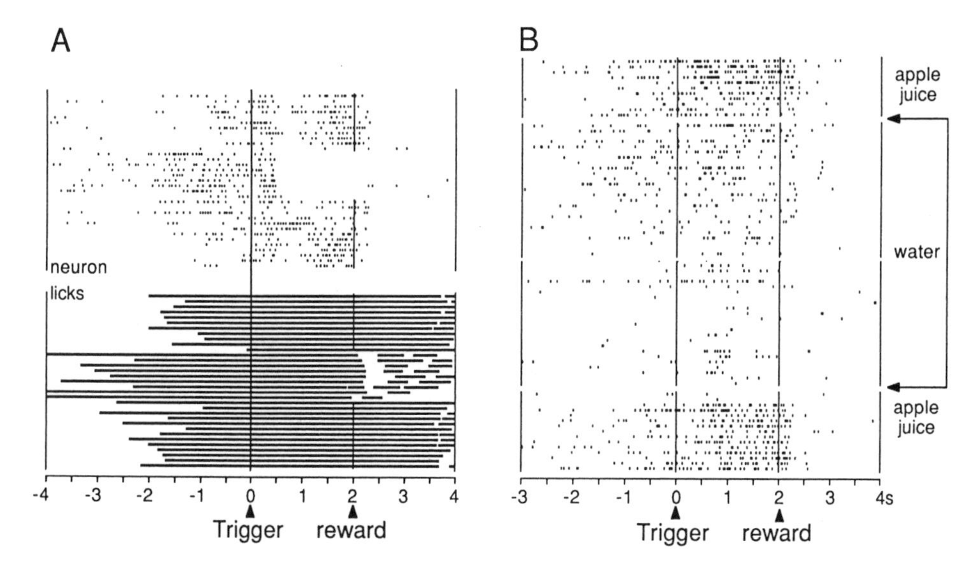

Figure 2.7 Neuronal activity in the primate ventral striatum related to the expectation of reward. *A,* Both onset and offset of activity are displaced when the temporal occurrence of reward changes. Delivery at the usual 2-second interval between a conditioned stimulus (Trigger) and a drop of liquid reward is marked by a vertical line in rasters, whereas this line is absent when reward is presented simultaneously with the stimulus. Bottom rasters show tongue contact with the liquid-delivering spout. An instruction stimulus precedes the trigger stimulus by randomly varying intervals of 2.5 to 3.5 seconds without indicating the changed trigger-reward interval. Note how the animal's behavior (*bottom*) and the neuronal activity refer to the expected onset of reward. *B,* Influence of different reward liquids on reward expectation–related activity of a primate ventral striatal neuron. The change from the regularly used apple juice to plain water (upper arrow) gradually reduces activation over successive trials. Reinstatement of apple juice (lower arrow) leads to increased activity after two trials. Trials in *A* and *B* are shown in original order downward. (From Schultz et al., 1992.)

each of these classes only engages a small fraction of striatal neurons, rarely exceeding 10% to 15% of the total striatal population, although considerable regional variations are sometimes observed. This does not suggest a low impact of these activities but reflects the coding of a much larger spectrum of behavior related information than hitherto assumed.

Context Dependency

The large number of different activity classes is surprising in the light of the relatively simple behavioral situations studied, which largely comprised ocular and skeletal reaction time tasks, tracking movements, and very basic delay and short-term memory tasks. Nevertheless, all studies noted a high incidence of striatal neurons that were not activated in any task component of a given experiment. It might be that these neurons are activated in more demanding tasks or that they are engaged

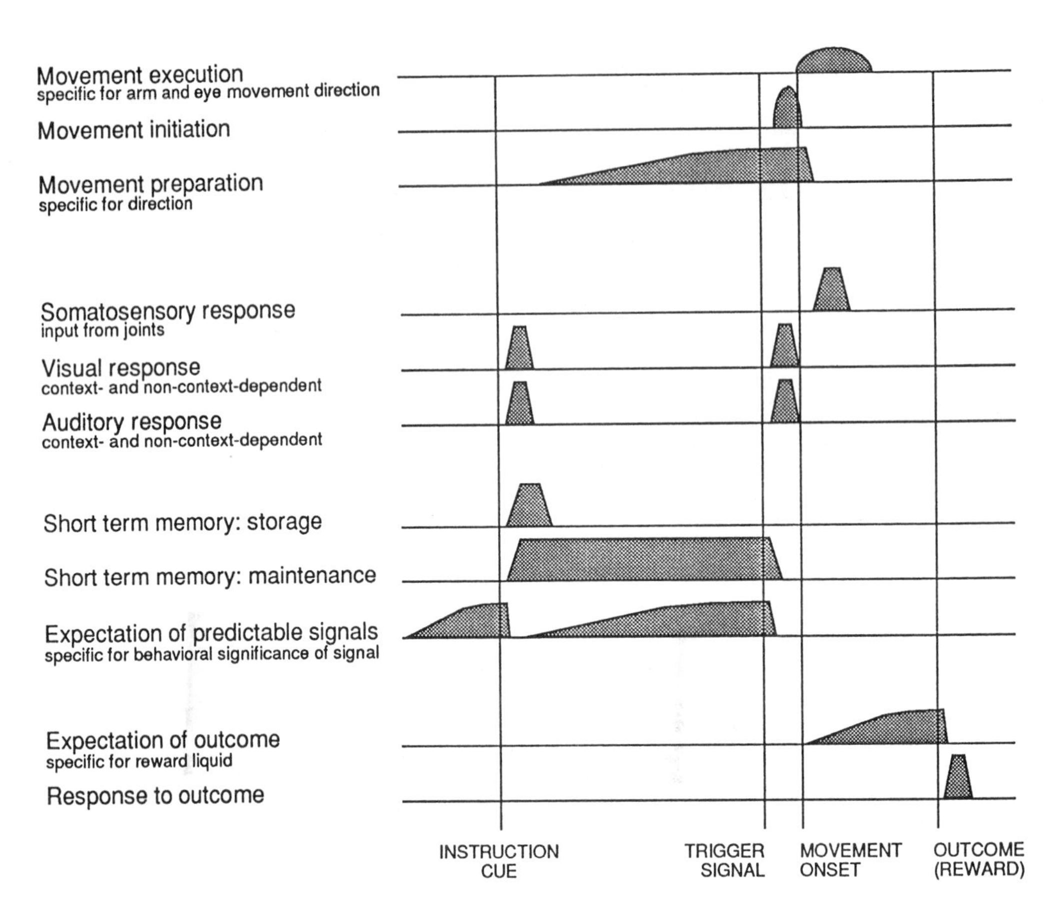

Figure 2.8 Schematic overview of changes in neuronal activity in the striatum during sensorimotor tasks involving movements, instruction cues, short-term memory, and rewards. There are both neuronal responses to stimuli and events, such as an initial instruction cue, a movement-triggering signal, or a drop of liquid reward, as well as activity preceding these events, including movements, which are being expected or prepared following extended experience in the task. There are also indicators for short-term memory–related activity, although necessary control tests are lacking so far. A given striatal neuron, if modulated in the task, usually shows only one or occasionally two of these changes.

in events that simply did not exist in the tested situations. Nevertheless, because of the limited numbers of neurons and the nearly unlimited number of possible behavioral situations with highly meaningful stimuli, we may assume that the same neurons would show rather specific activities in different tasks. A neuron specifically related to the expectation of a particular stimulus in one task might well show a similarly specific expectation-related activity in another task in which another stimulus with a similar behavioral significance is used or occurs at a similar point in time or space. Just as neurons in sensory systems are driven by specific physical stimulus parameters and thus are activated in largely different behavioral situations in which stimuli with these parameters occur, one might wonder what precisely are the parameters responsible for the activation of expectation-related neurons. With the high heterogeneity of striatal activities in mind, the observed activities may not be as exclusive to the task as they might appear. They might simply show only one of several possible neuronal solutions for solving the problem of appropriately treating the information underlying task performance (Fetz, 1992). It might then be that the same neurons perform differently in totally different contexts, an intriguing possibility suggesting experiments that remain to be done.

Ventral Striatum and Reward

Another surprising finding in recent years is the high incidence of reward-related activity and its distinctive forms, comprising both activity preceding reward delivery and responses signaling its reception. These activities are more frequent in the ventral striatum but are also found to surprising extents in more dorsal regions. Several findings from primate anatomical and rat psychopharmacological studies have already suggested an involvement of the basal ganglia in reward-driven learning mechanisms. These concepts received less attention from neurophysiologists focusing on the motor role of the basal ganglia because of the human movement disorders associated with their dysfunction. In particular, the ventral striatum receives heavy inputs from several cortical and subcortical limbic structures (e.g., see Russchen et al., 1985; Selemon and Goldman-Rakic, 1985), and numerous studies showed that reward-driven learning requires intact limbic and dopaminergic inputs to that structure (for review, see Robbins and Everitt, 1992).

Conceptually, the two main types of reward-related activity should have considerably different functions. The expectation of reward evoked by a learned stimulus develops when the intrinsically neutral stimulus has been repeatedly paired with a primary reward, such as food or liquid. The central tenet of the largely pavlovian incentive learning theory is that the conditioned stimulus elicits an approach response because it evokes the expectation of reward. It is unclear how this conditioned expectation of reward could be represented in the brain. One possibility is the observed sustained activity of striatal neurons preceding reward, although this might also be a correlate for a more explicit knowledge of reward occurrence in the present context. Responses to the reception of reward may be involved in the mechanism in which a drop of liquid is used for reinforcing behavior in a particular task. This signal should be maintained during continuing task performance, as it is in the striatum, lest the conditioned behavior undergo extinction. This kind of striatal response to reward contrasts with reward responses of dopamine neurons which occur only when reward is unexpected or uncertain, as during learning, and which disappear with established task performance (Ljungberg et al., 1992).

Function of the Striatum as Goal Selector as Opposed to Performance Controller

The presence and form of context-dependent responses to various signals suggest that the striatum is informed about the occurrence and behavioral significance of individual events in the current environmental situation. On the other hand, the presence of expectation-related activity shows that striatal neurons have access to stored knowledge about upcoming events of behavioral significance which span an entire task—from the initial cue through the intermediate events to the outcome of the task. This knowledge has been gained through the experience in the specific context of the behavioral task and can be used for predicting the next events, planning the appropriate actions, and predicting their outcome. In combining the responses to past events and the activity related to the expectation of learned events, the striatum obtains coherent information about the behavioral situation and can use it for integrating the instantaneous situation into the learned context. The presence of movement preparatory activity, including that preceding self-initiated movements, suggests that the striatum may use the knowledge about the behavioral situation for appropriately planning and adapting behavioral actions beyond the instantaneous moment, rather than just reacting in a conditioned manner to the current stimuli.

The large spectrum of behavior-related striatal activity reflecting both past and future events and spanning an entire behavioral act from the intention to the outcome, the particular conditions of reward-related striatal activity, the cited responses of dopamine neurons to unexpected rewards, and the intimate connections of the basal ganglia with the frontal cortex and the limbic system may give us an idea about the role of the basal ganglia in goal-directed behavior. Results from lesions and some neurophysiological studies suggest that the prefrontal cortex, and to a certain extent also the premotor cortex, may serve to define the goals of future actions on the basis of explicit representations of a multitude of actions and by acquiring new such representations from the continuing experience of the subject and by inference from previous knowledge. Which of these processes exactly are transmitted to the far smaller striatum is unknown, and it is rather probable that the basal ganglia use only a part of this information. So far, we see a surprising amount of highly differentiated sustained activity consistent with goal-directed neuronal processes in the monkey striatum which very much resembles activity found in the frontal cortex. Maybe the tasks used up to now did not test sufficiently complex behaviors to reveal clear behavioral neurophysiological distinctions between the two structures. By assuming that the striatum does not contain the same complexity of cognitive representations, we may assume that its role in goal-directed behavior is less global than that of the frontal cortex and could be limited to the selection of goals that are initially defined by the frontal cortex. The described range of behavior-related striatal activity suggesting knowledge on received signals and expectation of future events would be compatible with such a goal selector role. The selection process may contain both short-term and long-term adaptive processes. The short-term processes could be based on explicit knowledge about the behavioral situation reaching the striatum predominantly from cortical inputs and the current demands of the subject entering from limbic inputs. The long-term goal selection involves learning processes in which the outcome of an action influences the next behavioral act. One learning mechanism could be behavioral conditioning using reward and punishment. A biological correlate for reward learning may be found in the activity of dopamine neurons which form distinctive synapses with literally all striatal neurons. Dopamine neurons respond rather uniformely to unexpected rewards which are particularly effective for learning (see chapter 12). The selected action is maintained by the reinforcer which striatal neurons register in the mentioned way without involving dopamine responses, thus suggesting multiple reward inputs to the striatum at different times of learning and performance. These distinctions between cortical and striatal operations may be too schematic because the frontal cortex presumably receives the same dopamine reward signal, thus allowing reward-dependent selection also to occur at the cortical level. Other structures could be involved in the execution and supervision of performance for reaching the selected goal, such as the premotor and the primary sensorimotor cortex, which may operate partly in closed and somatotopically segregated loops with the lateral putamen (see chapter 6). A particular control system for skeletal, ocular, and vestibular motor functions during task initiation and execution appears to be the cerebellum, operating again in closed loops with the premotor and sensorimotor cortex. Adaptive learning in this structure should use an error code of movement indicating the deviation between the obtained and the desired outcome, as when manipulating a new object or moving against a different load (Gilbert and Thach, 1977), and should not be based on reinforcement signals, as in the basal ganglia.

ACKNOWLEDGMENTS

This work was supported by the Swiss National Science Foundation, the Fyssen Foundation (Paris), the Fondation pour la Recherche Médicale (Paris), the United Parkinson Foundation (Chicago), and the Italian National Research Council.

REFERENCES

Aldridge, J. W., Anderson, R. J. and Murphy, J. T. (1980) The role of the basal ganglia in controlling a movement initiated by a visually presented cue. *Brain Res.* 192:3–16.

Alexander, G. E. and Crutcher, M. D. (1990a) Preparation for movement. Neural representations of intended direction in three motor areas of the monkey. *J. Neurophysiol.* 64:133–150.

Alexander, G. E., and Crutcher, M. D. (1990b) Neural representations of the target (goal) of visually guided arm movements in three motor areas of the monkey. *J. Neurophysiol.* 64:164–178.

Apicella, P., Ljungberg, T., Scarnati, E., and Schultz, W. (1991a) Responses to reward in monkey dorsal and ventral striatum. *Exp. Brain Res.* 85:491–500.

Apicella, P., Scarnati, E., and Schultz, W. (1991b) Tonically discharging neurons of monkey striatum respond to preparatory and rewarding stimuli. *Exp. Brain Res.* 84:672–675.

Apicella, P., Scarnati, E., Ljungberg, T., and Schultz, W. (1992) Neuronal activity in monkey striatum related to the expectation of predictable environmental events. *J. Neurophysiol.* 68:945–960.

Cools, A. R., Van den Bercken, J. H. L., Horstink, M. W. I., Van Spaendonck, K. P. M., and Berger, H. J. C. (1984) Cognitive and motor shifting aptitude disorder in Parkinson's disease. *J. Neurol. Neurosurg. Psychiatry* 47:443–453.

Crutcher, M. D., and Alexander, G. E. (1990) Movement-related neuronal activity selectively coding either direction or muscle pattern in three motor areas of the monkey. *J. Neurophysiol.* 64:151–163.

Crutcher, M. D., and DeLong, M. R. (1984) Single cell studies of the primate putamen. II. Relations to direction of movement and pattern of muscular activity. *Exp. Brain Res.* 53:244–258.

Fetz, E. E. (1992) Are movement parameters recognizably coded in the activity of single neurons? *Behav. Brain Sci.* 15:679–690.

Funahashi, S., Chafee, M. V., and Goldman-Rakic, P. S. (1993) Prefrontal neuronal activity in rhesus monkeys performing a delayed anti-saccade task. *Nature* 365:753–756.

Gilbert, P. F. C., and Thach, W. T. (1977) Purkinje cell activity during motor learning. *Brain Res.* 128:309–328.

Hikosaka, O., and Wurtz, R. H. (1983) Visual and oculomotor functions of monkey substantia nigra pars reticulata. III. Memory-contingent visual and saccade responses. *J. Neurophysiol.* 49:1268–1284.

Hikosaka, O., Sakamoto, M., and Usui, S. (1989a) Functional properties of monkey caudate neurons. I. Activities related to saccadic eye movements. *J. Neurophysiol.* 61:780–798.

Hikosaka, O., Sakamoto, M., and Usui, S. (1989b) Functional properties of monkey caudate neurons. II. Visual and auditory responses. *J. Neurophysiol.* 61:799–813.

Hikosaka, O., Sakamoto, M., and Usui, S. (1989c) Functional properties of monkey caudate neurons. III. Activities related to expectation of target and reward. *J. Neurophysiol.* 61:814–832.

Ilinsky, I. A., Jouandet, M. L., and Goldman-Rakic, P. S. (1985) Organization of the nigrothalamocortical system in the rhesus monkey. *J. Comp. Neurol.* 236:315–330.

Johnstone, S., and Rolls, E. T. (1990) Delay, discriminatory, and modality specific neurons in striatum and pallidum during short-term memory tasks. *Brain Res.* 522:147–151.

Kimura, M. (1986) The role of primate putamen neurons in the association of sensory stimuli with movement. *Neurosci. Res.* 3:436–443.

Kimura, M. (1990) Behaviorally contingent property of movement-related activity of the primate putamen. *J. Neurophysiol.* 63:1277–1296.

Kimura, M., Rajkowski, J., and Evarts, E. (1984) Tonically discharging putamen neurons exhibit set-dependent responses. *Proc. Natl. Acad. Sci. U. S. A.* 81:4998–5001.

Kimura, M. Kato, M., and Shimazaki, H. (1990) Physiological properties of projection neurons in the monkey striatum to the globus pallidus. *Exp. Brain Res.* 82:672–676.

Liles, S. L. (1985) Activity of neurons in putamen during active and passive movements of the wrist. *J. Neurophysiol.* 53:217–236.

Ljungberg, T., Apicella, P., and Schultz, W. (1992) Responses of monkey dopamine neurons during learning of behavioral reactions. *J. Neurophysiol.* 67:145–163.

Niki, H., Sugita, S., and Watanabe, M. (1990) Modification of the activity of primate frontal neurons during learning of a go/no-go discrimination and its reversal. In E. Iwai and M. Mishkin (eds.), *Vision, Memory and the Temporal Lobe.* New York: Elsevier, pp. 295–304

Nishino, H., Ono, T., Sasaki, K., Fukuda, M., and Muramoto, K. I. (1984) Caudate unit activity during operant feeding behavior in monkeys and modulation by cooling prefrontal cortex. *Behav. Brain Res.* 11:21–33.

Robbins, T. W., and Everitt, B. J. (1992) Functions of dopamine in the dorsal and ventral striatum. *Semin Neurosci.* 119–128.

Rolls, E. T., Thorpe, S. J., and Maddison, S. P. (1983) Responses of striatal neurons in the behaving monkey. 1. Head of the caudate nucleus. *Behav. Brain Res.* 7:179–210.

Romo, R., Scarnati, E., and Schultz, W. (1992) Role of primate basal ganglia and frontal cortex in the internal generation of movements: Comparisons in striatal neurons activated during stimulus-induced movement initiation and execution. *Exp. Brain Res.* 91:385–395.

Russchen, F. T., Bakst, I., Amaral, D. G., and Price, J. L. (1985) The amygdalostriatal projections in the monkey. An anterograde tracing study. *Brain Res.* 329:241–257.

Schell, G. R., and Strick, P. L. (1984) The origin of thalamic inputs to the arcuate premotor and supplementary motor areas. *J. Neurosci.* 2:539–560.

Schultz, W., and Romo, R. (1992) Role of primate basal ganglia and frontal cortex in the internal generation of movements: Comparison with instruction-induced preparatory activity in striatal neurons. *Exp. Brain Res.* 91:363–384.

Schultz, W., Apicella, P., Scarnati, E., and Ljungberg, T. (1992) Neuronal activity in monkey ventral striatum related to the expectation of reward. *J. Neurosci.* 12:4595–4610.

Selemon, L. D., and Goldman-Rakic, P. S. (1985) Longitudinal topography and interdigitation of corticostriatal projections in the rhesus monkey. *J. Neurosci.* 5:776–794.

3 The Contribution of Cortical Neurons to the Firing Pattern of Striatal Spiny Neurons

Charles J. Wilson

One of the most striking features of striatal spiny projection neurons is their peculiar pattern of spontaneous activity, which is characterized by long periods of silence interrupted by episodes of activity lasting tenths of seconds to seconds. This firing pattern was originally described in anesthetized animals, but was later recognized as typical of striatal neurons under a variety of circumstances (for review, see Wilson, 1993). At the time of the early descriptions of this firing pattern from the microelectrode studies of Albe-Fessard and her associates (1960), the prevailing anatomical view considered the neostriatum to consist of a large number of medium-sized spiny interneurons and a much smaller number of larger projection cells (Ramón y Cajal, 1911; Vogt and Vogt, 1920). By analogy with the spinal cord and peripheral nervous system, it was presumed that the larger and more rare cells were probably excitatory, while the smaller but more numerous interneurons were assumed to be inhibitory (e.g., Kemp and Powell, 1971; Fox et al., 1971.). Because most of the neurons fire in the phasic pattern at very low rates, it was concluded that this firing pattern was probably that of the most common neuron, the spiny cell. The large number of presumed inhibitory interneurons and the low rate of firing of these cells were thus for many years believed to be causally related. The informally stated but commonly held view maintained that most striatal neurons were silent because of the strong inhibition exerted by the large population of interacting interneurons (e.g., Hull et al., 1973; Bernardi et al., 1976). It is ironic that this pattern originally came to be attributed to inhibition owing to a misidentification of the spiny neurons as interneurons. When this anatomical error was corrected with the discovery that the spiny neurons were inhibitory projection neurons, their intensive local collateral arborizations were offered as a functionally equivalent substitute. This allowed the new anatomical facts to be incorporated into the conventional view that massive amounts of tonic inhibition were the dominant synaptic force within intrastriatal circuits. This corresponded in time to the discovery that the spiny projection neurons

Supported by ONR grant N00014-922-J-111-3 and NIH grant NS20743.

synthesize and release γ-aminobutyric acid (GABA), and are powerfully inhibitory in their effects on their target neurons in the globus pallidus and substantia nigra (e.g., Precht and Yoshida, 1971; Fonnum et al., 1974; Ribak et al., 1976).

IS INHIBITION RESPONSIBLE FOR THE SILENCE OF SPINY NEURONS?

The identity of striatal spiny neurons as GABA-containing and -releasing cells was confirmed by a wide variety of anatomical, biochemical, and pharmacological studies. Electron microscopic studies likewise confirmed that the main target of the local axonal arborizations of spiny neurons were other spiny neurons. The extremely low rate of activity combined with the large number of excitatory inputs received by these neurons (see below) requires that the inhibition must be extremely powerful if it were to be responsible for enforcing that same low average rate of firing. This could be achieved if either each inhibitory synapse were extremely effective or if there were a tremendous convergence of inhibitory inputs from spiny neurons onto one another. The latter case would be effective because it would insure that each neuron was receiving some inhibitory input all the time even if most of the neurons surrounding it were silent. However, subsequent neurophysiological studies have failed to reveal the powerful inhibition that should result from the release of GABA from the local collaterals of these neurons (Wilson et al., 1989; Pennartz and Kitai, 1991). A more complete analysis of the firing patterns of identified striatal spiny neurons has shown that the relative silence of striatal neurons is not primarily due to inhibition (Wilson and Groves, 1981; Wilson et al., 1983a; Wilson, 1986, 1993; Calabresi et al., 1990). Among the data supporting this view are: (1) local application of GABA antagonists does not convert striatal spiny cells to tonically active neurons (although their firing rate is increased when they do fire) (Nisenbaum and Berger, 1992); (2) the membrane potential fluctuations responsible for the episodes of firing do not reverse upon experimental hyperpolarization of the membrane to levels that reverse inhibitory postsynaptic potentials (IPSPs) (Wilson et al., 1983a; Wilson, 1986); (3) Input resistance of the neurons is increased during the periods of silence (not decreased as expected for postsynaptic potentials [PSPs]) (Wilson, 1993); and (4) Experimentally produced pauses in excitatory afferent input can mimic the naturally occurring silent periods of striatal neurons (Wilson et al., 1983a; Wilson, 1986, 1993).

An alternative mechanism for the generation of firing patterns in striatal neurons is available from recent findings on the nonlinear properties of the spiny cell membrane (e.g., Kita et al., 1984; Calabresi et al.,

1987, 1990; Bargas et al., 1989; Kawaguchi et al., 1988; Surmeier et al., 1991; Jiang and North, 1991). While the characterization of the voltage-dependent conductances present in the spiny neuron membrane is not complete, several critical mechanisms have been uncovered that can help to explain the behavior of the cells in the subthreshold membrane potential range. A very powerful inwardly rectifying potassium current dominates the behavior of the cell over a wide range of membrane potentials, including that maintained during the silent periods (Kita et al., 1985; Kawaguchi et al., 1989; Jiang and North, 1991) This current, which accounts for most of the resting input resistance of the spiny cell, acts as an inhibitory shunt on weak and uncorrelated excitatory synaptic inputs on dendritic spines and distal dendrites (Wilson, 1992). It is proposed that this current is distributed on the dendrites, as well as the soma of the spiny neuron, where it can alter the effective dendritic length of the cell. Local powerful synaptic input located at one region of the dendritic tree is ineffective because it is shunted by potassium currents located on the rest of the dendritic tree. Only relatively coherent synaptic input located over a large portion of the dendritic tree is capable of depolarizing enough of the neuron to deactivate the inward rectifier. Once this level of overall depolarization is attained, however, the neuron's electrotonic structure collapses explosively. In addition to the reduction of the effective electrotonic distance from each excitatory input to the soma, the time constant of the dendritic membrane increases as the membrame resistance is increased, and the requirement of temporal coherence of inputs for cooperative action is relaxed. If inputs are maintained under these conditions, the cell will depolarize to a limit imposed by a set of outwardly rectifying potassium conductances that are activated by depolarization (Bargas et al., 1989; Surmeier et al., 1991) These conductances limit the depolarization to a level set by the voltage sensitivity of their activation and their state of inactivation. This new level will be maintained until excitatory synaptic input drops to a level that allows reactivation of the inward rectifier. At this voltage, the process reverses itself and forces the neuronal membrane back to a level close to the potassium reversal potential.

The existence of the potassium conductances responsible for this bistable-like behavior have been established in a set of voltage clamp studies conducted using dissociated striatal spiny neurons, and in intracellular recording experiments using spiny neurons in striatal slices (Kawaguchi et al., 1989; Jiang and North, 1991; Surmeier et al., 1991; Nisenbaum et al., 1994). The currents turn on and off in the subthreshold range for action potentials, so that both the polarized state and the depolarized state predicted by the properties of the currents are subthreshold. Thus they predict two subthreshold states, one more depolarized and one more polarized, with sharp transitions depending

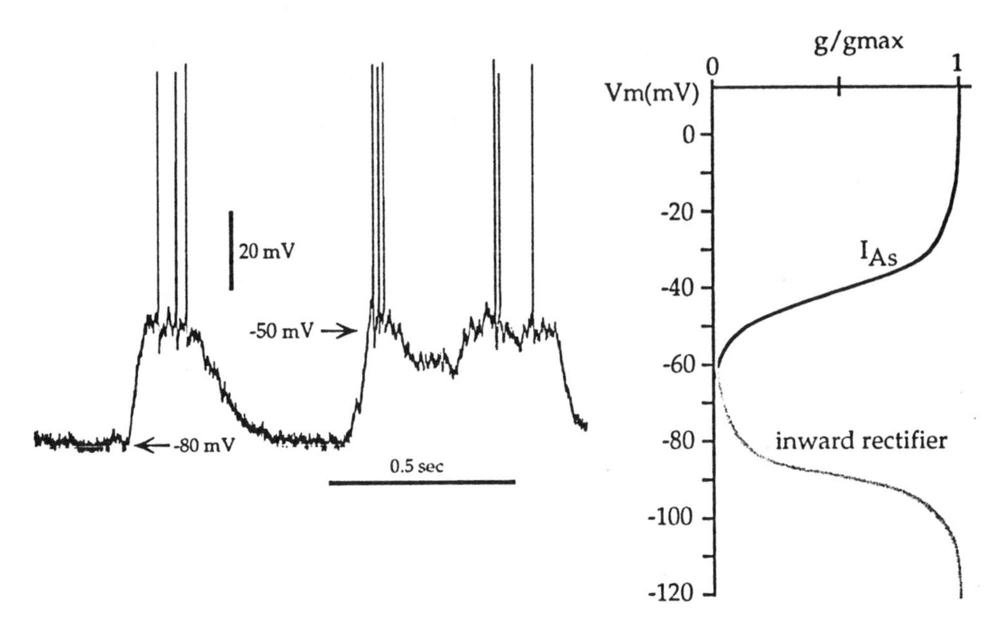

Figure 3.1 Spontaneous membrane potential shifts and associated firing of a spiny striatal projection neuron in the urethane-anesthetized rat. At the right a schematic representation of the voltage dependence of activation of two potassium currents proposed to contribute to the existence of preferred membrane potentials in these neurons is shown at the same voltage scale. The high-input resistance region over the intermediate range of membrane potentials makes this an unstable region, while more depolarized and hyperpolarized potentials are stabilized by the strong potassium conductances.

upon the spatial and temporal pattern of excitatory input rather than the activity of inhibitory neurons (Wilson, 1993).

This scheme, based on the theoretical properties of striatal spiny cells, predicts the intracellularly recorded behavior of the neurons seen in vivo (figure 3.1). In intact, anesthetized, animals, striatal neurons show abrupt transitions between two preferred membrane potentials. One, called "up," is a depolarized (−50 to −55 mV) state only a few millivolts from threshold. Small fluctuations of the membrane potential in this state can trigger action potentials, and action potentials are generally (but not always) observed during this state. Transitions to the up state are responsible for the episodes of firing seen in striatal neurons in vivo. The other state, called "down," is a very hyperpolarized state, negative to the reversal potential for IPSPs, and near the potassium equilibrium potential. Experiments intended to measure the input resistance of the neurons in the up and down states in a voltage-independent manner show that synaptic conductances are higher during the up state (Wilson et al., 1983a). Lesions or deactivation of the cerebral cortex result in spiny neurons remaining constantly in the down state (Wilson, 1993).

INHIBITION- AND EXCITATION-BASED MODELS CAN BE DISTINGUISHED BY PREDICTIONS ON SPATIAL PATTERNING

Both the inhibition-based explanation for the firing patterns of spiny neurons and the excitation-based explanation predict that the neurons will tend to have two relatively stable membrane potential states, and both mechanisms could be working simultaneously. Of course, inhibition is certainly present in the striatum. The issue is not the presence or absence of inhibition among spiny neurons but the extent to which it controls the overall pattern of striatal spiny cell firing. It is therefore useful to consider the similarities and differences in the predicted behavior of the system under control of each process. Mutual inhibition creates membrane potential states by competitive interaction between groups of neurons in a winner-take-all fashion (Wickens et al., 1991). In this scheme, the temporal pattern of activity seen in any one neuron is a reflection of spatial patterning of activity within the intrinsic circuitry of the striatum. The excitation-based explanation requires no mutual interactions between neurons to produce the states, and therefore no particular spatial pattern associated with membrane potential transitions. Although direct visualization of the spatial patterns of striatal activity is not currently feasible, it is possible to predict the scale of local interaction that must be operating if mutual inhibition by spiny neurons is the crucial factor in temporal patterning of spiny cell activity. This scale is set by the local axon collateral fields and the dendritic fields of the spiny cells (figure 3.2). Studies of intracellularly stained spiny neurons have shown that these cells can be divided into two categories based on the spatial extent of their axonal arborizations (Kawaguchi et al., 1989, 1990). The more common type of spiny neuron has a more dense but compact axonal field, about the same size as, but not always completely overlapping with, the cell's own dendritic field. A less common cell type has a much larger but sparser axonal field that ramifies in the region around the dendritic tree of the cell of origin, but does not overlap much with it. The dendritic tree of both cell types is 200 to 500 μm in diameter. These two cells are likely to produce very different patterns of mutual inhibition, with the more common type producing highly focused inhibition between neurons with overlapping dendritic trees, and so perhaps receiving similar afferent input (see below). The fewer cells having larger axonal fields could produce a more global but more diffuse and less powerful inhibition among spiny neurons.

By way of their local effects, inhibition-based models produce spatial grouping of excited neurons, and predict correlations in the activity of nearby cells. Positive correlations in firing will occur for nearby neurons receiving similar synaptic input, because they come to be excited, inhibited, and disinhibited together (Wickens, et al., 1991; see also Woodward et al., chapter 16). Self-organizing groups of neurons form under

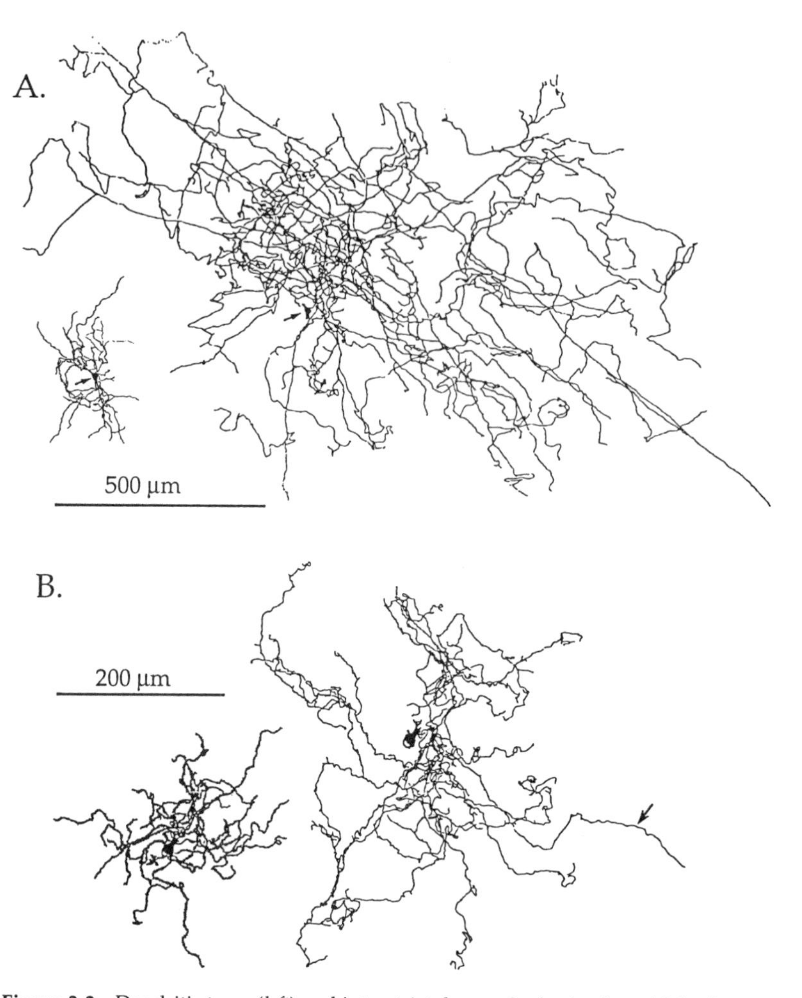

Figure 3.2 Dendritic trees (*left*) and intrastriatal axonal arborizations (*right*) for two spiny striatal projection neurons in parasagittal projections. The two cells (*A* and *B*) have similar-sized dendritic trees (note the difference in magnification in *A* and *B*). Their axonal arborizations within the striatum are very different, one covering a region of about 0.4 mm in diameter, while the other spans most of the rostrocaudal extent of the striatum and is over 1.0 mm long. (Reconstructions from Kawaguchi et al., 1990.)

such conditions and compete with one another, often producing anti-correlated activity across cell groups. Because these groups tend to organize around their mutual interactions, inhibition-based models do not require that the pattern of input to any one striatal neuron be particularly coherent, or have any special resemblance to the striatal cell's output pattern. A continuously varying input can be converted actively to two states by interactions among competing spatially defined groups of striatal neurons, with the temporal pattern of the inhibition largely determining the pattern of firing within each group. In the excitation-based scheme, the pattern of activation depends entirely upon the pattern of afferent inputs. In such schemes, the excitatory input must contain periods of activity and silence that define the overall temporal structure of the state transitions in striatal cells. Thus the spatial pattern of afferent fibers and the temporal structure of their firing patterns are critical to the functioning of the excitation-based model. A comparison of the features of these two schemes for explaining the temporal patterning of the firing of striatal neurons requires some consideration of the spatial organization of afferent terminations.

The issue is related to, but not identical with, the usual problems of determining the topographical mapping of afferents in the striatum. For the operation of mutually inhibitory networks, it is useful to assume that nearby neurons receive related inputs, and that the input pattern is distributed continuously as a function of distance in the postsynaptic network. For the neostriatum, this means that nearby spiny neurons would receive inputs from a common set of input fibers, and that spiny neurons located at greater and greater distances from one another would have more and more disjoint sets of input fibers. If the gradient of change in input fibers is shallow over the range of action of the mutual inhibition, interacting cells have effectively the same input and the dynamic groups of synchronously firing neurons are created mainly by the intrinsic connections of the network. If the inputs are changing greatly over the range of the mutual inhibition (so that two neurons within local axonal range have substantially different inputs), then the interconnections among spiny neurons will act primarily to accentuate differences in the input pattern over the small spatial scale. This effect is analogous to the well-known effects of lateral inhibition in the visual system. The spatial gradient of similarity in input patterns depends on both the across-fiber pattern of activity in the input pathway and the anatomical overlap between the axonal fields of corticostriatal input. Similarity of input to two neurons based on anatomical features (i.e., actual shared inputs as opposed to cross-fiber correlation) is a relatively static feature of the pathway with no dependence on moment-to-moment changes in the signal carried by the axons. It is determined by the shape and size of the postsynaptic dendritic trees, the degree of overlap of dendritic trees of nearby neurons, and by the structure of the axonal arborization of the presynaptic axon. While the shapes and sizes of dendritic trees of striatal spiny neurons have been known in detail for many years, structural information on single afferent axons has become available only recently. Because it is the substrate of only the anatomical contributor to the correlation of inputs converging onto spiny striatal neurons, it can only set a lower limit on the degree to which these neurons receive correlated input. If there is little divergence in the single axons (i.e., each corticostriatal axon connects to few striatal spiny neurons), correlated patterns of input across neurons can still be achieved by correlations among corticostriatal axons. If there is a great deal of divergence, with each axon contacting a large number of spiny cells, the pattern generated by that will be a feature of the processing of all corticostriatal input patterns. The converse problem, the number of corticostriatal axons contacting any one striatal neuron, will determine the degree to which single axons projecting to more than one spiny cell can correlate the activity of those neurons.

HOW MANY CORTICAL SYNAPSES DOES A SPINY CELL RECEIVE?

To interpret the spontaneous firing patterns of striatal spiny cells it is necessary to know whether it is driven by the action of a few very powerful afferent neurons or is instead the result of the combined effort of many afferents. Indirect evidence from electrophysiological studies has suggested that the membrane potential state changes and the evoked excitatory postsynaptic potentials (EPSPs) seen in striatal neurons in vivo must both arise from the action of a large number of afferent synapses. The basis for these conclusions was the inability to see underlying elementary response components within the spontaneous or evoked activity. For example, by reducing the strength of an electrical stimulus to a level at which an EPSP is evoked only part of the time, one might expect to see that the EPSP is composed of small all-or-nothing components representing the action of single synapses. Such unitary response components are visible in the EPSPs evoked in giant aspiny neostriatal neurons by cortical stimulation (Wilson et al., 1990), but they have not been seen in spiny cells (Wilson et al., 1982). Similarly, the spontaneous membrane potential fluctuations of giant aspiny neurons have a structure imposed by the presence of a relatively small number of unitary synaptic events. The smooth nature of the up and down membrane potential transitions of spiny neurons suggests instead a large number of synaptic events all of approximately equal and almost negligible importance (Wilson, 1993). If this suggestion turns out to be true, it implies that there must be both a large number of afferent synapses and that these must arise from a large number of different axons. The synapses must arise from different axons because axons making many weak synaptic inputs would be functionally equivalent to a single powerful synaptic input from the point of view of the physiological experiments described above. It is useful, therefore, to attempt to get anatomical estimates of both the number of synapses made upon a spiny cell by a single afferent fiber, and the total number of synapses formed on the cells. The second answer is easier to obtain. Nearly all synapses from the cerebral cortex and from the intralaminar thalamic nuclei are formed on dendritic spines of spiny cells (the thalamic parafascicular-centromedian complex is an exception; see Dube et al., 1988, Xu et al., 1991). Because individual spines almost always possess exactly one of these synapses, it is possible to count the number of dendritic spines and so get an estimate of the total number of synaptic inputs to the neurons. Such counts show considerable variability from cell to cell, mostly because spiny neurons vary in the number of spiny dendrites and to a lesser degree because they vary in spine density (Wilson et al., 1983b). Integration of the high-voltage electron microscopy–corrected spine density curve measured for spiny neurons yields estimates of about 500 spines per spiny dendrite. The number of spiny dendrites varies from about 20 to 60 per cell, yielding between 10,000 and 30,000 spines per cell. Of these, about half are expected to arise from the corticostriatal projection (Kemp and Powell, 1971). Of course, these 5000 to 15,000 synapses need not arise from the same number of corticostriatal cells, as one afferent axon could make more than one synapse with the same spiny cell (see below).

CELLS OF ORIGIN

Studies of individual corticostriatal axons have been undertaken to examine the sizes and patterns of arborizations of these cells within the striatum (Wilson, 1987; Cowan and Wilson, 1994). The cells of origin of the corticostriatal projection have been the subject of controversy and conjecture since the time of Cajal. He reported observing axon collaterals arising from descending internal capsule fibers and arborizing in the neostriatum in rats and mice (Ramón y Cajal, 1911), and concluded that the projection to the neostriatum was primarily collateral in nature and arose from neurons projecting to the brainstem or beyond. More recent studies employing retrograde axonal tracing in the 1970s and 1980s seemed to contradict this view, as they revealed that, at least in some cortical regions, most corticostriatal neurons were located in a laminar pattern that did not correspond to that of any of the other known corticofugal pathways (e.g., Jones et al., 1977; Hedreen, 1977; Schwab et al., 1977). It should be noted that this experiment with retrograde tracing was made in rats and monkeys with the same result, and so species differences could not be offered as an explanation for the observations made by Cajal. It seemed necessary to conclude that Cajal was simply mistaken. This view was reinforced by the observation that in neurophysiological experiments, few cortical neurons could be antidromically excited by stimulation of both the cerebral peduncle and the neostriatum (e.g., Kitai et al., 1976). The corticostriatal projection was a difficult one to study using retrograde tracing with horseradish peroxidase (HRP) as was employed in those early studies, however, and Nauta et al. (1974) suggested that this difficulty might be due to dilution of the retrogradely transported label in the large axonal arborizations of the neurons. Moreover, some antidromic stimulation experiments suggested that there were brainstem projecting neurons with axonal branches in the neostriatum, but these experiments were not considered conclusive because of the possibility of spread of current to the internal capsule (Oka and Jinnai, 1978; Jinnai and Matsuda, 1979).

More recent studies using retrograde tracing in cats, rats, and monkeys have shown corticostriatal neurons distributed over several laminae in patterns that vary among cortical regions (e.g., Royce, 1982; Arikuni and Kubota, 1986; Wilson, 1987). Double retrograde labeling has shown that corticostriatal neurons located in the same cortical area but in different laminae have specific patterns of collateralization (Royce and Bromley, 1984; Fisher et al., 1984). In addition, recent orthograde tracing studies have shown that injections centered in different laminae produce variations in the patterns of labeling in the striatum, suggesting that corticostriatal cells in different laminae arborize differently in the striatum (Gerfen, 1989). Intracellular staining experiments, starting with Donoghue and Kitai (1981), have shown that in rat, neurons antidromically activated from the brainstem give off fine axon collaterals that arborize in the neostriatum. In early experiments employing HRP staining it could not be determined how common such collaterals were, or how widely they arborized in the striatum. The collaterals, being much finer than the main axon in the internal capsule, were very faintly stained. They could not be followed to their terminations in the neostriatum, and they were so faint that when they were not seen it could not be said with certainty that they were absent (Donoghue and Kitai, 1981; Landry et al., 1984). Recent experiments employing intracellular staining with biocytin have shown that these axons are much more common than previously expected, and may be a consistent feature of pyramidal tract cells in the premotor cortex (Cowan and Wilson, 1994). The failure to detect these axon collaterals in antidromic stimulation studies is probably due to their extremely fine caliber in comparison to the main axon in the internal capsule, which may cause failure of propagation of the antidromic spike at the branch point. Brainstem projecting neurons make up a very small proportion of the total population of corticostriatal cells, however. The vast majority of these neurons are more superficially situated, smaller pyramidal neurons. Antidromic activation (figure 3.3) and intracellular staining studies of these cells in rats have shown that most of these neurons have primarily corticocortical connections, in addition to their projection to the neostriatum (Wilson, 1987; Cowan and Wilson, 1994). In the premotor cortex of rats, most or all of these neurons are bilaterally projecting, with connections to the neostriatum and neocortex of both sides. In cortical regions with less prominent bilateral projections, it is likely that these cells are primarily ipsilaterally projecting corticocortical cells (Wilson, 1987). These results show that the corticostriatal projection arises from several different cell populations, each collateralized and participating in some other corticocortical or corticofugal pathway, as suggested by Royce and Bromley (1984).

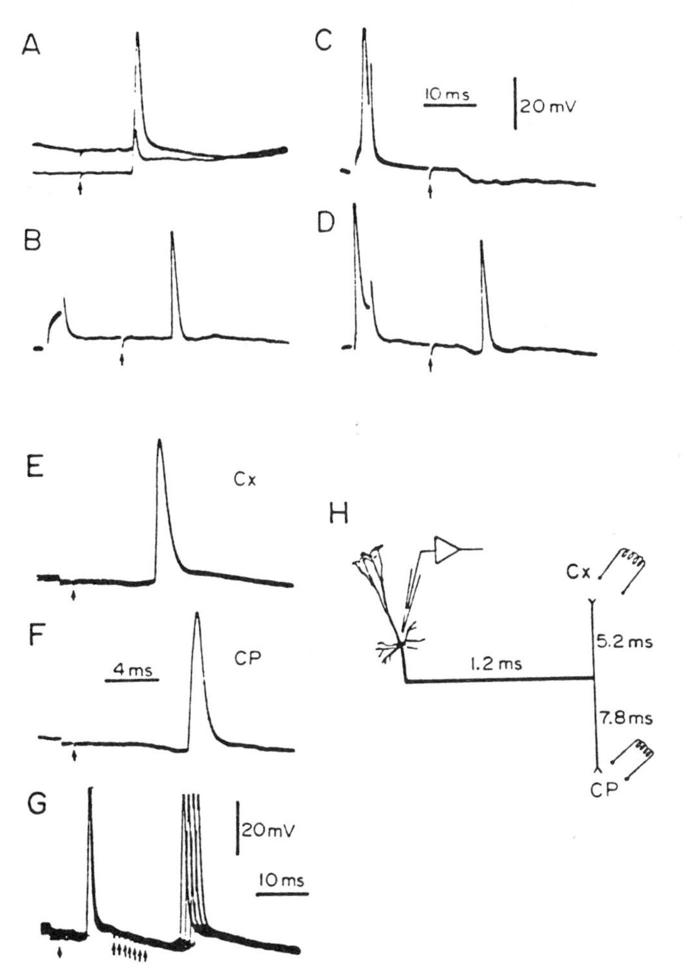

Figure 3.3 Antidromic identification of crossed corticostriatal neurons *A,* Antidromic action potential evoked in a cortical neuron by stimulation in the contralateral neostriatum. Hyperpolarization caused spike fractionation but a constant latency is maintained. *B–D,* Collision of the antidromic spike with an orthodromic action potential evoked by an intracellular current pulse. *E* and *F,* Antidromic activation of the same neuron from stimulation of the contralateral cortex (Cx) and contralateral neostriatum (CP). *G,* Sequential stimulation of the two sites produces collision at short interstimulus intervals. *H,* From the interstimulus interval over which collision occurs and the conduction time of the two antidromic spikes, it is possible to calculate the conduction time in the common part of the axon and in each of the two branches, one to the striatum and one to the contralateral cortex. (From Wilson, 1987.)

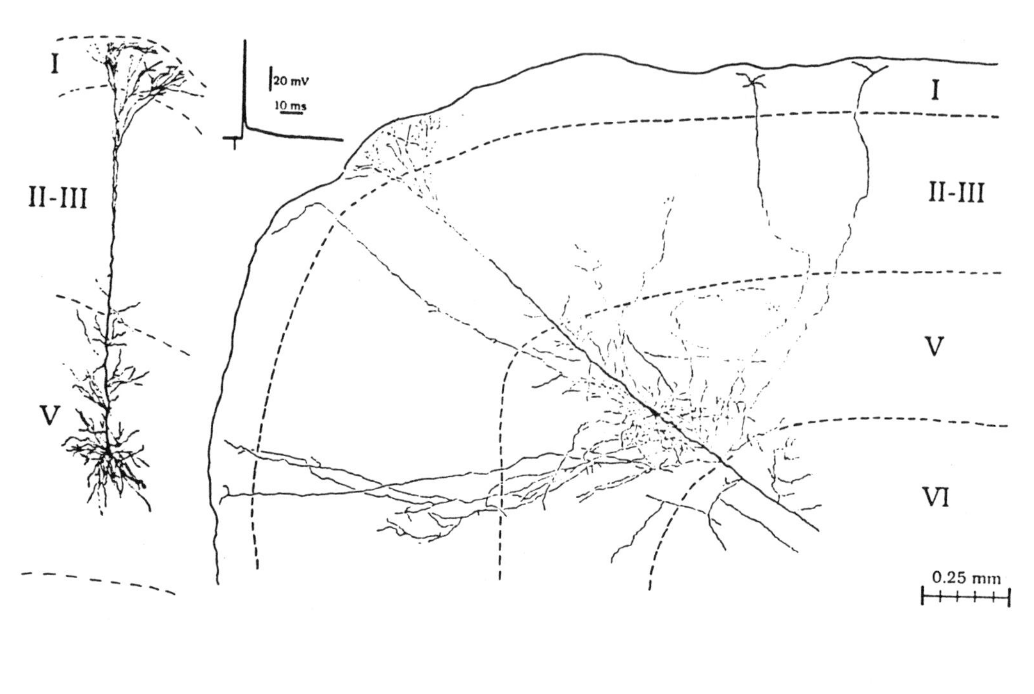
I
II-III
V
20 mV
10 ms
I
II-III
V
VI
0.25 mm

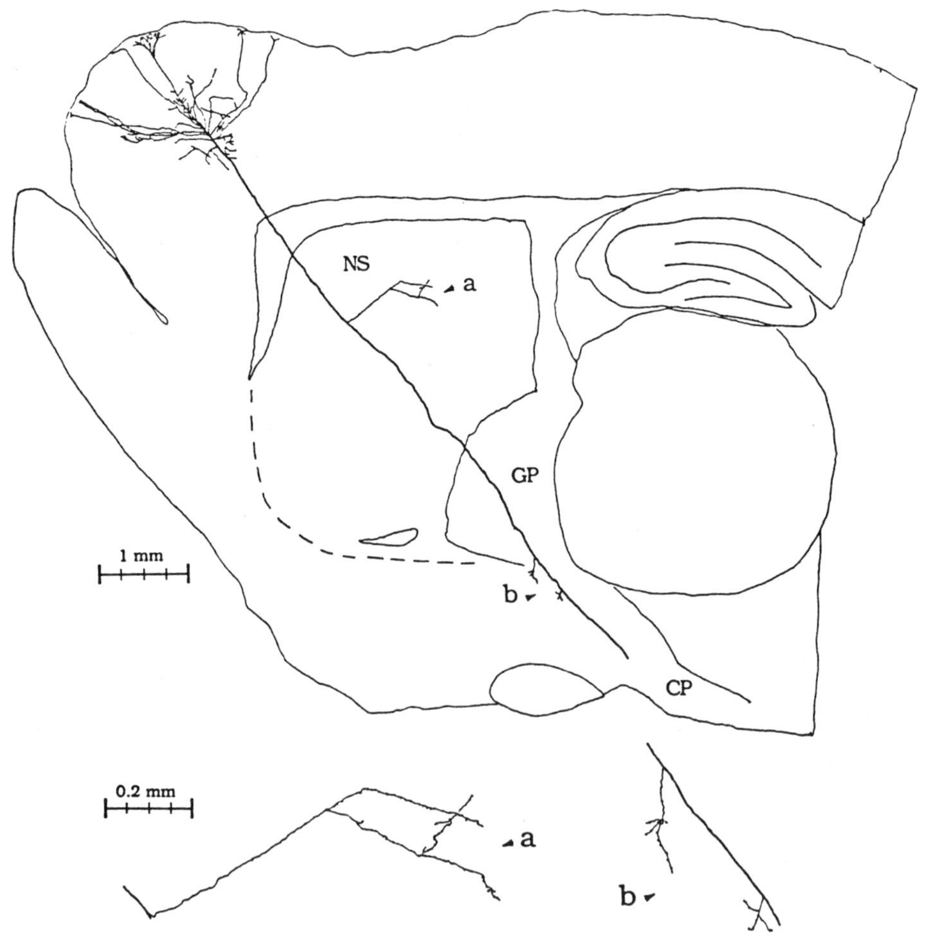
NS
a
GP
1 mm
b
CP
0.2 mm
a
b

The intrastriatal axonal arborizations of two of these cell populations, the bilaterally projecting corticocortical-corticostriatal cells and the pyramidal tract–corticostriatal cells of the premotor cortex have been described by Cowan and Wilson (1994) using intracellular staining with biocytin. An example showing the axonal projections of a pyramidal tract neuron is shown in figure 3.4. As expected from the small EPSP component contributed by this part of the corticostriatal pathway (Wilson et al., 1982), the intrastriatal collateral arborizations of these cells were modest in size and contributed few boutons to the striatal neuropil. Less predictable was the compact nature of these arborizations. The small volume of the arborization suggests the possibility that it might participate in a topographically simple and precise representation in the cortex. Although the small sample of intracellularly injected neurons of this type and the restricted cortical area we have sampled preclude a systematic study of topography, the arborizations we have seen so far are consistent with this interpretation. Within the arborization, however, the axons branched sparsely and boutons were formed en passant on axonal branches that took a predominantly straight course through the neuropil. This arrangement, in conjunction with the geometry of the dendritic field of the spiny neuron, prevents any one axon from making more than a few contacts with any one spiny cell. This is illustrated in figure 3.5, in which a drawing of a spiny neuron has been superimposed upon the intrastriatal axonal arborization of the pyramidal tract–corticostriatal axon shown in figure 3.4. Given the sparse volume filling of both the axon and the dendritic field, there is no way that the cortical axon could contact more than a tiny proportion of the total number of spiny cells whose dendrites enter the volume occupied by the axonal arborization, and no one neuron could possiblly receive more than a handful of synapses. If the axon were specifically addressed to one spiny cell and all of the synapses it made within that cell's dendritic tree were made with the one neuron, it could make as many as ten synapses. Even this would be a small proportion of the 10,000 to 30,000 synapses received by that spiny cell. It is much more likely that it forms synapses with ten different cells, each of which receives one synapse.

Figure 3.4 Dendritic and axonal arborizations of a pyramidal tract neuron in the rat medial agranular cortex. (*Top left*) A reconstruction of the dendritic tree is shown relative to the cortical lamination. The antidromic response to pyramidal tract stimulation is also shown. (*Top right*) The intracortical axonal arborization of the cell is shown superimposed on the dendritic tree. In addition to the arborizations in layers V and VI in the region of the basilar dendritic tree, there are longer projections to more caudal and rostral regions of the cortex. (*Middle*) Drawing shows the entire telencephalic axonal arborization of the cell, including collateral arborizations in the neostriatum (*a*) and basilar forebrain (*b*). At *bottom*, these two collateral arborizations are shown at higher magnification. GP, globus pallidus; NS, neostriatum; CP, cerebral peduncle. (From Cowan and Wilson, 1994.)

Figure 3.5 The intrastriatal collateral axonal arborization of the same neuron shown in figure 3.4, at higher magnification and with a spiny neuron dendritic tree superimposed in gray. The spiny neuron was injected in a different rat, and is superimposed simply to illustrate the dimensions of the axonal arborization. Because they are at different mediolateral levels, the two main branches of the cortical axon could not both cross the dendritic tree of the spiny neuron. No matter how it is placed within the axonal arborization, this cell could not receive more than three or four contacts from the axon.

Intracellular staining studies show that individual crossed corticostriatal neurons also have too sparse connections to allow one to one control of striatal neurons. An example of the dendritic and axonal arborizations of one of these cells is shown in figure 3.6. The various kinds of corticostriatal neurons (identified by cortical layer and axon collateralization) have different-sized axonal arborizations in the striatum, but none of them has a dense enough arborization to allow it to make large numbers of contacts on individual neurons. The number of corticostriatal axons innervating any one striatal cell must therefore be measured in the thousands. Although the striatal cell can successfully filter out input patterns that do not produce synchronous uniform excitation of the dendrites, such a large number of afferent inputs, if independently activated, could produce a relatively constant excitatory barrage even if the cortical neurons fired randomly. The excitation-based model thus requires that the cell accentuate a pattern of correlated input activity already well established in a large number of afferent fibers. The complex shapes of the single axonal arborizations brings into question an assumption of the inhibition-based model as well, the similarity of input to nearest neighbors. Because individual corticostriatal axons synapse so sparsely, the striatal neurons receiving input from any one corticostriatal axon (and so having similar input because of it) are spread out in a complex pattern over a very large region of the neostriatum. They are not nearest neighbors, and most are not within inhibitory range of one another. Making sense of this odd arrangement will require a systematic understanding of the relationship between afferent axons. Do nearby cortical neurons have axons that branch similarly and so describe complex branching shapes of striatal neurons receiving related input? Or would the next corticostriatal neuron arborize in the same general volume but in an entirely different pattern within it? Studies of axonal staining after small injections of tritiated amino acids in the neostriatum show a fine organization to the axonal labeling caused by small groups of nearby cortical neurons (e.g., Flaherty and Graybiel, 1993). This grouping of labeled axons, called matrisomes, raises the possibility of across-fiber similarities in the fine organization of single axonal arborizations. This problem is, of course, an extension of the well-studied issue of topography in the corticostriatal projection. Discussions of the topographical organization of this projection are usually concentrated on the question of whether multiple cortical areas converge upon the same striatal neuron or not. The results with single corticostriatal axons show that there are a very large number of cortical neurons converging onto each striatal neuron, but they leave open the question of the distribution of those neurons within the cortex.

CORTICOSTRIATAL NEURONS SHOW STATE TRANSITIONS SIMILAR TO THOSE OF STRIATAL NEURONS

Because there are a large number of different afferent inputs that converge upon each neuron, regardless of their origins in the cortex, it becomes important to determine how many of these are required to make the striatal spiny neuron undergo the transition to the up state, and at what rate they must continue to fire to maintain a spiny cell in the up state. Certainly for the second question, and perhaps for the first, the temporal summation of inputs from single fibers must be taken into consideration. If one axon fires at a very high rate, is this equivalent to having several different axons firing at about the same time? If so, perhaps a small number of rapidly firing afferents could control the behavior of spiny neurons as effectively as the convergence of a much larger number of more slowly firing cortical cells.

A number of different input patterns could function to produce the known behavior of spiny neurons. Because there are so many different afferent neurons, a large-scale patterning of inputs, with each afferent contact participating in only a small way, can be considered. In this case, corticostriatal neurons would not necessarily fire in a pattern resembling that of the striatal neuron. For example, imagine there were 5000 participating cortical cells divided into hypothetical nonoverlapping groups of 100 that fired synchronously and were each able to trigger a brief transition to the up state in the striatal neuron. Imagine further that these 50 groups fired in volleys spaced at intervals adjusted to maintain a constant up state in the striatal neuron (about 25 ms or

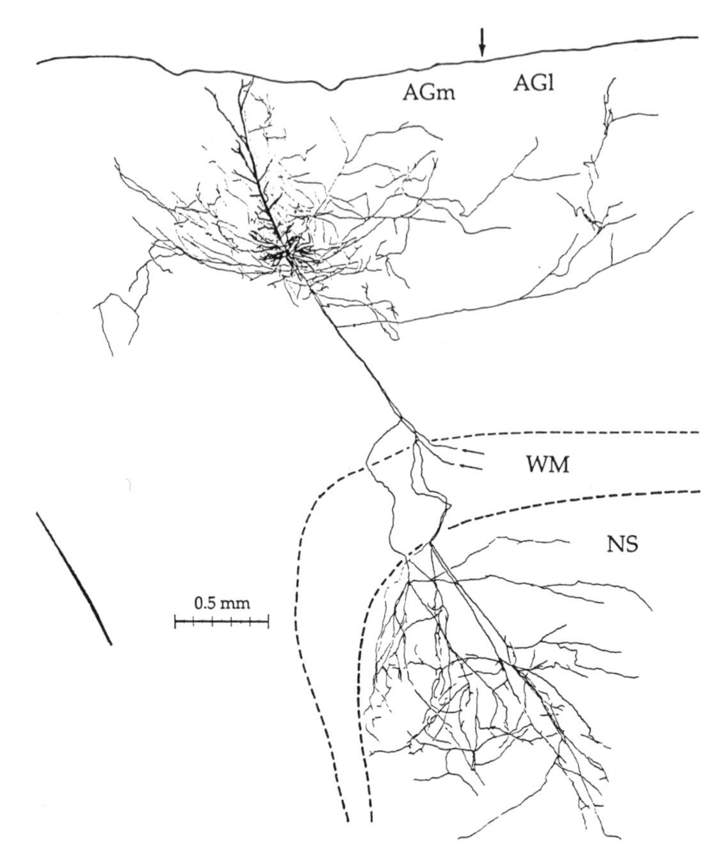

Figure 3.6 The dendritic tree and ipsilateral axonal arborization of a bilaterally projecting corticocortical-corticostriatal neuron like the one whose activation is shown in figure 3.3. The ipsilateral intracortical axonal arborization extends widely in the region around the cell (AGm), and also into the neighboring lateral agranular area (AGl). Four main axonal branches are formed before the axon enters the white matter (WM). Two of them, marked with arrows, were followed through the corpus callosum to the contralateral hemisphere (not shown). These probably correspond to the two axonal branches shown in figure 3.3. The others enter the ipsilateral neostriatum (NS), where they make a large and extremely sparse axonal arborization. While much larger than the collateral arborization of the pyramidal tract cell shown in figure 3.5, this arborization is just as sparse. (From Cowan and Wilson, 1994.)

less). If we recorded from any one of these hypothetical corticostriatal neurons, we would observe it to fire only once during a typical 125-ms episode of depolarization in the striatal neuron. The pattern of firing we would see in individual corticostriatal neurons would reveal little about their role in controlling any one striatal cell. Of course, this scheme would produce very strictly tuned corticostriatal function, with each striatal neuron detecting very complex and specific patterns of activity in a large number of corticostriatal cells. The opposite view would have cortical neurons firing in a pattern similar to that of the membrane potential fluctuations seen in striatal cells. In this scheme, each afferent neuron would be contributing over a longer time window. If corticostriatal neurons converging onto a single striatal cell were firing episodically, and fired in episodes that began at the same time, fewer of them would be required to impose an up-state transition in the striatal cell. The number required would depend on their firing rates and coherence in the short (millisecond) time scale, as well as the strength of their synapses in striatum and their spatial distribution on the postsynaptic neuron. In the extreme case, if all the converging cortical cells were firing rhythmically and in phase during their episodes of activity, the 5000 cells described above could be replaced by only one group of 100. The other 49 (hypothetical) groups of converging corticostriatal neurons could represent alternative cortical patterns, any of which might produce identical activity in the striatal cell.

Some progress toward sorting out the possibilities between these extremes has been made by recording the spontaneous activity of corticostriatal neurons in vivo (Cowan and Wilson, 1994). The striatal projections of these same cells were analyzed by reconstructing them in serial sections after intracellular staining with biocytin. These corticostriatal neurons showed spontaneous membrane potential fluctuations on the same time scale as those of striatal cells, and fired in episodes matching the periods of depolarization in striatal cells. An example is shown in figure 3.7. Thus groups of corticostriatal neurons working together to depolarize striatal cells may be correlated in two different time scales of importance to the striatal cell. One is the time scale of the depolarizing episodes (0.1–3.0 seconds). Corticostriatal neurons whose episodes of firing do not overlap cannot cooperate to generate the up-state transition in striatal neurons. The second time scale is that relevant to the synaptic summation in the postsynaptic cell, the time window during which two synapses can summate. Although this depends upon membrane potential (becoming long as the cell is depolarized and short when hyperpolarized), it is measured in milliseconds

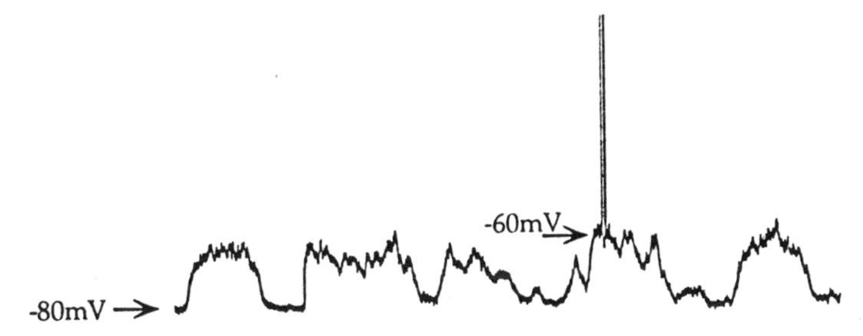

Figure 3.7 Spontaneous membrane potential shifts of a corticostriatal neuron like the one shown in figures 3.3 and 3.6. Note the similarity between this cell and the striatal neuron shown in figure 3.1. Sweep duration is 5 sec.

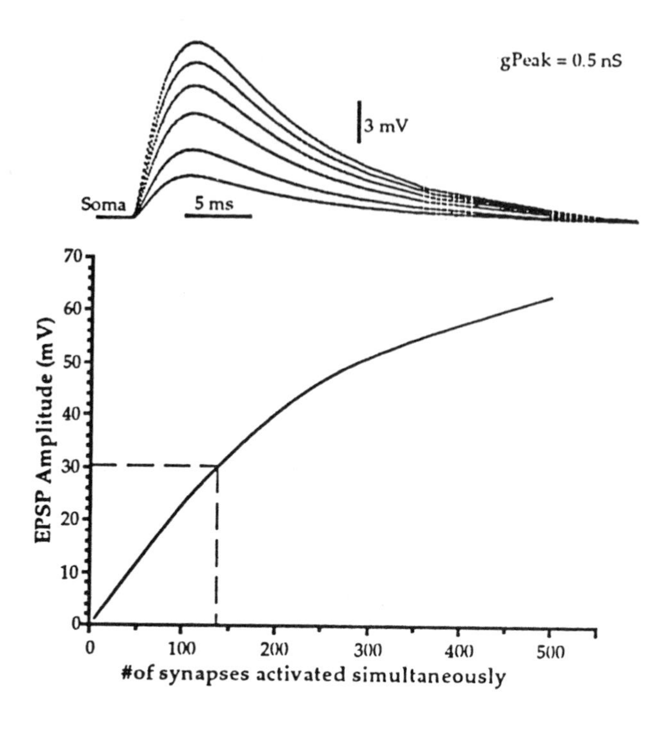

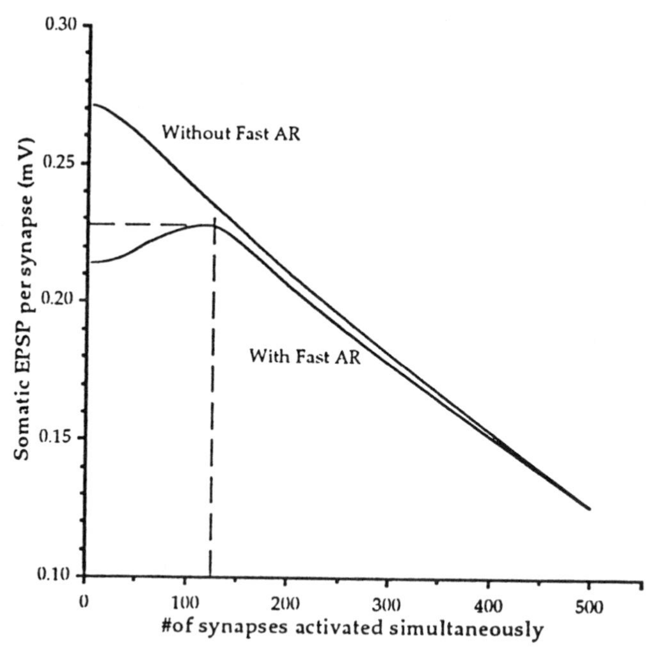

Figure 3.8 Computer-simulated excitatory postsynaptic potentials (EPSPs) generated by synchronous uniformly distributed corticostriatal stimulation in a model of the spiny neuron. This simulation shows the minimum number of synaptic inputs required to achieve a somatic depolarization comparable to the up state of the spiny neuron. (*Top*) The somatic peak EPSP amplitude is shown as a function of the number of activated synapses. Dotted lines show the number of synapses requires to achieve a 30-mV depo-

and is roughly the duration of the evoked corticostriatal EPSP (approximately 25 ms). Corticostriatal neurons firing in a correlated fashion on this time scale can cooperate on a local scale on the postsynaptic neuron. Some evidence for correlated activity in both time scales is available. Corticostriatal cells generating episodes of activity independently and combining their output onto individual striatal neurons would not produce depolarizations of the same duration in the postsynaptic cell. The resemblance between the cortical and striatal episodes of depolarization indicates that there is a correlation in generation of episodic activity. Corticostriatal neurons were also observed to fire in phase with a subthreshold high-frequency (35–45 Hz) membrane potential oscillation that, if it is coherent across a group of corticostriatal neurons (as it is, at least sometimes, in the visual cortex), would synchronize their activity on the short time scale.

MULTIPLE ALTERNATIVE PATTERNS OF INPUT MAY EXCITE SINGLE STRIATAL NEURONS

Computer simulations of synaptic activation of striatal neurons by afferent inputs firing in patterns similar to those of corticostriatal neurons indicate that the number of correlated input fibers required to fire a striatal neuron may be numbered in the hundreds, not the thousands. Of course, this depends critically on the strength (and time scale) of the correlation between fibers and their distribution on the neuron. Using the low innervation density per fiber suggested by single axon arborization data, we assume that each fiber makes only one synapse on any one striatal spiny cell. Low estimates of synaptic conductance (0.1–0.5 nS) were used to ensure that the simulations were not biased by the effects of unrealistically large synaptic inputs. Simulations using stronger synapses gave comparable results, but required even fewer synaptic inputs (Wilson 1992). These results, shown in figure 3.8, suggest that although the spiny neurons each receive input from thousands of cortical and thalamic neurons, they do not require the cooperative effort of all of them to make the transition to firing. Thus different sets of inputs may, at different times, control a single spiny cell. Their control of the cell would be by virtue of a correlation of their activity maintained over tens to hundreds of milliseconds, and from the point of view of

larization (approximately the size of the up-state transition). (*Bottom*) The number of synapses achieving an up-state transition are shown to correspond to the peak in the synaptic effectiveness generated by the presence of the anomalous rectification (AR). While the absolute number of synapses achieving this effect depends upon the peak synaptic conductance (0.5 nS in this simulation, but unknown for corticostriatal synapses), this correspondence between the up state and the peak in the synaptic effectiveness curve does not. (From Wilson, 1992.)

the individual spiny cell, each of these possible patterns of input would be indistinguishable. Individual afferent fibers could, at different times, participate in different controlling ensembles of corticostriatal and thalamostriatal neurons.

REFERENCES

Albe-Fessard, D., Rocha-Miranda, C., and Oswaldo-Cruz, E. (1960) Activités évoqueés dans le noyau caude du chat en réponse à des types divers d'afférences. *Electroencephalogr. Clin. Neurophysiol.* 12:649–661.

Arikuni, T., and Kubota, K. (1986) The organization of prefrontocaudate projections and their laminar origin in the macaque monkey: A retrograde study using HRP-gel. *J. Comp. Neurol.* 244:492–510.

Bargas, J., Gallarraga, E., and Aceves, J. (1989) An early outward conductance modulates the firing latency and frequency of neostriatal neurons of the rat brain. *Exp. Brain Res.* 75:146–156.

Bernardi, G., Marciani, M. G., Morocutti, C., and Giacomini, P. (1976) The action of picrotoxin and bicuculline on rat caudate neurons inhibited by GABA. *Brain Res.* 102:379–384.

Calabresi, P., Misgeld, U., and Dodt, H. U. (1987) Intrinsic membrane properties of neostriatal neurons can account for their low level of spontaneous activity. *Neuroscience* 20:293–303.

Calabresi, P., Mercuri, N. B., Stefani, A., and Bernardi, G. (1990) Synaptic and intrinsic control of membrane excitability of neostriatal neurons. I. An in vivo analysis. *J. Neurophysiol.* 63:651–662.

Cowan, R. L., and Wilson, C. J. (1994) Spontaneous firing patterns and axonal projections of single corticostriatal neurons in the rat medial agranular cortex. *J. Neurophysiol.* 71:17–32

Donoghue, J. P., and Kitai, S. T. (1981) A collateral pathway to the neostriatum for corticofugal neurons of the rat sensory-motor cortex: An intracellular HRP study. *J. Comp. Neurol.* 212:76–88.

Dube, L., Smith, A. D., and Bolam, J. P. (1988) Identification of synaptic terminals of thalamic or cortical origin in contact with distinct medium-size spiny neurons in the rat neostriatum. *J. Comp. Neurol.* 267:455–471.

Fisher, R. S., Shiota, C., Levine, M. S., Hull, C. D., and Buchwald, N. A. (1984) Interhemispheric organization of corticocaudate projections in the cat: A retrograde double-labelling study. *Neurosci. Lett.* 48:369–373.

Flaherty, A. W., and Graybiel, A. M. (1993) Two input systems for body representations in the primate striatal matrix: Experimental evidence in the squirrel monkey. *J. Neurosci.* 13:1120–1137.

Fonnum, F., Grofova, I., Rinvik, E., Storm-Mathisen, J., and Walberg, F. (1974) Origin and distribution of glutamate decarboxylase in substantia nigra of the cat. *Brain Res.* 71:77–92.

Fox, C. A., Andrade, A. N., Hillman, D. E., and Schwyn, R. C. (1971) The spiny neurons in the primate striatum: A Golgi and electron microscopic study. *J. Hirnforsch.* 13:181–201.

Gerfen, C. R. (1989) The neostriatal mosaic: Striatal patch-matrix organization is related to cortical lamination. *Science* 246:385–388.

Hedreen, J. C. (1977) Corticostriatal cells identified by the peroxidase method. *Neurosci. Lett.* 4:1–7.

Hull, C. D., Bernardi, G., Price, D. D., and Buchwald, N. A. (1973) Intracellular responses of caudate neurons to temporally and spatially combined stimuli. *Exp. Neurol.* 38:324–336.

Jiang, Z.-G., and North, R. A. (1991) Membrane properties and synaptic responses of rat striatal neurones in vitro. *J. Physiol. (Lond.)* 443:533–553.

Jinnai, K., and Matsuda, Y. (1979) Neurons of the motor cortex projecting commonly on the caudate nucleus and the lower brainstem in the cat. *Neurosci. Lett.* 4:1–7.

Jones, E. G., Coulter, J. D., Burton, H., and Porter, R. (1977) Cells of origin and terminal distribution of corticostriatal fibers arising in the sensori-motor cortex of monkeys. *J. Comp. Neurol.* 173:53–80.

Kawaguchi, Y., Wilson, C. J., and Emson, P. C. (1989) Intracellular recording of identified neostriatal patch and matrix spiny cells in a slice preparation preserving cortical inputs. *J. Neurophysiol.*, 62:1052–1068.

Kawaguchi, Y., Wilson, C. J., and Emson, P. C. (1990) Projection subtypes of rat neostriatal matrix cells revealed by intracellular injection of biocytin. *J. Neurosci.* 10:3421–3438.

Kemp, J. M., and Powell, T. P. S. (1971) The synaptic organization of the caudate nucleus. *Philos. Trans. R. Soc. Lond.*, [*Biol.*] 262:403–412.

Kita, T., Kita, H., and Kitai, S. T. (1984) Passive electrical properties of rat neostriatal neurons in in vitro slice preparation. *Brain Res.* 300:129–139.

Kita, H., Kita, T., and Kitai, S. T. (1985) Active membrane properties of rat neostriatal neurons in in vitro slice preparation *Exp. Brain Res.* 60:54–62.

Kitai, S. T., Kocsis, J. D., and Wood, J. (1976) Origin and characteristics of the cortico-caudate afferents: An anatomical and electrophysiological study. *Brain Res.* 191:137–141.

Landry, P., Wilson, C. J., and Kitai, S. T. (1984) Morphological and physiological characteristics of pyramidal tract neurons in the rat. *Exp. Brain Res.* 57:177–190.

Nauta, H. J. W., Pritz. M. B., and Lasek, R. J. (1974) Afferents to the caudato-putamen studied with horseradish peroxidase. An evaluation of a retrograde neuroanatomical research method. *Brain Res.* 67:219–238.

Nisenbaum, E. S., and Berger, T. W. (1992) Functionally distinct subpopulations of striatal neurons are differentially regulated by GABAergic and dopaminergic inputs. 1. In vivo analysis. *Neuroscience* 48:561–578.

Nisenbaum, E. S., Xu, Z. C., and Wilson, C. J. (1994) Contribution of a slowly-inactivating potassium current to the transition to firing of neostriatal spiny projection neurons. *J. Neurophysiol.*, in press.

Oka, H., and Jinnai, K. (1978) Common projection of the motor cortex to the caudate nucleus and cerebellum. *Exp. Brain Res.* 31:31–42.

Pennartz, C. M. A., and Kitai, S. T. (1991) Hippocampal inputs to identified neurons in a in vitro slice preparation of the rat nucleus accumbens: Evidence for feed-forward inhibition. *J. Neurosci.* 11:2838–2847.

Precht, W., and Yoshida, M. (1971) Blockage of caudate-evoked inhibition of neurons in the substantia nigra by picrotoxin. *Brain Res.* 32:229–233.

Ramón y Cajal, S. (1911) *Histologie du système nerveux de l'homme et des vertèbres.* 1911 Maloine, Paris; [Chapter 23 trans. by Haycock, J. W., and Bro, S. (1975) *Behav. Biol.* 14:387–402].

Ribak, C. E., Vaughn, J. E., Saito, K., Barber, R., and Roberts, E. (1976) Immunocytochemical localization of glutmate decarboxylase in rat substantia nigra. *Brain Res.* 116:287–298.

Royce, G. J. (1982) Laminar origin of cortical neurons which project upon the caudate nucleus: A horseradish peroxidase investigation in the cat. *J. Comp. Neurol.* 205:8–29.

Royce, G. J., and Bromley, S. (1984) Fluorescent double labelling studies of thalamostriatal and corticostriatal neurons. In J. S. McKenzie, R. E. Kemm, and L. N. Wilcock (eds.), *The Basal Ganglia.* New York: Plenum Press, pp. 131–146.

Schwab, M., Agid, Y., Glowinski, L., and Thoenen, H. (1977) Retrograde axonal transport of I-tetanus toxin as a tool for tracing fiber connections in the central nervous system: Connections of the rostral part of the rat neostriatum. *Brain Res.* 126:211–224.

Surmeier, D. J., Foehring, R., Stefani, A., and Kitai, S. T. (1991) Developmental expression of a slowly-inactivating voltage dependent potassium current in rat neostriatal neurons. *Neurosci. Lett.* 122:41–46.

Vogt, C., and Vogt, O. (1920) Zu Lehre der Erkrankungen des striaren Systems. *J. Psychol. Neurol.* 25:628–846.

Wickens, J. R., Alexander, M. E., and Miller, R. (1991) Two dynamic modes of striatal function under dopaminergic-cholinergic control: Simulation and analysis of a model. *Synapse* 8:1–12.

Wilson, C. J. (1986) Postsynaptic potentials evoked in spiny neostriatal projection neurons by stimulation of ipsilateral or contralateral neocortex. *Brain Res.* 367:201–213.

Wilson, C. J. (1987) Morphology and synaptic connections of crossed corticostriatal neurons in the rat. *J. Comp. Neurol.* 263:567–580.

Wilson, C. J. (1992) Dendritic morphology, inward rectification, and the functional properties of neostriatal neurons. In T. McKenna, J. Davis, and S. F. Zornetzer (eds.), *Single Neuron Computation.* San Diego: Academic Press, pp. 141–171.

Wilson, C. J. (1993) The generation of natural firing patterns in neostriatal neurons. In G. W. Arbuthnott and P. C. Emson (eds.), Chemical Signaling in the Basal Ganglia *Prog. Brain Res.*, in press.

Wilson, C. J., and Groves, P. M. (1981) Spontaneous firing patterns of identified spiny neurons in the rat neostriatum. *Brain Res.* 220:67–80.

Wilson, C. J., Chang, H. T., and Kitai, S. T. (1982) Origins of post-synaptic potentials evoked in identified neostriatal neurons by stimulation in substantia nigra. *Exp. Brain Res.* 45:157–167.

Wilson, C. J., Chang, H. T., and Kitai, S. T. (1983a) Disfacilitation and long-lasting inhibition of neostriatal neurons in the rat. *Exp. Brain Res.* 51:227–235.

Wilson, C. J., Groves, P. M., Kitai, S. T., and Linder, J. C. (1983b) Three-dimensional structure of dendritic spines in the rat neostriatum. *J. Neurosci.* 3:383–398.

Wilson, C. J., Kita, H., and Kawaguchi, Y. (1989) GABAergic interneurons, rather than spiny cell axon collaterals, are responsible for the IPSP responses to afferent stimulation in neostriatal spiny neurons. *Soc. Neurosci. Abstr.* 15:907.

Wilson, C. J., Chang, H. T., and Kitai, S. T. (1990) Firing patterns and synaptic potentials of identified giant aspiny interneurons in the rat neostriatum *J. Neurosci.* 10:508–519.

Xu, Z. C., Wilson, C. J., and Emson, P. C. (1991) Restoration of thalamostriatal projections in rat neostriatal grafts: An electron microscopic analysis. *J. Comp. Neurol.* 303:22–34.

4 Elements of the Intrinsic Organization and Information Processing in the Neostriatum

Philip M. Groves, Marianela Garcia-Munoz, Jean C. Linder, Michael S. Manley, Maryann E. Martone, and Stephen J. Young

Early conceptions of the neostriatum typically placed this cell mass in a functional path originating in neocortex and eventually impinging on motor systems, by way of pallidum, thalamus, and motor cortex, but without much consideration of the intrinsic organization of the neostriatum or the potential input-output transformation that it performs (e.g., see Denny-Brown, 1962; Laursen, 1963; Penney and Young, 1983). The neostriatum appeared cytoarchitectonically uniform when viewed with conventional histologic stains; it possessed no clear laminar organization such as that found in the neocortex, nor could it be easily parcellated based on differences in cell density or size. Early attempts to develop schemata for the intrinsic organization of the neostriatum did not adequately reflect the proportions of neostriatal neuron types, at least as revealed with the Golgi method (Féger et al., 1979; Pasik et al., 1979). A schematic diagram which more accurately depicts this prior issue relevant to the intrinsic organization of elements within the neostriatum is shown in figure 4.1. This figure illustrates that the vast majority of neostriatal neurons are common spiny neurons (S1 in the classification of Pasik et al., 1979) and a small remainder comprise other cell types. While this classification is now being refined and expanded by additional anatomical and neurochemical differentiations of neostriatal neurons, this early view provided an initial step toward considering the intrinsic organization and the basis for information processing in the basal ganglia (Groves, 1983).

Knowledge regarding neostriatal intrinsic organization and function has evolved rapidly. A remarkable insight was derived from the evidence reported by Grofová (1975) using the then new method of retrograde transport of horseradish peroxidase. She observed that the medium-sized, common spiny neuron of neostriatum was actually a neostriatal efferent, in contrast to the previous belief that these cells were interneurons (e.g., Kemp and Powell, 1971). This was followed by the revelation that the neostriatum was heterogeneous when viewed during development with dopamine histofluorescence and later with acetylcholinesterase histochemistry in the adult animal (Olson et al.,

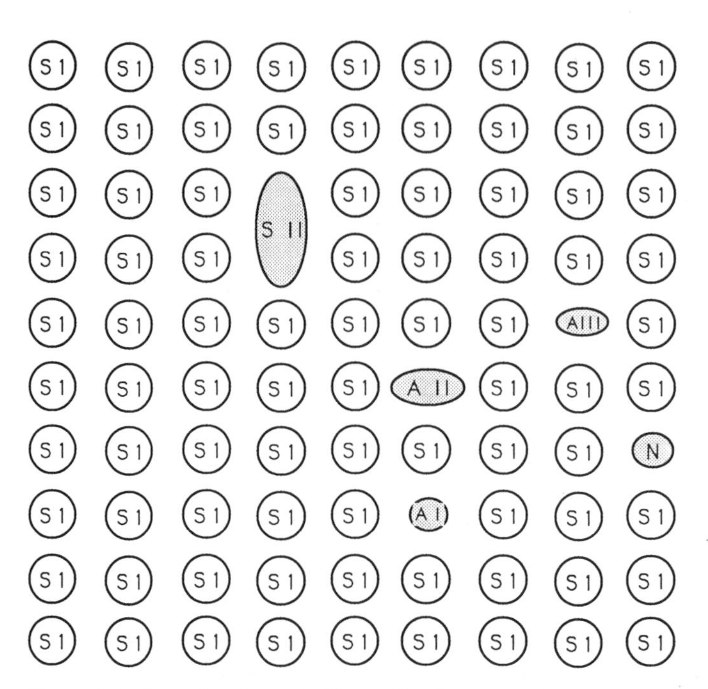

Figure 4.1 An early schematic illustrating the preponderance of the common spiny type 1 neuron (S1) in the mammalian neostriatum with a relative minority of other cell types including the spiny type 2 (S11) and the aspiny interneurons (A1, A11 and A111). It is uncertain whether the neurogliform cell (N) is neuron or glia.

1972; Graybiel and Ragsdale, 1978), in contrast to the then prevalent view that the neostriatum was a homogeneous collection of small to medium-sized interneurons recalcitrant to understanding. Other discoveries followed, including the fact that the large neurons of neostriatum were not efferent neurons but rather were interneurons containing acetylcholine or other neurotransmitters (Kimura et al., 1981; DiFiglia and Aronin, 1982)

These pioneering studies provided a basic framework for an explosion of experimental studies during the next decades. During this fruitful period of research, the use of an increasing array of anatomical and physiological approaches has led to a refinement in the concept of compartmental organization, the identification of major neurochemical classes of striatal neurons, and the emergence of new conceptions regarding the ordering of striatal afferent and efferent systems. Currently, the neostriatum is conceived of in terms of macroscopically organized parallel channels that participate in processing information from functionally related cortical and thalamic areas. Information flowing through these channels is processed in striatum, passed to basal ganglia output stations, and sent out directly to the brainstem as well as back through direct and indirect pathways formed by these output stations to distinct regions of thalamus and cortex (e.g., Alexander et al., 1986; Alexander and Crutcher, 1990).

Afferents from parietal, occipital, frontal, temporal, and limbic cortex have been found to terminate in distinct longitudinal zones within the neostriatum (Selemon and Goldman-Rakic, 1985). The majority of these inputs are to the dendrites of the common spiny output neurons, which constitute up to 95% of striatal cells in some species. These termination zones are further differentiated into small subregions which are associated with functionally related cortical areas (Yetarian and Van Hoesen, 1978; Parthasarathy et al., 1992) These observations have suggested that cortical signals from different regions are processed in parallel via separate modules within the striatum.

In addition to their specific cortical inputs and different output sites, some of these modules appear to be associated with a compartment within the striatum with a distinct neurochemistry (reviewed in Graybiel, 1990; Gerfen, 1992). Afferent inputs terminating mainly within these patchy compartments or striosomes include the projections from prefrontal cortex, the substantia nigra pars reticulata, the ventral tier of pars compacta, and from a variety of limbic structures. In contrast, inputs to the surrounding extrastriosomal regions or the "matrix" include the sensorimotor cortex, the centromedian and parafascicular nuclei of the thalamus, the dorsal tier of the substantia nigra pars compacta, and the retrorubral and ventral tegmental areas. Evidence indicates that the local dendritic and axonal fields of many spiny input-output neurons are confined within these compartments, at least within the rodent neostriatum (Penny et al., 1988; Kawaguchi et al., 1989). It has been hypothesized that certain classes of interneurons, for example the somatostatin-containing neuron, may serve to communicate between compartments (Gerfen, 1984).

While we are developing a broad view of the organization of parallel processing systems within the striatum, our understanding of the specific computations performed within the striatum on the information flowing through this structure is also evolving. Though knowledge regarding intrinsic striatal organization and its function is increasing, our information concerning the interactions between groups of striatal neurons and afferent systems is still limited. Much is still to be learned about the connectivity and processing operations of striatal neurons both within and across processing channels. It is only with this information that we will be able to understand the computational functions performed within the striatal processing modules.

In this chapter we present work from our laboratory on structural and physiological aspects of striatal processing. First, we discuss studies examining the relation between striosomal organization and the distribution of peptides and their association with cholinergic interneurons. This work has revealed additional principles of organization and interaction between neuronal systems within the striatum which refine and extend the dichotomous striosomal and matrix partitioning of the striatum. Next we examine observations from ultrastructural studies focused on the dopaminergic innervation of striatal neurons, employing serial section three-dimensional reconstruction techniques. These studies provide insight into the morphology of dopaminergic axons and postsynaptic interactions involving this important neurotransmitter. Finally, we discuss electrophysiological studies which demonstrate that, in addition to the integration of information performed by striatal neurons, an initial level of computation is performed via presynaptic interactions between striatal afferents. Our work suggests that these interactions can lead to long-term modifications at this level of processing as a result of the pattern of information arriving within afferent systems.

COMPARTMENTAL ORGANIZATION

In the early 1970s, a new conception about the intrinsic organization of the neostriatum began to emerge based on staining patterns observed in histochemical, immunocytochemical, receptor-binding, and tract-tracing studies. The consistent finding that neuroactive substances and afferent and efferent connections were distributed in discrete areas of differentially high or low density in a variety of species led to the now familiar concept of the compartmental organization of the mammalian neostriatum. In its simplest form, the neostriatum is typically divided into two compartments, a smaller patch or striosomal compartment and a larger surrounding matrix (reviewed in Graybiel, 1990; Gerfen, 1992), each possessing unique patterns of staining and afferent and efferent connections. While the presence of a compartmental organization within most species is not disputed, the nature and number of these compartments is still in question. Part of this confusion arises from the definition of a "patch." The striosomal compartment is usually defined in terms of its staining characteristics. Most striosomal markers generally coincide with patches of differentially low acetylcholinesterase (AChE) staining or increased mu opiate receptor binding, the first markers to be used to define these compartments (Graybiel and Ragsdale, 1978; Herkenham and Pert, 1981). Thus the patch compartment is considered to stain intensely for the peptides enkephalin and substance P and poorly for tyrosine hydroxylase, somatostatin, or calbindin. However, all of these markers exhibit regional differences within a species. For example, calbindin-poor patches are poorly defined within the dorsal-lateral rat neostriatum (Gerfen et al., 1985; Kawaguchi et al., 1989) while substance P–rich patches are most clearly visible in this area in the cat (Beckstead and Kersey, 1985; Malach & Graybiel, 1986; Martone et al., 1993). The staining characteristics of a putative striosome can also change across the dorsal-ventral extent of the caudate nucleus (Graybiel et al., 1981; Martin et al., 1991b), across species (Johnson et al., 1990) or as a result of technical factors (Graybiel and Chesselet, 1984; Bolam et al., 1988).

The intra- and interspecies variability exhibited by various patch markers raises the question of whether the patch compartment represents a single compartment or is perhaps represented by two or more separate or partially overlapping compartments. Our laboratory has employed computer-assisted three-dimensional reconstruction to examine the organization of peptide-immunoreactive patches in the cat neostriatum. We have shown that the patches of intense enkephalin-like immunoreactivity, generally considered to represent the striosomal compartment (Graybiel et al., 1981), form a highly organized, three-dimensional lattice extending throughout the caudate nucleus of the cat (Groves et al., 1988). A recent study on the three-dimensional organization of the compartmental markers enkephalin and calbindin in the human striatum revealed a similar latticelike structure (Manley et al., 1991). In both the cat and the human, a remarkable similarity in the structure of the network was seen from brain to brain, despite differences in age, sex, sectioning, and staining (figure 4.2).

Although there was a high degree of connectivity among the individual patches forming the network, it was not clear in these and other studies (Desban et al., 1989) whether the patches were contained within a single network. The presence of discontinuous patch networks raises the question of whether some of the regional variability observed in the striosomal compartment reflects the presence of different systems, each with a unique neurochemical and afferent character. The regional differentiation is seen most dramatically in the cat and primate neostriatum where patches in the dorsal and ventral caudate nucleus can show opposite staining patterns. For example, in the cat neostriatum, enkephalin-rich patches line up with substance P–rich patches in the dorsal caudate nucleus and substance P–poor patches in the ventral caudate nucleus (Graybiel et al., 1981). In the primate, we and others (Martin et al., 1991b; Manley et al., 1991) have observed that both enkephalin and substance P show similar reversals along the dorsal-ventral axis. In more dorsal regions, substance P– and enkephalin-rich zones are coextensive with calbindin- and AChE-poor striosomes, while at more ventral levels, peptide-poor zones correspond to the striosomes. Dorsal-ventral reversals have also been noted in the cat caudate nucleus for afferents from certain frontal cortical regions (Ragsdale and Graybiel, 1981; 1990) and from subregions of the substantia nigra (Beckstead, 1987; Jiménez-Castellanos and Graybiel, 1987).

In order to determine whether the cat caudate nucleus is characterized by two separate patch networks, a substance P–rich network in the dorsal caudate nucleus and a substance P–poor network in the ventral caudate nucleus, we compared the three-dimensional distribution of enkephalin-rich patches with that of substance P–rich and —poor zones

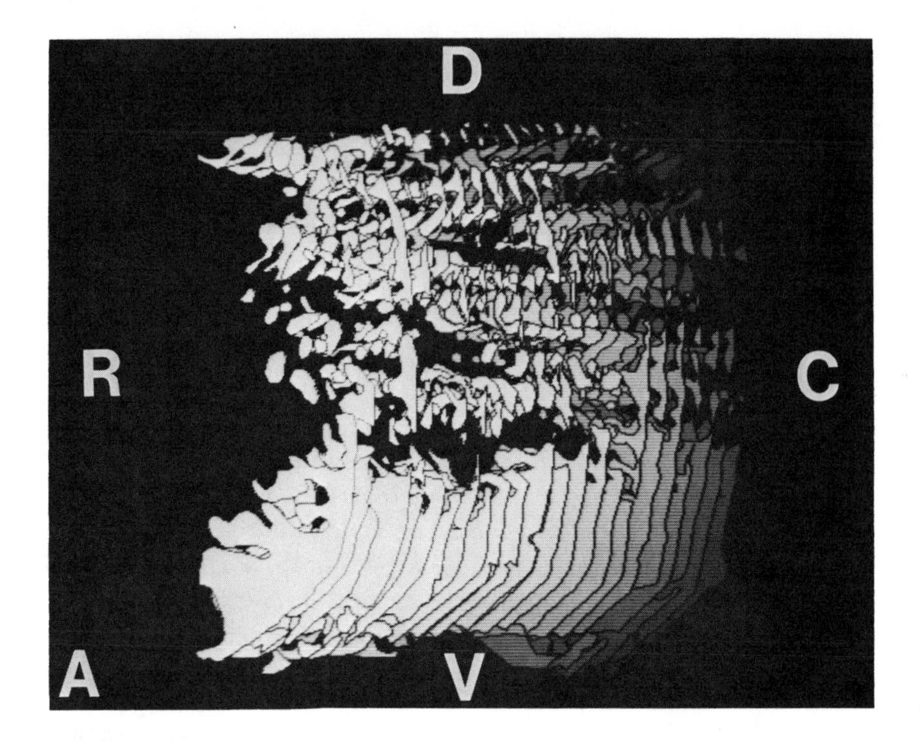

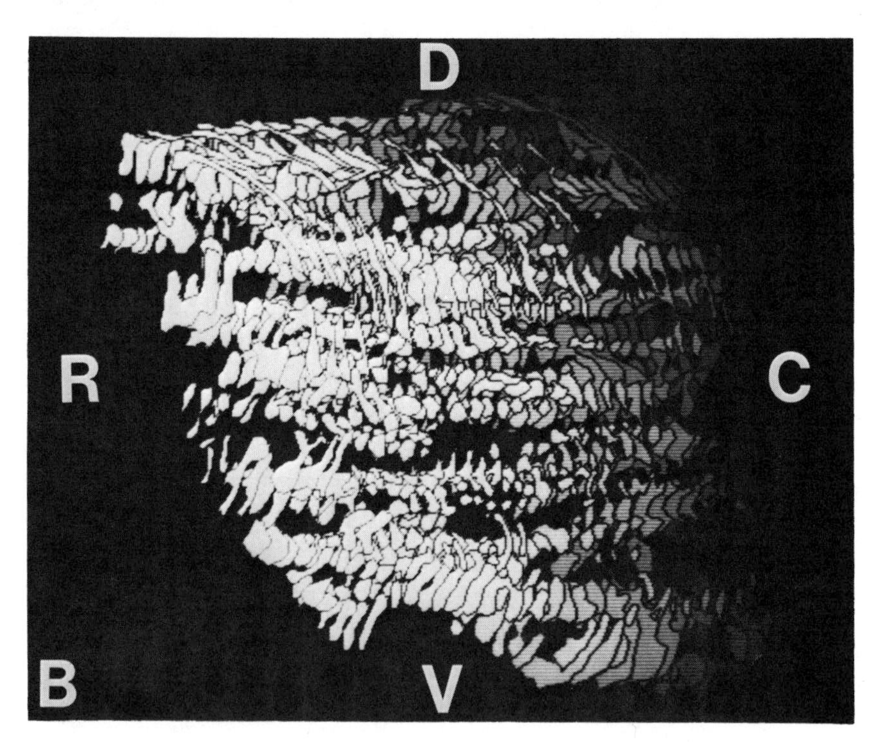

Figure 4.2 Images of normal human right caudate nuclei traced with the camera lucida, digitized and displayed using an IBM-PC–based three-dimensional reconstruction program (Young et al., 1987). D, dorsal; V, ventral; R, rostral, C, caudal. Each caudate

visualized in adjacent sections. We observed that a continuous enkephalin-rich patch network in the middle of the caudate nucleus overlapped with both substance P–rich and –poor zones. In fact, in fortuitous pairs of sections, a single enkephalin-rich patch could be seen to overlap with a substance P–poor patch medially and a substance P–rich zone laterally (figure 4.3). The transition between substance P–rich and –poor areas was not gradual but occurred abruptly. A similar pattern within an individual patch was also observed in the cat caudate nucleus by Jiménez-Castellanos and Graybiel (1989) following injections of retrograde tracers into the densicellular zone of the substantia nigra pars compacta. In that study, patches of retrogradely labeled neurons were sometimes observed to coincide with only part of a single AChE-poor patch observed in an adjacent section (see figure 6 in Jiménez-Castellanos and Graybiel, 1989).

The results of these analyses suggest that the same patch network within the cat caudate nucleus contains subregions that can be distinguished on the basis of neurochemistry and perhaps afferent connections. These transitions between rich and poor zones could reflect a differential distribution of neuronal subtypes within certain patches or perhaps the influence of an afferent system on peptide expression. For example, numerous studies have demonstrated that dopamine agonists and antagonists exert opposite effects on the amount of substance P and enkephalin expressed by striatal neurons (reviewed in Gerfen, 1992). Large injections of anterograde tracers into the cat substantia nigra pars compacta result in patches of labeling that coincide with substance P–rich patches dorsally and leave fields with gaps of labeling ventrally (Beckstead, 1987). Since dopamine is thought to increase levels of substance P expression in striatal cells (reviewed in Gerfen, 1992), substance P–rich areas within the striatum may be characterized by increased dopaminergic influence.

These studies do not answer the question of whether a single striosomal compartment exists or whether functionally distinct networks are present which may overlap in certain regions of the caudate nucleus. For example, there may be an enkephalin-rich network and a substance P–rich network that are coextensive in the dorsal caudate nucleus but are out of register at more ventral regions. The resolution of this issue will require additional studies on the anatomical basis of the patch compartment. Although the patch compartment exhibits neurochemical

nucleus is displayed as though the viewer were sitting in the lateral ventricle, looking at the ventricular edge of the nucleus. *A*, Outlines of intense leucine-enkephalin immunoreactivity from 24 coronal sections taken from the head of the nucleus. The large contours at the ventral limit of the caudate represent matrix enkephalin immunoreactivity. *B*, Calbindin-D–poor patches from a different individual. This sample of caudate includes a portion of the body of the caudate as well as the head.

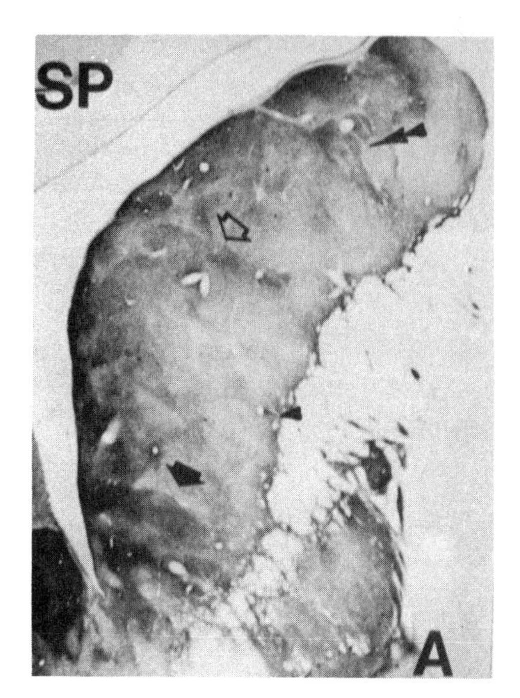

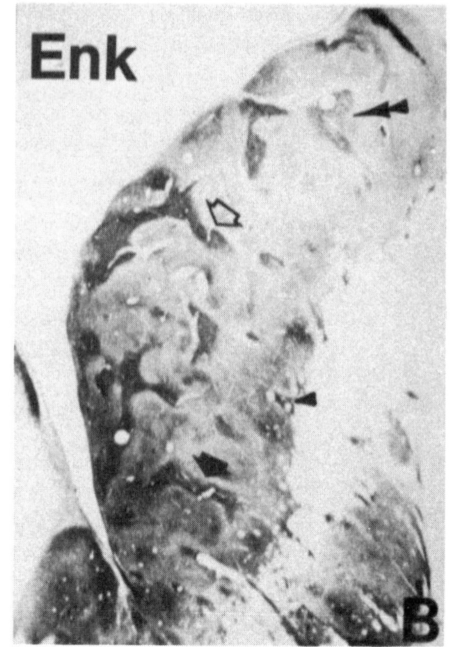

Figure 4.3 Adjacent coronal sections labeled for (*A*) substance P (SP) and (*B*) enkephalin (Enk) showing the correspondence between the patchy distributions of these two peptides in the cat caudate nucleus. In dorsal and lateral regions of the caudate, there is a general correspondence between substance P–rich and enkephalin-rich patches (double arrowheads and small arrowheads). In the ventral-medial caudate nucleus, substance P–poor patches are seen which correspond to enkephalin-rich patches (thick arrows). At mid-dorsal ventral levels, a single enkephalin patch is present (open arrow) which overlaps with a substance P–poor patch medially and a substance P–rich patch laterally.

differentiation, are there defining characteristics which identify patches throughout the neostriatum? Early studies in the rodent neostriatum suggested that patches could be distinguished on the basis of their input from prelimbic cortex (Gerfen, 1984; Donoghue and Herkenham, 1986). Within the cat neostriatum, however, prefrontal cortex projects to striosomes in some regions of the caudate nucleus and to the matrix in others (Ragsdale and Graybiel, 1981, 1990). A recent study in the rat found that it is the cortical layer from which projections arise rather than the cortical area that relates to whether cortical afferents terminate within patch or matrix (Gerfen, 1989). Patches receive input from deep layer V from cortical areas which project to the particular striatal region while complementary matrix areas receive input from superficial layer V and layer III. Thus, in the dorsal rat neostriatum, the patch compartment may be defined across its extent by projections from deep layer V. Additional features which argue for a single patch compartment in the rat are the demonstration that neurons which come to lie within the patch compartment are born at the same time (as determined by thymidine labeling) (van der Kooy and Fishell, 1987) and that they are partitioned off from the matrix in developing rats by glycoconjugate-rich boundaries (Steindler et al., 1988; O'Brien et al., 1992).

It is not known whether a patch compartment exists within cats and primates that can be identified by a single anatomical criterion. The laminar organization of cortical projections to patch and matrix compartments has not yet been examined in these species. Clusters of neurons within the developing cat caudate nucleus that come to lie in the striosomal compartment in the adult are born concurrently, suggesting that the patch neurons share a common birth date, similar to the rat (Graybiel and Hickey, 1982; Graybiel, 1984). however, Graybiel and Hickey (1982) noted that thymidine-labeled cell clusters were rare in the medial part of the caudate nucleus, except in animals injected at later embryonic stages. As seen in figure 4.3, this region is characterized by substance P–poor zones in the adult. If neurons that will come to lie in the substance P–poor zones exhibit a different developmental time course than patch neurons in more dorsal and lateral regions, it may warrant their consideration as a distinct compartment. Clearly, the relationship of substance P–rich and –poor patches to developing neurons and nigral and cortical afferents will need to be investigated more directly before the nature of these zones is understood.

Although the complex staining patterns of substance P and enkephalin within the caudate nucleus of the cat may be difficult to relate to a dichotomous striosome-matrix organization, the patterns of staining may be useful in examining the intrinsic organization of the neostriatum. The majority of substance P and enkephalin staining within the neostriatum is believed to arise from subpopulations of spiny neurons that contain one or both of these peptides (Penny et al., 1986; Gerfen and Young, 1988; Izzo et al., 1987; Besson et al., 1990). Thus, the relationship between heterogeneities in these peptides and other striatal systems may yield clues to functional interactions within the neostriatum. In a series of studies, we examined the organization of cholinergic interneurons, as identified by immunolabeling for choline acetyltransferase (ChAT) in the cat neostriatum, and how their distribution related to enkephalin and substance P patches (Martone et al., 1991, 1993). We first examined the three-dimensional distribution of cholinergic perikarya within the neostriatum. These reconstructions showed that ChAT+ neurons did not appear to be evenly distributed throughout the cat caudate nucleus but were distributed in loose clusters. In some cases, the size and shape of these clusters were similar in size and shape to enkephalin-rich patch networks reconstructed from adjacent sections. Double-labeling studies indicated that a higher density of ChAT+ cells was found within the enkephalin-rich patches than in the matrix, but only within the dorsal-lateral sector of the caudate nucleus. At more ventral regions, no difference in density between patch and matrix was observed. Since substance P is also concentrated within patches in the dorsal-lateral caudate nucleus, we then compared the distribution of ChAT+ cells to the distribution of substance P staining using both three-dimensional reconstruction techniques and double-label immunocytochemistry (Martone et al., 1993). Consistent with the previous results, we found that increased densities of cholinergic neurons were associated with substance P–rich patches in the dorsal-lateral caudate nucleus. However, in addition, increased densities of ChAT+ neurons appeared to be associated with substance P–rich areas throughout the caudate nucleus and ventral striatum (figure 4.4).

Although these studies indicated that the density of ChAT+ neurons was increased in the dorsal striosomal compartment compared to matrix, it was not clear that these neurons were specifically associated with the striosomal compartment in this area. Unlike clusters of substance P– and dynorphin-containing neurons seen in this same region of the cat caudate nucleus (Bolam et al., 1988; Besson et al., 1990), these neurons did not appear to be constrained by patch borders but were rather loosely clustered in the vicinity of a patch. This lack of a strong association between patch borders and ChAT+ cells, along with the observation that ChAT+ neurons were found in substance P–rich areas regardless of whether they corresponded to patch or matrix, led us to conclude that the presence of cholinergic perikarya in areas exhibiting intense substance P immunoreactivity likely reflected a strong anatomical relationship of ChAT+ neurons with substance P fibers rather than with the striosome or matrix compartment. Cholinergic neurons have been shown in ultrastructural studies to be contacted by substance P–containing boutons in the rat neostriatum (Bolam et al., 1986; Martone et al., 1992). Cholinergic neurons appear to be the only striatal cell class that expresses detectable levels of messenger RNA (mRNA) for the tachykinin receptor (Gerfen, 1991) and also exhibit the strongest staining for the substance P receptor protein (Kowall et al., 1993). Pharmacological evidence has also shown that tachykinins, including substance P, can cause release of acetylcholine from striatal slices (Arenas et al., 1991).

The light microscopic analyses suggested that cholinergic neurons were more strongly associated with substance P–rich areas than with enkephalin-rich regions in the cat caudate nucleus. In order to determine whether ChAT+ neurons also receive synaptic input from enkephalin terminals, we examined the ultrastructural relationship between enkephalin- and substance P–containing terminals and cholinergic perikarya in the rat neostriatum (Martone et al., 1992). Both enkephalin+ and substance P+ synapses onto cholinergic perikarya and dendrites were observed. However, while we were able to demonstrate numerous substance P+ synapses onto ChAT+ cell bodies and den-

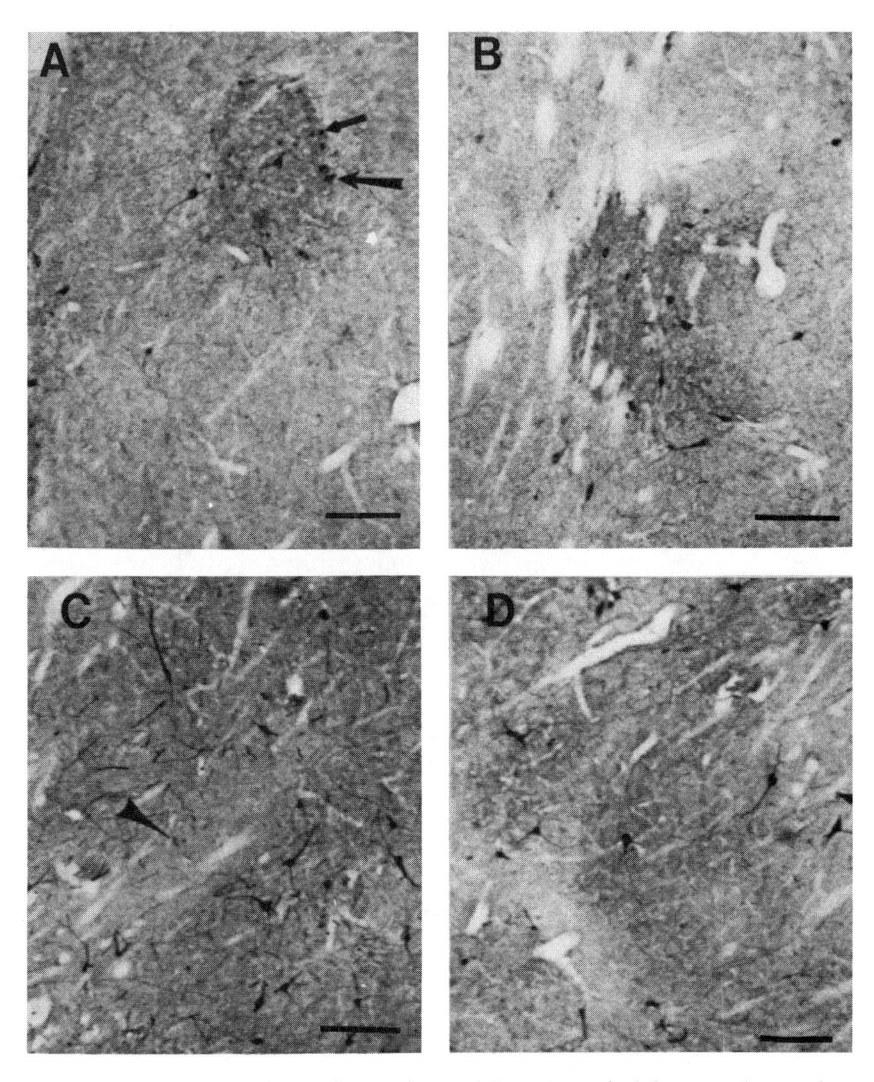

Figure 4.4 Dorsal (*A* and *B*) and ventral (*C* and *D*) regions of adult cat caudate nucleus double-labeled for choline acetyltransferase (ChAT) and substance P. ChAT-like immunoreactivity was localized to a small population of large cells (long large arrow in *A*) and dendrites, while substance P labeling was seen in clusters of small cells (short small arrow in *A*) in the dorsal caudate nucleus and fine-caliber puncta and fibers in the ventral caudate nucleus. ChAT+ cells tended to lie within substance P-rich patches in the dorsal caudate nucleus (*A* and *B*) and to avoid substance P–poor regions (arrowhead in *C*) in the ventral caudate nucleus. Note the high concentration of ChAT+ dendrites outside of the substance P–poor zones in *C* and *D*. Data and details of immunolabeling protocol are given in Martone et al., 1993. Scale bars: 200 μm.

drites, we found relatively few examples of enkephalin+ synapses onto ChAT-labeled profiles. These results were consistent with a study by Chang et al. (1987) in the basal forebrain reporting few enkephalin+ synapses on basal forebrain cholinergic neurons.

Although quantitative conclusions cannot be drawn from this study, the results suggest that perhaps ChAT+ neurons receive a relatively stronger synaptic input from substance P–containing striatal neurons compared to enkephalin-containing neurons. These results are of interest given that enkephalin- and substance P–containing spiny neurons appear to differ in their efferent projections. Enkephalin-containing medium spiny neurons project mainly to the globus pallidus while substance P–containing neurons project mainly to the entopeduncular nucleus and substantia nigra (Gerfen and Young, 1988; Flaherty and Graybiel, 1993). Medium spiny neurons also possess an extensive local axonal plexus that arborizes within the neostriatum. The differences observed in the electron microscopic study between the interaction of enkephalin and substance P boutons with cholinergic neurons suggest that neurons containing these peptides may participate to different degrees in intrastriatal circuits in addition to their distinct efferent projections.

The results of studies from our laboratory and many others are consistent with the view of multiple levels of organization within the neostriatum (Alexander et al., 1986; Gerfen, 1992). At least one level of organization is represented by the presence of one or more patch compartments. Another level is represented by differential extrinsic and perhaps intrinsic connections of subpopulations of striatal cells. Further work is necessary not merely to catalogue what is inside and outside of the patches but to determine whether the differential distribution of various striatal systems reflects fundamental differences in information processing between compartments.

MORPHOLOGY AND INTERACTIONS OF DOPAMINERGIC AFFERENTS

The differential control of neuropeptide expression in substance P– and enkephalin-containing spiny neurons discussed above represents one important example of dopaminergic control of neostriatal information processing and function. We have been investigating the anatomical basis for dopaminergic actions in the striatum by examining the ultrastructure of the dopamine axons, their synapses, and the local postsynaptic environment in which they occur. The morphology and extent of the nigrostriatal dopamine projection has been examined using many histological techniques, including catecholamine histofluorescence (Carlsson et al., 1962; Andén et al., 1966; Fuxe, 1965; Arluison et al., 1982) uptake of 5-hydroxydopamine (5-OHDA) (Tennyson et al., 1974;

Arluison et al., 1978; Groves, 1980), autoradiography of labeled amino acids (Hattori et al., 1973, 1991) or dopamine (Doucet et al., 1986), anterograde transport of *Phaseolus vulgaris*–leucoagglutinin (PHA-L) (Gerfen and Sawchenko, 1985), and immunocytochemistry for tyrosine hydroxylase (Pickel et al., 1981; Arluison et al., 1984; Bouyer et al., 1984a; Freund et al., 1984) and dopamine (Voorn et al., 1986). Some of the early studies led to the conclusion that dopamine axons had conspicuous varicosities that were associated either with asymmetric synapses or a lack of defined synaptic specializations. In contrast, the more recent, mostly immunocytochemical work indicates that dopamine afferents to the neostriatum generally lack large varicosities and make tiny, symmetric synapses with spines and distal dendrites. We have recently attempted to resolve these opposing views of dopaminergic axonal and synaptic morphology in a reexamination of 5-OHDA-labeled rat striatum using three-dimensional serial reconstruction techniques (Groves et al., 1994). Our analysis generally confirms the conclusions of recent immunocytochemical studies and extends our knowledge of the various types of interactions involving dopamine afferents.

Striatal axons labeled with 5-OHDA are small (0.06–1.5 μm in diameter). Variations in diameter are generally minor and occur at irregular intervals. Synaptic sites are located along the axon (en passant) rather than at terminal boutons. The locations of synapses are not correlated with dilated portions of the axon. Dilations are more commonly associated with the presence of mitochondria. Dopamine axon diameters measured at 50 synapses ranged from 0.1 to 0.77 μm with a mean of 0.26 μm. Of these, the larger axon diameters were more likely to be associated with synapses onto dendritic shafts than those onto spine heads or necks. These characteristics are generally in agreement with those based on axons immunolabeled for tyrosine hydroxylase, which also show no association of synapses with regions of large diameter (Pickel et al., 1981; Freund et al., 1984). The lack of correlation between dopaminergic varicosities and synaptic sites is significant because there is a persisting view that dopamine synapses are associated with varicosities and that therefore the number and location of synapses can be determined by identifying varicosities (e.g., Doucet et al., 1986). Unfortunately, an accurate estimate of the extent of the dopaminergic innervation will not be so easy to obtain. We have some measures of dopamine innervation, but these are based on very small volumes of tissue and must still be considered preliminary. For one 230-μm^3 block cut into a series of 70 sections, we measured the total volume of 5-OHDA-labeled dopamine axons and found that they occupied 6.6 μm^3, or about 3% of the total volume. For three other series, we analyzed all labeled and unlabeled synapses, and found that of 381 synapses, 9% were dopaminergic.

Dopamine synapses are very small, usually appearing in only two to three sections. The delicate symmetric synaptic specializations average 0.039 μm^2. The area of the synaptic specialization varies with the type of postsynaptic target: synapses onto dendritic shafts are larger than those onto spine heads or necks. Synaptic vesticles are relatively large, sparse, and clustered near the contact zone. Our description of dopamine synapses is very similar to a recent characterization of cortical dopamine synapses (Smiley and Goldman-Rakic, 1993). The tiny dimensions of the dopamine synapses and their relatively small numbers have probably contributed to the difficulties with their identification over the years. They are very hard to recognize without examining serial sections. Additionally, all labeling methods with any significant background will sometimes mistakenly mark the much more common nondopaminergic boutons, which are large and asymmetric. It is not surprising that the identity of the dopamine synapse has been controversial for so long, since they are so much harder to recognize than those larger and more frequently encountered nondopaminergic synapses.

The features of dopamine synapses are in complete contrast to the most common type of unlabeled synapse in the striatum which is large, asymmetric, and contains small round vesicles. This class of synapse composed 72% of our sample of 381 labeled and unlabeled synapses. Most of these synapses contacted spine heads (67% of our entire sample), the majority of which probably represent glutamatergic corticostriatal afferents (Frotscher et al., 1981; Somogyi et al., 1981; Dubé et al., 1988). Some of these asymmetric spine synapses may also arise from medial geniculate thalamic projections (Moriizumi and Hattori, 1992), but in rats the parafascicular nucleus of the thalamus seems to send similar asymmetric synapses only to dendritic shafts of both cholinergic cells and a class of less spiny, medium spiny cells (Dubé et al., 1988; Lapper and Bolam, 1992). In our sample, asymmetric shaft synapses composed 5% of the total. In other species (Kemp and Powell, 1971; Chung et al., 1977; Sadikot et al., 1992), a higher proportion of thalamic afferents may synapse with spines, with the ratio of spine to shaft synapses depending on which particular thalamic nucleus is the source of the projection.

Unlabeled symmetric synapses in our study usually contacted spine necks or dendritic shafts. These synapses could be divided into two classes. One type (7%–8%) possessed large round vesicles, resembling γ-aminobutyric acid (GABA)-containing and peptide-containing local axon collaterals of the neostriatal medium spiny cells (Wilson and Groves, 1980; Somogyi et al., 1981; DiFiglia et al., 1982; Bolam et al., 1983; Bolam and Izzo, 1988). The other class (7%–8%) contained small pleomorphic vesicles. A small percentage of synapses in our sample could not be definitively categorized. Synapses of serotonergic afferents from the dorsal raphe (Calas et al., 1976; Arluison and De La Manche, 1980; Pasik and Pasik, 1982; Pasik et al., 1984; Soghomonian et al., 1989) and synapses from several types of striatal interneurons (Kita et al., 1990; Takagi et al., 1983; 1984; Phelps et al., 1985; Vuillet et al., 1989; DiFiglia and Aronin, 1982; Bennett and Bolam, 1993) also contribute to the neostriatal innervation, but their ultrastructural identities are not yet well established.

Three-dimensional reconstructions of labeled dopamine axons and some of their postsynaptic targets revealed important features relating to possible pre- and postsynaptic interactions. It was clear from the three-dimensional reconstructions that most of the postsynaptic dendrites were portions of spiny cells. These cells are probably the medium-sized common spiny neurons that account for about 95% of the neurons in the rat neostriatum. The spiny neurons have been shown by Freund et al. (1984) to receive tyrosine hydroxylase–labeled synapses. In agreement with previous reports (Freund et al., 1984; Triarhou et al., 1988), the most common postsynaptic target of our 5-OHDA-labeled dopamine synapses (56%) was the head or neck of a dendritic spine, always with a second, unlabeled, asymmetric synapse located on the same spine head (figure 4.5A). Almost all other dopamine synapses (42%) were found along small dendritic shafts, often near the base of spines. In our study, a few contacts were also observed in the vicinity of branch points of dendrites where spines were not present even when followed in long series. These could be proximal spine-free portions of medium spiny cell dendrites (Wilson et al., 1983), but their relatively small diameter and the extent of the aspiny region suggest that they represent one of the aspiny cell types of the neostriatum.

A single short dopaminergic axonal segment can synapse with multiple dendrites. The 10-μm portion of labeled axon shown in a reconstruction including only two of its postsynaptic targets in figure 4.5B actually makes six synapses with six different dendrites. It contacts a spine head, two spine necks, a spiny dendritic shaft, and branch points of two aspiny dendritic segments. Other labeled axons were found to make multiple synapses with the same dendrite, and many dendrites receive synapses from multiple dopamine axonal segments. It is also clear that dopamine axons may remain in close, but not necessarily synaptic, contact with dendrites, spine necks, unlabeled terminals, and other dopamine axons. Other researchers have observed this type of nonsynaptic apposition between dopamine and cortical terminals (Bouyer et al., 1984b) and have recently reported extensive nonsynaptic contacts between dopamine axons and cell bodies of the large striatal cholinergic neurons (Dimova et al., 1993).

Dopaminergic synaptic contacts appeared to be distributed nonuniformly even within a relatively small volume of neuropil as suggested

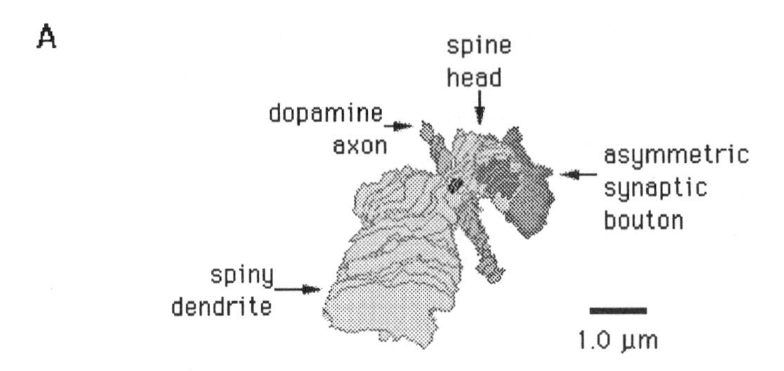

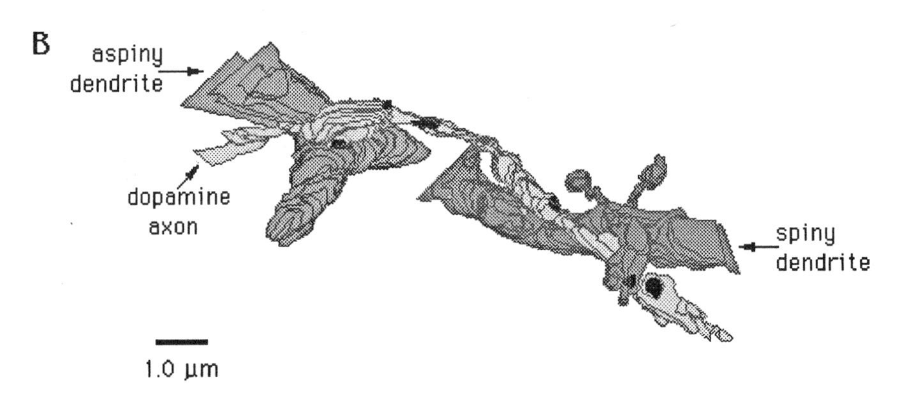

Figure 4.5 Dopamine axons illustrated with computer-assisted three-dimensional reconstructions from serial electron micrographs. *A,* A 5-hydroxydopamine-labeled dopamine axon makes a symmetric synapse with a spiny dendrite on the neck of a spine (synapse in black). A second synapse (large, asymmetric, and unlabeled) occurs more distally on the spine head (dark gray). This unlabeled synapse most closely resembles those made by excitatory glutamatergic cortical boutons. *B,* A 10-µm-long segment of dopamine axon shows slight variations in diameter uncorrelated to the locations of its six synapses onto six different dendrites (synapses are marked in *black*). Two of the postsynaptic dendrites are also reconstructed: a spiny dendrite (with a dopamine synapse on one of the spine necks) and an aspiny dendritic segment (with a dopamine synapse at the branch point). (Data from Groves et al., 1994.)

by the reconstruction (figure 4.6) of all 27 labeled axon segments contained within one series. Labeled axons appeared to form many synapses in one region, but none as they traversed an adjacent area that appeared to contain potential synaptic sites. This distribution suggests that dopaminergic input might be directed to specific dendritic processes or neurons, or be associated with a larger pattern of striatal inhomogeneity such as the striosomal compartment (Gerfen et al., 1987a,b; Jiménez-Castellanos and Graybiel, 1987).

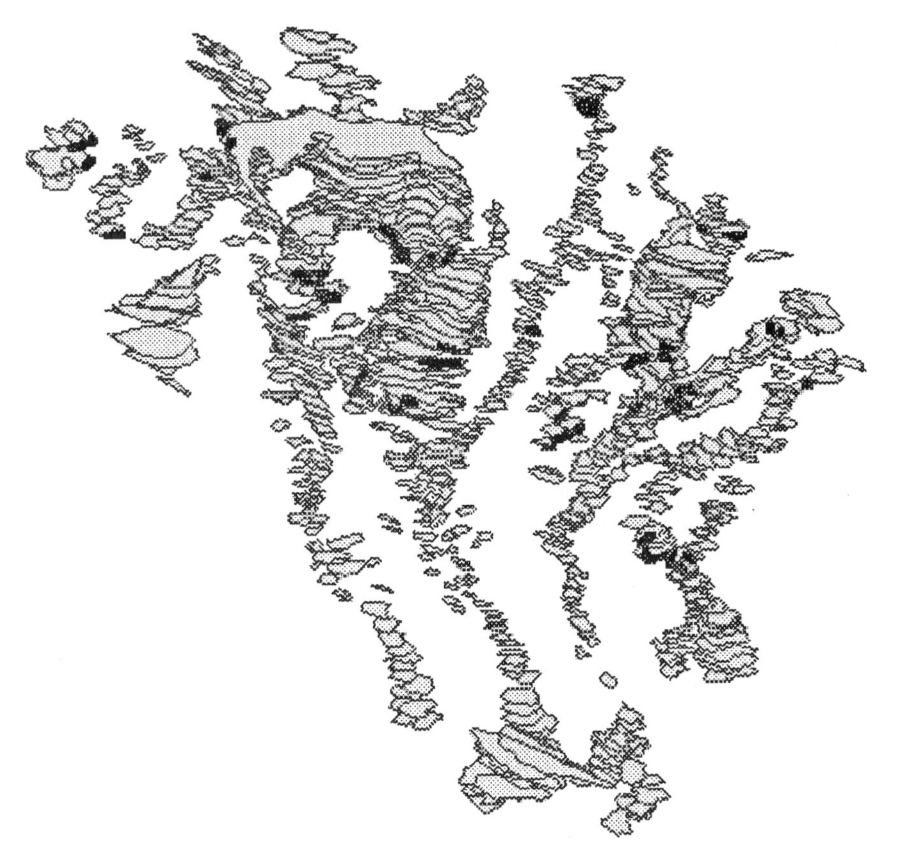

Figure 4.6 All 27 dopaminergic axon segments (gray) in this block of tissue have been reconstructed to illustrate their roughly parallel arrangement and the locations of their synapses (black), which seem to cluster in one region rather than being distributed uniformly. (Data from Groves et al., 1994.)

POSTSYNAPTIC INTERACTIONS INVOLVING DOPAMINE AND OTHER AFFERENTS

As previously described by Freund et al. (1984), spines with labeled symmetric synapses always possessed an unlabeled asymmetric synapse located on the spine head. Most spines were not contacted by labeled synapses, though in many cases labeled or unlabeled symmetric synapses were located near the base of the spine on the dendritic shaft. Asymmetric synapses are invariably found on the spine heads of the most common type of neostriatal neuron (Wilson and Groves, 1980; Wilson et al., 1983). These synapses are most likely made by the excitatory afferents from cortex or thalamus (Kemp and Powell, 1971; Bouyer et al., 1984b). In our previous studies, analyses of images of the spiny neurons of rat neostriatum revealed that the density of spines varies between one and four spines per micron of dendritic length (figure 4.7A). These studies indicate that there are 300 to 500 spines per dendrite and between 25 and 30 dendrites per neuron in the rat, and similar estimates have been made for other species. Thus, a single spiny neuron receives around 10,000 excitatory inputs via its spines (Wilson and Groves, 1980; Wilson et al., 1983; Wilson, 1990). It has been estimated that spiny neurons in the neostriatum receive over 80% of their synaptic input through the dendritic spines (Groves, 1980). Wilson (1984, 1992) has modeled the electrotonic properties resulting from this configuration of synaptic input to common neostriatal spiny neurons using parameters derived from morphological measurements of spines and their distribution on these neurons, as well as a consideration of the effects of the marked anomalous rectification observed in these cells. The model suggests that when a small number of excitatory synapses are active, the neuron is relatively hyperpolarized, and excitatory events produce small somatic signals due to the low input resistance of the neuron as a consequence of the combined effect of the anomalous rectifier and the large surface area afforded by the spiny dendrite. A sufficient increase in synchronous excitatory activity can depolarize the membrane leading to a decrease in the action of the anomalous rectifier. Under this condition, neuronal input resistance increases and dendritic synaptic events produce larger somatic signals. Thus under low levels of uncorrelated input, the neuron remains in a polarized state. Sufficient correlated input leads to a positive feedback effect tending to produce an abrupt transition to a depolarized state.

The synaptic arrangement observed in the neostriatum, in which an excitatory synapse is present on the spine head with labeled and unlabeled symmetric contacts in the immediate vicinity, resembles configurations seen in many other brain regions (Shepherd, 1990) in which presumed inhibitory synapses appear located to strongly influence excitatory signals by hyperpolarizing the dendritic membrane or increasing its membrane conductance to shunt excitatory currents. In addition to its role as the striatal output neuron, the common spiny neuron ramifies an axon collateral field within the striatum which is approximately coextensive with its dendritic field, as illustrated in figure 4.7. These collaterals make symmetric synapses onto other spiny neurons, as illustrated in figure 4.8. The neurotransmitter released at these synapses is believed to be GABA, and immunocytochemical studies (Ribak et al., 1980) have revealed that these GABA synapses are similar in morphology to those labeled by intracellular injection of horseradish peroxidase into medium spiny cells. The discovery that spiny neurons make symmetric synapses onto each other, including synapses on the axon initial segments, led to the conception that the neostriatum is a large lateral inhibitory network. Several models of striatal processing have suggested that this network serves to filter spatial and temporal patterns of input (e.g., Groves, 1983; Groves et al., 1988; Wickens et al.,

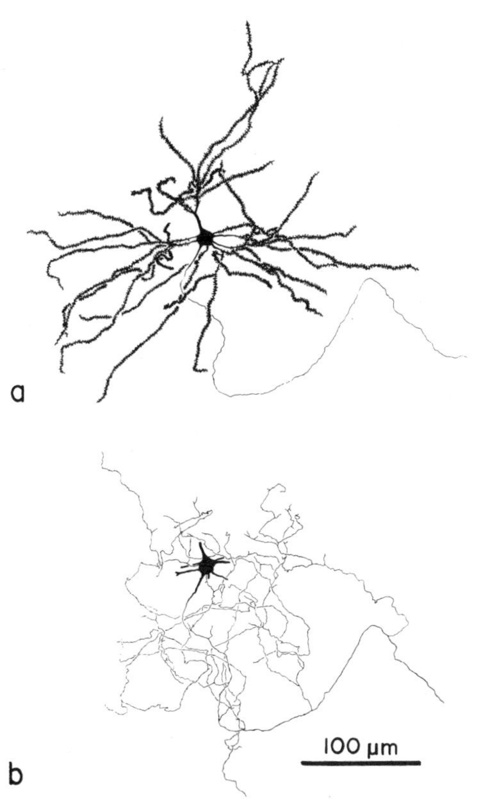

Figure 4.7 Camera lucida drawings of the common spiny neuron (spiny type I neuron) of neostriatum. *a,* The spiny neuron is drawn to illustrate the very heavy investment of dendritic spines on its dendrites, beginning at 20 μm from the cell body and continuing throughout the entire length of the dendrite. *b,* The extensive axon collaterals of this neuron are illustrated without the dendrites in order to reveal their extensive length and circuitous path within the domain of the dendritic field. (From Wilson and Groves, 1980.)

1991). Physiological evidence for a recurrent inhibitory effect has proved elusive. While a short-acting GABA inhibition involving a chloride conductance has been observed in striatal neurons (Park et al., 1980; Misgeld et al., 1982; Lighthall and Kitai, 1983; Kita et al., 1985 Mercuri et al., 1991), it is not clear at the present time whether this inhibition represents the action of recurrent collaterals or an effect due to a GABA interneuron (Wilson et al., 1989).

The function of dopamine input on the responses of striatal neurons has also proved to be frustratingly difficult to establish. In anesthetized and immobilized animals, dopaminergic agonists decrease the firing rate of most neurons (Groves et al., 1975; Siggins, 1978); however, some of these studies and experiments on freely moving animals have given more varied results (e.g., Trulson and Jacobs, 1979; Rolls et al., 1984; Haracz et al., 1989; Ryan et al., 1989; Williams and Millar, 1990). Recent studies of neostriatal neurons employing intracellular slice recording

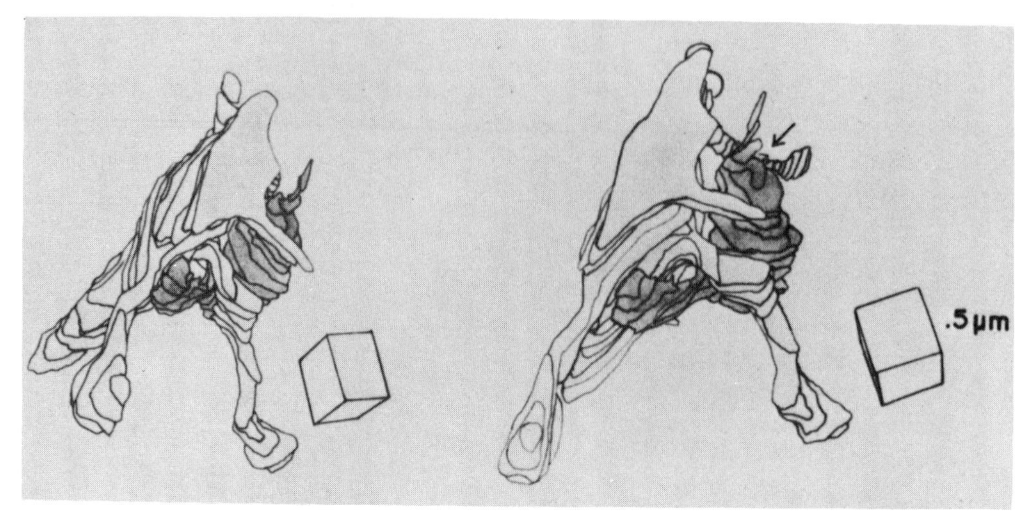

Figure 4.8 Part of a local axon collateral (dark gray) of a common spiny neuron filled with horseradish peroxidase has been reconstructed from serial electron micrographs to illustrate its association with a dendrite of a second, unlabeled common spiny neuron. Two different rotations are presented that show one synapse onto the dendritic shaft near the base of a dendritic spine and a second onto a spine neck (arrow). (From Wilson and Groves, 1980.)

techniques have established that dopamine does not produce the changes in resting membrane potential and resistance typically associated with a conventional inhibitory action (Akaike et al., 1987; Calabresi et al., 1987, 1988; Rutherford et al., 1988; Garcia-Munoz et al., 1989). Instead, these experiments have revealed that dopamine alters the voltage gating and inactivation characteristics of several ion currents which are significant over different ranges of membrane potential in the striatal neuron (Calabresi et al., 1988; Rutherford et al., 1988; Surmeier et al., 1992; Kitai and Surmeier, 1993). These studies are beginning to reveal that dopamine input may produce different effects which depend on the membrane potential range in which the neuron is operating.

The spine may form a relatively isolated chemical compartment thereby localizing the effects of dopamine or other inputs on voltage-sensitive mechanisms (Koch et al., 1992; Koch and Zador, 1993). In support of this possibility, recent experiments using microfluoremetry (Guthrie et al., 1991; Müller and Connor, 1991, 1992) suggest that spines may serve as separate compartments for calcium accumulation. This possible function of spines suggests that excitatory afferents arriving on those spines which also have dopamine inputs could belong to a specific class subject to selective control by dopamine. However, dopaminergic input is relatively indiscriminate. The same axon may contact several locations on the same dendritic segment as well as on multiple dendritic segments. As discussed above and illustrated in figure 4.5B, a short stretch of axon can make synaptic contacts on spines, dendritic shafts, and near branch points. Some of the dendritic segments contacted by this axon may arise from the same neuron; however, considering the typical distribution of the dendrites of common spiny neurons and the aspiny character of branch contacts, it is more likely that they stem from different neurons. Dopaminergic input to dendritic shafts and branch points, which was observed to be associated with larger synaptic appositions, could influence excitatory signals arriving from entire distal dendritic branches. Thus, while it is conceivable that a signal from a dopamine neuron traveling through this short segment of axon could selectively influence excitatory input to a specific spine, it is also likely to regulate signals from more distal dendritic regions on several different neurons.

STRIATAL INTERACTIONS DEPENDENT ON THE ACTIVATION OF PRESYNAPTIC RECEPTORS

In addition to the postsynaptic interactions resulting from dopaminergic and glutamatergic signals to striatal neurons, discussed above, there is also evidence for presynaptic interactions within the striatum involving these two neurotransmitter systems. Studies have indicated that the terminal axons of these two projections possess receptors for their own transmitter (autoreceptors) which provide a substrate for feedback regulation of neurotransmitter release. There is also evidence for presynaptic glutamate receptors on dopamine axons (Leviel et al., 1990; Moghaddam et al., 1990; Wang, 1991; Westerink et al., 1992; Desce et al., 1992) and dopamine receptors on the cortical terminals (Rowlands and Roberts, 1980; Maura et al., 1989; Earle and Davies, 1991; Naudon et al., 1992; Yamamoto and Davy, 1992). Though axoaxonic synapses are extremely rare in the striatum (Kornhuber and Kornhuber, 1983), the close proximity of cortical glutamate, nigral dopamine, and other afferent terminals as they form synapses on spiny neurons (figure 4.9) provides a potential locus for nonsynaptic interactions between different classes of afferents resulting from the activation of presynaptic receptors by diffusion of transmitter from synaptic sites (Cuello, 1966; Fuxe and Agnati, 1991).

We have been employing an electrophysiological technique that provides a means for examining, in an in vivo preparation, the consequences of the activation of receptors on striatal axon terminals of a single neuron over brief time intervals. Though it is not possible at present to obtain intracellular electrophysiological recordings of events underlying presynaptic auto- and heteroreceptor stimulation in mammalian nerve terminals owing to their very small size, a method based on the measurement of the electrical excitability of the axonal terminal field may be used to infer electrophysiological events in the terminal. Excitability is evaluated by determining the stimulating current neces-

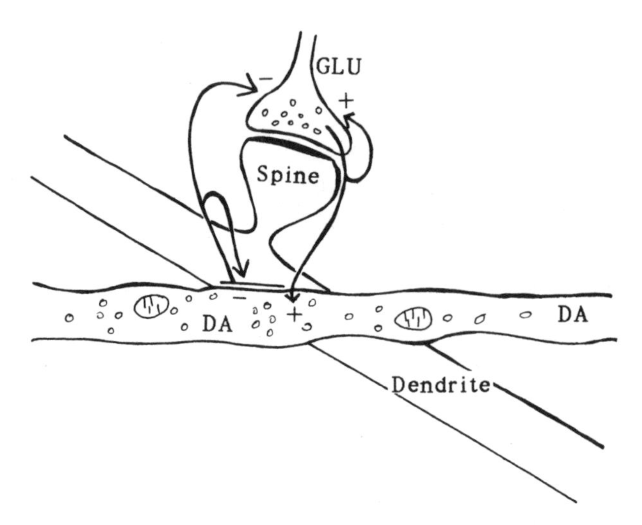

Figure 4.9 Even in the absence of striatal axoaxonic synapses, glutamatergic (GLU) and dopaminergic (DA) terminals may interact to modify excitability and transmitter release owing to the presence of presynaptic heteroreceptors and the diffusion of the neurotransmitters a small distance away from their release site. Feedback loops are also possible owing to the presence of autoreceptors on both terminals. A positive feedback is observed in glutamatergic terminal axons and a negative feedback in dopaminergic terminal fields.

sary to elicit antidromic action potentials from the terminal field of the neuron with a fixed probability over a series of antidromic stimulus presentations, or alternatively, by determining the probability of the occurrence of antidromic potentials to a fixed stimulating current (Groves et al., 1981; Tepper and Groves, 1990).

A variety of studies using in vitro and in vivo preparations have provided evidence that the activation of autoreceptors on dopaminergic terminals decreases release from these terminals (for references, see Kalsner and Westfall, 1990). Consistent with these biochemical studies, we have found that enhancement of extracellular dopamine by local striatal infusion of amphetamine produces a dose-dependent decrease in the excitability of the nigrostriatal terminal. This reduction in excitability, as well as that observed following direct-acting dopaminergic agonists, can be reversed by subsequent administration of dopamine receptor antagonists such as haloperidol or sulpiride (Tepper et al., 1984, 1985). The decrease in terminal excitability induced in these experiments does not occur following depletion of endogenous dopamine with α-methyl-*p*-tyrosine (Groves et al., 1981; Tepper et al., 1984). Local neurons do not mediate these excitability effects since they are still observed following kainic acid lesions (Tepper and Groves, 1990). When administered by themselves, autoreceptor antagonists lead to increases in terminal excitability (Groves et al., 1981; Tepper et al., 1984), suggesting that autoreceptors on these terminals are tonically activated by exogenous extracellular dopamine. A decrease in excitability is also observed following an increase in impulse activity resulting from stimulation of dopaminergic axons in the medial forebrain bundle. Importantly, terminal excitability is found to vary inversely with increases or decreases in the spontaneous firing rate of the dopamine neuron, indicating that the modulation of dopaminergic excitability, and presumably release, occurs under physiological conditions (figure 4.10A). These observations indicate that the negative feedback on the terminal resulting from an increase in impulse activity operates with a relatively short time constant (Takeuchi et al., 1982; Tepper et al., 1984). As discussed in the previous section, a dopaminergic axon makes multiple synapses within a small region. Release from these several synapses could contribute to the observed variation in excitability associated with changes in impulse traffic. In addition, the terminal could also be affected by dopaminergic overflow from nearby synapses of other nigral neurons producing a possibly uncorrelated background level of autoreceptor activation. The spatial extent over which these interactions occur is difficult to evaluate. However, the sensitivity of the neuron's terminal field to changes in its impulse activity may indicate that, under normal physiological conditions, feedback is predominantly local.

In addition to the negative feedback produced by dopamine autoreceptor activation on terminal excitability and dopamine release, there is evidence from excitability studies that striatal dopaminergic terminals may also be influenced by presynaptic interactions due to activation of heteroreceptors by neurotransmitter released from other classes of afferents forming synapses nearby. Chavez-Noriega et al. (1986) found that the excitability of the striatal dopaminergic terminal field decreased following cortical stimulation and that this effect did not occur in dopamine-depleted animals (Garcia-Munoz and Chavez-Noriega, 1986). These studies suggest that an increase in glutamate release due to cortical stimulation results in greater activation of glutamatergic heteroreceptors on dopamine terminals and thus increases dopamine release. This increase in dopamine release leads in turn to increased stimulation of dopamine autoreceptors and a decrease in excitability. In support of this interpretation, studies have shown that cortical stimulation increases the level of extracellular glutamate in the striatum (Herrera-Marschitz et al., 1990; Galli et al., 1991), and that local striatal administration of glutamate, or elevations in endogenous glutamate induced by cortical stimulation, increase the release of dopamine (Leviel et al., 1990; Moghaddam et al., 1990; Wang, 1991; Desce et al., 1992).

The level of striatal dopaminergic activity also appears to act presynaptically to affect terminal excitability and release from the glutamate-containing corticostriatal terminals. Increased extracellular levels of dopamine following local striatal administration of amphetamine or electrical stimulation of nigrostriatal dopamine-containing cells led to a decrease in excitability of the corticostriatal terminal field. These de-

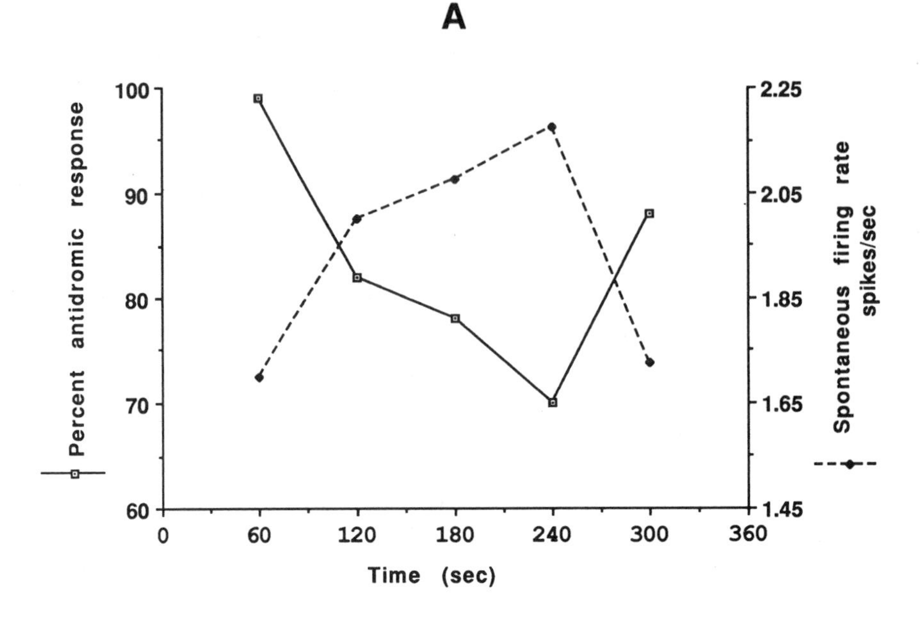

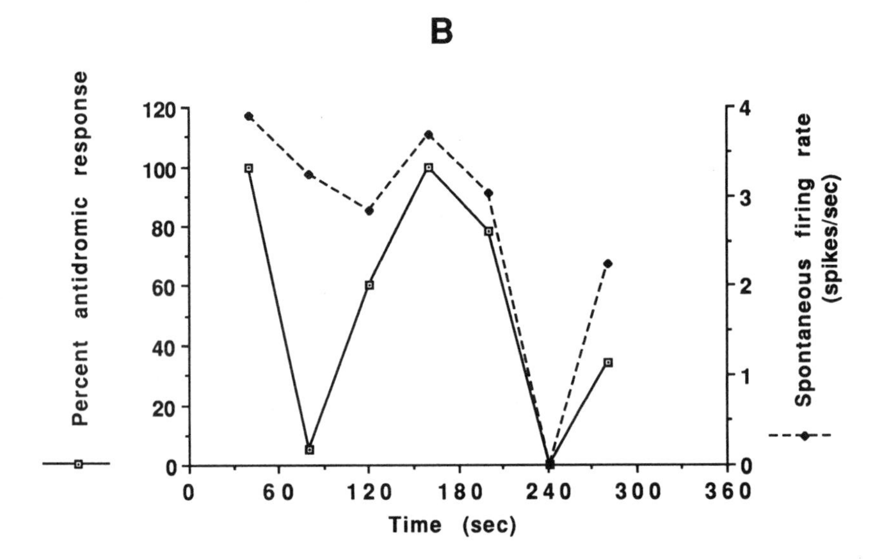

Figure 4.10 Correlation between spontaneous firing rate and the excitability of the striatal axon terminal field for (*A*) a nigrostriatal and (*B*) a corticostriatal neuron. Excitability is measured as the percentage of antidromic responses elicited by striatal stimulation with a fixed stimulating current. Note that nigrostriatal terminal field excitability decreases when spontaneous firing rate increases, whereas there is a positive correlation between excitability and spontaneous firing rate in corticostriatal neurons. (Data from Tepper et al., 1984, and Garcia-Munoz et al., 1991b.)

creases in excitability could be reversed by local infusion of haloperidol or L-sulpiride and did not occur following depletion of dopamine stores with α-methyl-*p*-tyrosine and reserpine (Garcia-Munoz et al., 1991a). An increase in excitability was observed following local striatal administration of dopamine antagonists alone, providing evidence that corticostriatal terminals are endogenously stimulated by dopamine in agreement with studies examining release from these terminals (Rowlands and Roberts, 1980; Maura et al., 1989; Earle and Davies, 1991; Naudon et al., 1992; Yamamoto and Davy, 1992). The observed modifications in corticostriatal terminal excitability to increased dopamine transmission do not appear to depend on postsynaptic actions since they were still seen in animals with kainic acid–induced lesions of the striatum (Garcia-Munoz et al., 1991a).

In contrast to the negative feedback control produced by autoreceptor regulation in dopaminergic neurons, we have found that, in corticostriatal terminals, excitability is positively correlated with variations in spontaneous firing rate; increases and decreases in firing rate are accompanied by corresponding changes in the percentage of antidromic responses elicited by a subthreshold stimulus (figure 4.10B). Local striatal administration of either the competitive *n*-methyl-D-aspartate (NMDA) antagonist D-α-aminoadipate (DAA) or D-2-amino-7-phosphonoheptanoate (AP-7), or the competitive non-NMDA antagonist 6-cyano-7-nitroquinoxaline-2,3-dione (CNQX) blocked the correlation between excitability and firing rate (Garcia-Munoz et al., 1991b). The increase in excitability seen with increased firing rate can also be observed following a single action potential. Following the arrival of an action potential, the corticostriatal terminal field is rendered more excitable for a period of 20 to 40 ms. This postimpulse facilitation of terminal excitability was attenuated after local application of AP-7 or CNQX. Doses of AP-7 or CNQX that were half the effective dose produced a nonsignificant effect, but when administered simultaneously, a significant attenuation was observed. The participation of interneurons in these excitability effects was ruled out since they were still seen following kainic acid lesions (Garcia-Munoz et al., 1991b).

The impulse-related facilitation of excitability suggests a positive feedback mechanism: an increase in firing rate increases glutamate release which in turn increases the activation of presynaptic autoreceptors to locally modify terminal membrane conductance or polarization, or both, and thereby increase electrical excitability and subsequent impulse-related release. In support of this hypothesis, there is evidence from in vitro studies that glutamate release is increased by presynaptic activation of autoreceptors with kainic acid, NMDA, and *trans*-1-amino-cyclopentyl-1,3-dicarboxylate (*t*-ACPD) (Collins et al., 1983; Ferkany and Coyle, 1983; Martin et al., 1991a). Presynaptic kainate-mediated glutamate release has been proposed as an important mechanism contributing to kainate neurotoxicity (Biziere and Coyle, 1978; McGeer et al., 1978).

These presynaptic feedback mechanisms on terminal excitability and neurotransmitter release may have important functional implications. Behavioral studies indicate that striatal dopamine receptor stimulation produces behaviors which are also induced by glutamate receptor blockade (i.e., enhanced locomotion, sniffing, rearing, and feeding) (Clark and White, 1987; Carlsson and Carlsson, 1989; Mehta and Ticku, 1990). Similarly, glutamate receptor stimulation produces behaviors that are also usually induced by dopamine receptor antagonists (akinesia, catatonia, and reduced locomotion) (Clark and White, 1987; Schmidt and Bury, 1988). Animals receiving dopamine receptor antagonists or the dopamine-depleting agents, reserpine and α-methyl-*p*-tyrosine, display virtually no locomotion. However if an NMDA-receptor antagonist (MK-801) is administered, a pronounced increase in locomotor activity is observed (Carlsson and Carlsson, 1989; Mehta and Ticku, 1990). A lesion of the corticostriatal pathway resulting from removal of the frontal cortex produces behavioral effects equivalent to those observed with an increase in dopamine transmission, i.e., a potentiation of stereotypies caused by apomorphine and a decrease in neuroleptic-induced catalepsy (Scatton et al., 1982).

In an animal model of Parkinson's disease (i.e., a lesion of the nigrostriatal pathway) there is an increase in the synthesis and release of dopamine in the striatum in the absence of an increase in the firing rate of the remaining intact cells in the substantia nigra (for review, see Zigmond et al., 1989). This observation raises the possibility of a modulatory influence exerted by cortical afferents on the remaining intact dopamine terminals as observed in the biochemical and electrophysiological experiments described above. From a clinical point of view, it may be hypothesized that an overdominant cortical influence due to a lack of inhibitory dopaminergic control, and therefore the exacerbation of the positive glutamate feedback, could result in rigidity and an impairment of the ability to move or in a slow execution of movements, as seen in Parkinson's disease (Marsden, 1989). Conversely, an imbalance in favor of dopamine not only leads to motor overactivity but has been related to the etiology of schizophrenia (Stevens, 1989). The ability of NMDA antagonists to precipitate a schizophrenic psychosis (Deutsch et al., 1989) strengthens the suggestion that glutamate and dopamine receptor stimulation produce opposite behavioral consequences, just as they result in opposite pre- and postsynaptic electrophysiological effects.

LONG-TERM PRESYNAPTIC CHANGES

The presence of a presynaptic positive feedback mechanism involving glutamate autoreceptors suggests that long-term changes in terminal excitability might be induced by brief high-frequency stimulation of the corticostriatal pathway, similar to the enduring effects observed in long-term potentiation (LTP) or depression (LTD). Indeed, we have recently observed that long-lasting changes in corticostriatal terminal excitability can be induced by a brief tetanizing stimulus to the cortex. As illustrated in figure 4.11, the tetanus typically initiates a long-term decrease in excitability lasting in most cases for as long as the recording is maintained (e.g., up to 140 minutes). A higher-current tetanus produced a brief increase in excitability, which was then followed by a long-lasting decrease. However, in a double tetanus procedure, in which a second tetanus was administered during the brief increase in excitability initiated by previous suprathreshold tetanic stimulation, a long-term increase in excitability was observed. Long-term increases in excitability could also be initiated by delivery of a single cortical tetanic stimulus following treatments which disrupted striatal dopamine and GABA transmission. These results were obtained in both intact animals and in rats in which postsynaptic striatal neurons had been destroyed with kainic acid. Thus, the induction and maintenance of these long-lasting effects may not depend on postsynaptic actions, but on alterations occurring in the presynaptic membrane (Garcia-Munoz et al., 1992).

Both LTP and LTD have been observed in the cerebellum (Crepel and Jaillard, 1991), in visual, somatosensory, motor, and prefrontal cortical areas (Bindman et al., 1988; Artola et al., 1990; Hirsch and Crepel, 1992), and in the hippocampus (Dunwiddie and Lynch, 1978; Levy and Steward, 1983). These long-term modifications in synaptic efficacy occur in neurons with synapses employing glutamate and their induction appears dependent on the stimulation of specific glutamate receptor types. The level of membrane polarization determined by the activity of the various classes of inputs to the cell during tetanization, including activation of kainate and AMPA (α-amino-3-hydroxy-5-methylisoxazole-4-propionic acid) glutamate receptors, appears to establish the conditions affecting the degree of response of NMDA glutamate receptors, and therefore the induction of LTP. It has been suggested that a rise in cytosolic calcium due to NMDA receptor activation leads to LTP, whereas an increase through voltage-dependent calcium channels or release from intracellular pools via activation of another type of glutamate receptor leads to LTD (Hirsch and Crepel, 1992). Accordingly, blockade of NMDA receptors prevents the induction of LTP, but does not suppress LTD (Artola et al., 1990). The work of Artola et al. (1990) suggests that cortical LTD occurs if postsynaptic depolarization exceeds a critical level, but remains below the level for activation of NMDA receptors, whereas LTP is induced if this second threshold is reached.

We have begun to explore the involvement of specific glutamate receptors in the long-lasting changes in corticostriatal excitability using intrastriatal administration of glutamate receptor antagonists and ago-

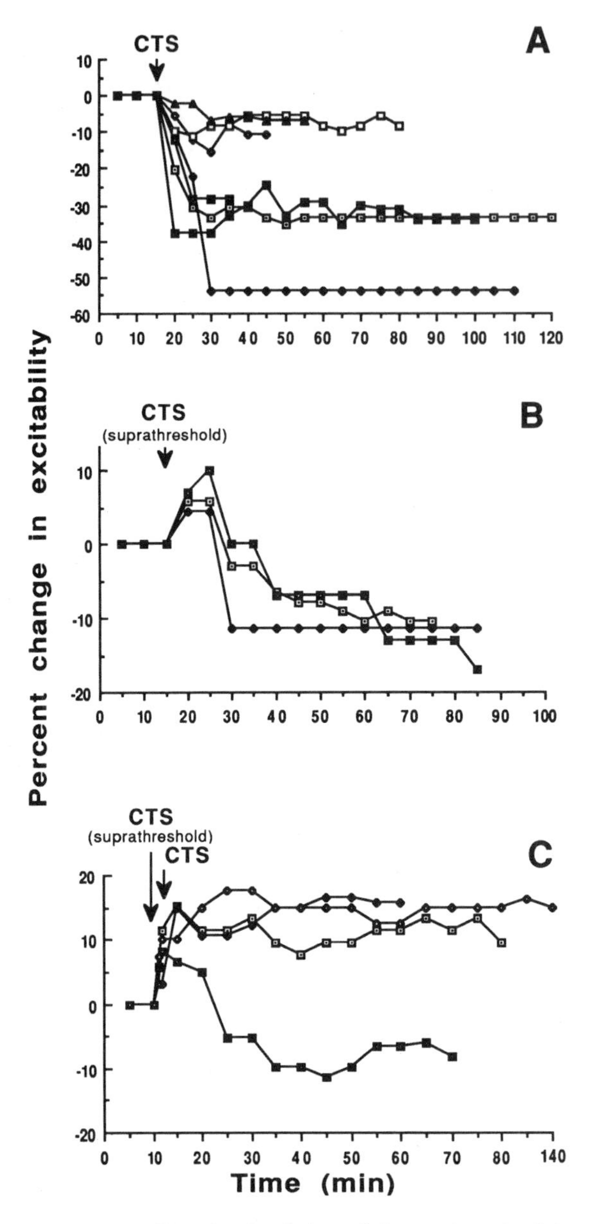

Figure 4.11 Tetanic stimulation of the corticostriatal terminal field induces long-term modifications in excitability. Curves are shown for individual cells. Excitability is expressed as the percent change in threshold current relative to control; positive and negative changes represent increased or decreased excitability. *A,* A long-lasting decrease in terminal excitability followed threshold cortical tetanic stimulation (CTS). *B,* Suprathreshold (2 × threshold) CTS initiated a brief increase in excitability followed by a long-term decrease. *C,* A second threshold CTS applied during the period of enhanced excitability initiated an enduring increase in excitability. (Data from Garcia-Munoz et al., 1992.)

nists. Both NMDA and non-NMDA receptors appear to mediate the induction of long-term increases in excitability, while participation of the *t*-ACPD receptor appears critical for the initiation of a long-term decrease in excitability (Garcia-Munoz et al., 1994).

LTP and LTD have recently been observed in the striatum. Tetanic stimulation of the corticostriatal pathway typically induces striatal LTD (Walsh, 1991; Calabresi et al., 1992a; Tyler et al., 1992; Walsh, 1993). LTD may predominate in the striatum owing to the actions of dopamine and GABA inputs to postsynaptic neurons. Consistent with this view, LTP has been observed using in vitro intracellular recording of striatal spiny neurons following high-frequency stimulation of the corpus callosum under conditions favoring disinhibition, i.e., under low magnesium or in the presence of a $GABA_A$ receptor antagonist, or both (Calabresi et al., 1992b; Walsh, 1991).

Our findings suggest that, similar to the induction of LTP or LTD, whether a persisting increase or decrease in presynaptic excitability is produced may depend on the level of depolarization of the terminal membrane achieved during the tetanization. The level of depolarization is presumed to depend on the balance of excitatory and inhibitory influences expressed via activation of auto- and heteroreceptors on the terminal. In addition to glutamate autoreceptors, there is evidence that corticostriatal terminals also possess heteroreceptors for dopamine and GABA (Maura et al., 1989; Calabresi et al., 1991) and, as discussed above, treatments which increase dopamine release in the striatum decrease corticostriatal terminal excitability (Garcia-Munoz et al., 1991a). In animals in which dopamine and GABA transmission have been disrupted, tetanic stimulation results in an enduring increase in excitability. We hypothesize that this occurs because inhibitory input in the vicinity of the terminal field is reduced so that the depolarization resulting from the glutamatergic positive feedback is sufficiently enhanced to initiate a long-lasting increase in excitability. According to this view, the particular pattern of increases and decreases in synaptic efficacy occurring at the synapses of a neuron or within a striatal module would depend not only on the pattern of cortical activity but also on the activity of dopaminergic, GABAergic, and other afferents operating in the local synaptic environment. Activity in these afferent systems would affect the likelihood that cortical signals would institute a long-term change and whether a long-term increase or decrease in synaptic efficacy would be established.

It is interesting that the induction and maintenance of the long-lasting presynaptic effects we have observed do not appear to depend on postsynaptic actions since they can be induced in animals with kainate striatal lesions. The initiation and maintenance of these long-lasting modifications in excitability may involve mechanisms similar to those operating in LTP and LTD to alter receptor affinity or conductance or to change the morphology of the synapse to maintain an altered level of excitability. It is conceivable that in some instances, depending on receptor distribution, synaptic activation, and state of membrane depolarization at the moment of the tetanus, pre- and postsynaptic changes could be opposite in direction or even occur only at the pre- or postsynaptic side. The hebbian principle of synaptic enhancement proposes a modification of both pre- and postsynaptic elements. Our view, that pre- and postsynaptic modifications in excitability may occur dissociated from each other, postulates a nonhebbian approach in which it is possible to either modify the presynaptic side and alter neurotransmitter release or to alter postsynaptic responsiveness to the tetanized pathway and possibly, through second messenger systems, modify the response to other classes of input as well. Such pre- and postsynaptic changes may provide a "switch" to allow the passage of preferential information. Such a focusing mechanism would not only allow a facilitated input to produce the strongest modulation but also permit, if needed, a substitution between alternative types of information. The idea that only a limited amount of the corticostriatal input may be facilitated while the rest may be suppressed has been proposed in reviews and theoretical papers on basal ganglia function several times in the last 20 years (e.g., see Denny-Brown and Yanizawa, 1976; Groves, 1983; Kimura et al., 1993).

LTP and LTD have been proposed as phenomena reflecting the action of mechanisms at the neuronal level underlying learning and memory in the hippocampus and cerebellum (e.g., Ito, 1989; Morris et al., 1988). The striatal pre- and postsynaptic changes in excitability could have important consequences for the integration of converging and intrinsic striatal activity which result, on the one hand, in regulation of motor activity such as locomotion and initiation of movement, and on the other hand, in more complex motor-related behaviors involving learning (Öberg & Divac, 1979; Carpenter and Jayaraman, 1987). Evidence suggests that different brain systems may be involved in learning and memory for different classes of information. For example, lesions of the striatum affect learning in several tasks which are not impaired by hippocampal lesions (O'Keefe and Nadel, 1978). These include a water maze task requiring a visual discrimination (Packard and McGaugh, 1992), different types of avoidance tasks (Prado-Alcalá et al., 1975; Giordano and Prado-Alcalá, 1986), and visual- and olfactory-conditioned emotional responses (Whishaw et al., 1987). Patients with hippocampal damage can acquire and demonstrate pattern-analysis skills necessary for mirror reading, sensorimotor skills required for rotary pursuit and bimanual tracking tasks, and some cognitive skills necessary to solve puzzles (Cohen et al., 1985). There is evidence indicating that the caudate may participate in these spared abilities (Martone et al., 1984; Cohen et al., 1985).

SUMMARY AND CONCLUSIONS

The studies reviewed in this paper describe several aspects of the intrinsic organization and function of the neostriatum. Our analysis of the distribution of peptides and their relation to cholinergic neurons suggests a preferential linkage between substance P–containing neurons and cholinergic interneurons within the striatum. In our immunocytochemical studies, we found that cholinergic neurons were more likely to be located in areas high in substance P than in enkephalin-rich regions. In agreement with these results, our studies and others suggest that cholinergic neurons receive a greater input from substance P neurons than from enkephalin-containing neurons. This preferential linkage may represent a component of the intrinsic organization within the striatum related to the functions of the differential projections of substance P neurons to the globus pallidus vs. the projections of enkephalin neurons to the entopeduncular nucleus and substantia nigra, and their participation in the direct and indirect output pathways (Alexander and Crutcher, 1990; Flaherty and Graybiel, 1993). This analysis emphasizes that in addition to the striosomal organization, there are critical intrinsic relationships represented by differential linkages between striatal neurons and their output stations. The analysis of peptide distributions also suggests differences between species which must be considered in developing models of striatal function.

Our ultrastructural studies of the dopamine innervation of striatal neurons provide additional evidence that dopamine synapses are symmetric and not necessarily present on axonal varicosities. Three-dimensional reconstructions of these axons and synapses along with synapses from other classes of afferents reveal the remarkable integration of information performed by common spiny output neurons. This arrangement is consistent with the multilayered connectionist network model of striatal information processing discussed in a recent paper by Alexander and colleagues (1992) who argue that a distributed network may best account for information processing in cortical-basal ganglia circuitry. In a connectionist model, information regarding specific motor sequences is not represented by separate discrete programs located within the motor circuitry. Instead, the information for multiple patterns of response is embedded within the pattern of modifiable and varying strength connections between neurons and in the properties of these neurons. Unique outputs result from the pattern of input signals and the resulting interactions within the network. This type of model appears to best represent the structure and functioning of cortical-basal ganglia circuitry where there is tremendous convergence and divergence at the neuronal level, and yet individual neurons can respond very specifically during phases of a motor task (Rolls et al., 1983; West et al., 1990; Apicella et al., 1992).

Within this framework, evidence from the electrophysiological studies described in this chapter suggests that the first layer of processing within the neostriatum occurs at the presynaptic level. The studies we have described suggest that presynaptic processes result in moment-to-moment alterations in afferent terminals which may alter impulse-related release. The effect of increased activation of autoreceptors on dopaminergic terminals can be represented as a negative feedback process which serves to reduce impulse-related release. In contrast, autoreceptor activation of corticostriatal terminals appears to operate as a positive feedback mechanism to enhance release from glutamatergic terminals. In addition to these mechanisms of autoregulation, our studies demonstrate presynaptic interactions between dopaminergic and corticostriatal glutamatergic axon terminals. Evidence suggests that these interactions occur as a result of the activation of heteroreceptors by neurotransmitter released from nearby sites. In support of this hypothesis, ultrastructural studies, including the work presented in this chapter, demonstrate the close proximity of corticostriatal and dopaminergic afferents as they form synapses on striatal neurons. Such interactions represent an important recurrent theme in the literature of dopamine since the introduction of levodopa therapy for Parkinson's disease. Recurrent issues have been the locations of sites where levodopa is decarboxylated and where dopamine is released and acts in the striatum to produce therapeutic gain in patients with Parkinson's disease, and whether or not dopamine can act in a hormone-like extrasynaptic mode to alter neostriatal information processing. Our results suggest that nonsynaptic effects operate at least over short distances within the local synaptic environment. While there is some consensus that dopamine can act at a distance from its site of synaptic release, precise information regarding the extent of the extracellular environment in which extrasynaptic dopamine action takes place is an issue of current conceptualization and investigation (Kawagoe et al., 1992; Zigmond et al., 1992).

In addition to the participation of these afferent systems in phasic computations, we have found evidence that corticostriatal afferents can undergo long-lasting alterations in function. Our experiments suggest that a persistent increase or decrease in synaptic efficacy can be instituted depending on the pattern of activity in corticostriatal fibers and coincident activity in other afferent systems in the local cortical terminal environment. Thus, a long-term decrease in the synaptic efficacy of a corticostriatal terminal is proposed to be initiated by an epoch of intense activity of a cortical neuron accompanied by activation of dopaminergic afferents in the vicinity of the terminal. This interaction between dopaminergic and cortical terminals, which can lead to a persistent modification in corticostriatal terminal excitability, may be relevant to studies showing that, in the primate, dopaminergic neurons respond to alerting stimuli and other cues critical for performance and learning (Jacobs et al., 1984; Schultz et al., 1993). While these studies suggest that dopamine neurons do not encode the learning required for the task, these neurons may be setting appropriate conditions at striatal terminal fields to institute long-term modifications in these afferents, thereby altering the connectivity of the net.

These presynaptic effects represent only the first of multiple layers of neostriatal processing. Another layer of processing is represented in the properties of postsynaptic interactions occurring within striatal neurons. As we have discussed, there is recent evidence for plasticity at postsynaptic sites in striatal neurons (Walsh, 1991, 1993; Calabresi et al., 1992a,b; Tyler et al., 1992). The nature of postsynaptic interactions will depend on the distribution of synapses formed by different classes of afferents (as illustrated by our ultrastructural studies), the types of receptors present on each neuron, the different intrinsic physiological properties of neuronal classes (Wilson 1990, 1992), and the patterns of connection between different classes of striatal neurons. Knowledge in these areas is rapidly evolving; although considerably more information is needed regarding the intrinsic structure and function of the neostriatum and the characteristics of its afferent systems, we are at least partly on the way to achieving this goal.

ACKNOWLEDGMENTS

This research was supported in part by grants from the Office of Naval Research (N00014-89-J-1254) and the National Institutes of Health (DA02854, DA00079, and MH18398).

REFERENCES

Akaike, A., Ohno, Y., Sasa, M., and Takaori, S. (1987) Excitatory and inhibitory effects of dopamine on neuronal activity of the caudate nucleus neurons in vitro. *Brain Res.* 418:262–272.

Alexander, G. E., and Crutcher, M. D. (1990) Substrates of parallel processing. *Trends Neurosci.* 13:266–271.

Alexander, G. E., DeLong, M. R., and Strick, P. L. (1986) Parallel organization of functionally segregated circuits linking basal ganglia and cortex. *Annu. Rev. Neurosci.* 9:357–381.

Alexander, G. E., DeLong, M. R., and Crutcher, M. D. (1992) Do cortical and basal ganglionic motor areas use "motor programs" to control movement? *Behav. Brain Sci.* 15:656–665.

Andén, N.-E., Fuxe, K., Hamberger, B., and Hökfelt, T. (1966) A quantitative study on the nigroneostriatal dopamine neuron system in the rat. *Acta Physiol. Scand.* 67:306–312.

Apicella, P., Scarnati, E., Ljungberg, T., and Schultz, W. (1992) Neuronal activity in monkey striatum related to the expectation of predictable environmental events. *J. Neurophysiol.* 68:945–960.

Arenas, E., Alberch, J., Perez-Navarro, E., Solsona, C., and Marsal, J. (1991) Neurokinin receptors differentially mediate endogenous acetylcholine release evoked by tachykinins in the neostriatum. *J. Neurosci.* 11:2332–2338.

Arluison, M., Agid, Y., and Javoy, F. (1978) Dopaminergic nerve endings in the neostriatum of the rat—1. Identification by intracerebral injections of 5-hydroxydopamine. *Neuroscience* 3:657–673.

Arluison, M., and De La Manche, I. S. (1980) High-resolution radioautographic study of the serotonin innervation of the rat corpus striatum after intraventricular administration of [^{3}H]5-hydroxytryptamine. *Neuroscience* 5:229–240.

Arluison, M., Javoy-Agid, F., Feuerstein, C., Tauc, M., Conrath-Verrier, M., and Mailly, P. (1982) Histofluorescence analysis of several systems of catecholaminergic nerve fibres within the rat neostriatum revealed by either restricted lesions of the substantia nigra or γ-hydroxybutyrate. *Brain Res. Bull.* 9:355–365.

Arluison, M., Dietl, M., and Thibault, J. (1984) Ultrastructural morphology of the dopaminergic nerve terminals and synapses in the striatum of the rat using tyrosine hydroxylase immunocytochemistry: A topographical study. *Brain Res. Bull.* 13:269–285.

Artola, A., Bröcher, S., and Singer, W. (1990) Different voltage-dependent thresholds for inducing long-term depression and long-term potentiation in slices of rat visual cortex. *Nature* 347:69–72.

Beckstead, R. M. (1987) Striatal substance P cell clusters coincide with the high density terminal zones of the discontinuous nigrostriatal dopaminergic projection system in the cat: A study by combined immunocytochemistry and autoradiographic axon-tracing. *Neuroscience* 20:557–576.

Beckstead, R. M., and Kersey, L. S. (1985) Immunohistochemical demonstration of differential substance P–, met-enkephalin-, and glutamic acid-decarboxylase–containing cell body and axon distribution in the corpus striatum of the cat. *J. Comp. Neurol.* 232:481–498.

Bennett, B. D., and Bolam, J. P. (1993) Characterization of calretinin-immunoreactive structures in the striatum of the rat. *Brain Res.* 609:137–148.

Besson, M. J. Graybiel, A. M., and Quinn, B. (1990) Co-expression of neuropeptides in the cat's striatum: An immunohistochemical study of substance P, dynorphin B and enkephalin. *Neuroscience* 39:33–58.

Bindman, L. J., Murphy, K. P. S. J., and Pockett, S. (1988) Postsynaptic control of the induction of long-term changes in efficacy of transmission at neocortical synapses in slices of rat brain. *J. Neurophysiol.* 60:1053–1065.

Biziere, K., and Coyle, J. T. (1978) Influence of cortico-striatal afferents on striatal kainic acid neurotoxicity. *Neurosci. Lett.* 8:303–310.

Bolam, J. P., and Izzo, P. N. (1988) The postsynaptic targets of substance P–immunoreactive terminals in the rat neostriatum with particular reference to identified spiny striatonigral neurons. *Exp. Brain Res.* 70:361–377.

Bolam, J. P., Somogyi, P., Takagi, H., Fodor, I., and Smith, A. D. (1983). Localizations of substance P–like immunoreactivity in neurons and nerve terminals in the neostriatum of the rat: A correlated light and electron microscopic study. *J. Neurocytol.* 12:325–344.

Bolam, J. P. Ingham, C. A., Izzo, P. N., Levey, A. I., Rye, D. B., Smith, A. D., and Wainer, B. H. (1986) Substance P–containing terminals in synaptic contact with cholinergic neurons in the neostriatum and basal forebrain: A double immunocytochemical study in the rat. *Brain Res.* 397:279–289.

Bolam, J. P., Izzo, P. N., and Graybiel, A. M. (1988). Cellular substrate of the histochemically defined striosome/matrix system of the caudate nucleus: A combined Golgi and immunocytochemical study in cat and ferret. *Neuroscience* 24:853–875.

Bouyer, J. J., Joh, T. H., and Pickel, V. M. (1984a). Ultrastructural localization of tyrosine hydroxylase in rat nucleus accumbens. *J. Comp. Neurol.* 227:92–103.

Bouyer, J. J., Park, D. H., Joh, T. H., and Pickel, V. M. (1984b) Chemical and structural analysis of the relation between cortical inputs and tyrosine hydroxylase–containing terminals in rat neostriatum. *Brain Res.* 302:267–275.

Calabresi, P., Mercuri, N. B., Stanzione, P., Stefani, A., and Bernardi, G. (1987). Intracellular studies on the dopamine-induced firing inhibition of neostriatal neurons *in vitro:* Evidence for D1 receptor involvement. *Neuroscience* 20:757–771.

Calabresi, P., Benedetti, M., Mercuri, N. B., and Bernardi, G. (1988). Endogenous dopamine and dopaminergic agonists modulate synaptic excitation in neostriatum: Intracellular studies from naive and catecholamine-depleted rats. *Neuroscience* 27:145–157.

Calabresi, P., Mercuri, N. B., De Murtas, M., and Bernardi, G. (1991) Involvement of GABA systems in feedback regulation of glutamate- and GABA-mediated synaptic potentials in rat neostriatum. *J. Physiol. (Lond.)* 440:581–599.

Calabresi, P., Maj, R., Pisani, A., Mercuri, N. B., and Bernardi, G. (1992a) Long-term synaptic depression in the striatum: Physiological and pharmacological characterization. *J. Neurosci.* 12:4224–4233.

Calabresi, P., Pisani, A., Mercuri, N. B., and Bernardi, G. (1992b) Long-term potentiation in the striatum is unmasked by removing the voltage-dependent magnesium block of NMDA receptor channels. *Eur. J. Neurosci.* 4:929–935.

Calas, A., Besson, M. J., Gaughy, C., Alonso, G., Glowinski, J., and Cheramy, A. (1976) Radioautographic study of in vivo incorporation of ^{3}H-monoamines in the cat caudate nucleus: Identification of serotoninergic fibers. *Brain Res.* 118:1–13.

Carlsson, M., and Carlsson, A. (1989) The NMDA antagonist MK-801 causes marked locomotor stimulation in monoamine-depleted mice. *J. Neural Transm. Gen. Sect.* 75:221–226.

Carlsson, A., Falck, B., and Hillarp, N. Å. (1962). Cellular localization of brain monoamines. *Acta Physiol. Scand. Suppl.* 196:1–28.

Carpenter, M. B., and Jayaraman, A. (1987) *The Basal Ganglia II.* London: Plenum Press.

Chang, H. T., Penny, G. R., and Kitai, S. T. (1987) Enkephalinergic-cholinergic interaction in the rat globus pallidus: A Pre-embedding double-labeling immunocytochemistry study. *Brain Res.* 426:197–203.

Chavez-Noriega, L., Patino, P., and Garcia-Munoz, M. (1986) Excitability changes induced in the striatal dopamine-containing terminals following frontal cortex stimulation. *Brain Res.* 379:300–306.

Chung, J. W., Hassler, R., and Wagner, A. (1977) Degeneration of two of nine types of synapses in the putamen after center median coagulation in the cat. *Exp. Brain Res.* 28:345–361.

Clark, D., and White, F. (1987) D1 dopamine receptor—the search for a function: A critical evaluation of the D1/D2 dopamine receptor classification and its functional implications. *Synapse* 1:347–388.

Cohen, N. J., Eichenbaum, H., De Acedo, H., and Corkin, S. (1985) Different memory systems underlying acquisition of procedural and declarative knowledge. *Ann. N. Y. Acad. Sci.* 444:54–71.

Collins, G. G. S., Anson, J., and Surtees, L. (1983) Presynaptic kainate and N-methyl-D-aspartate receptors regulate excitatory amino acid release in the olfactory cortex. *Brain Res.* 265:157–159.

Crepel, F., and Jaillard, D. (1991) Pairing of pre- and postsynaptic activities in cerebellar purkinje cells induces long-term changes in synaptic efficacy *in vitro. J. Physiol. (Lond.)* 432: 123–141.

Cuello, C. A. (1966) Nonclassical neuronal communications. *FASEB J.* 42:2912–2922.

Denny-Brown, D. (1962) *The Basal Ganglia and Their Relation to Disorders of Movement.* Oxford: Oxford University Press.

Denny-Brown, D., and Yanizawa, N. (1976) The role of the basal ganglia in the initiation of movement. In M. D. Yahr (Ed.), The Basal Ganglia. *Res. Publ. Assoc. Nerv. Ment. Dis.* 55:115–148.

Desban, M., Gauchy, C., Kemel, M. L., Besson, M. J., and Glowinski, J. (1989) Three-dimensional organization of the striosomal compartment and patchy distribution of striatonigral projections in the matrix of the cat caudate nucleus. *Neuroscience* 29:551–566.

Desce, J. M., Godeheu, G., Galli, T., Artaud, F., Cheramy, A., and Glowinski, J. (1992). L-Glutamate-evoked release of dopamine from synaptosomes of the rat striatum: Involvement of AMPA and N-methyl-D-aspartate receptors. *Neuroscience* 47:333–339.

Deutsch, S. I., Mastropaola, J., Schwartz, B. L., Rosse, R. B., and J. M. Morihisa (1989) A "glutamatergic hypothesis" of schizophrenia. Rationale for pharmacotherapy with glycine. *Clin. Neuropharmacol.* 12:1–13.

DiFiglia, M., and Aronin, N. (1982) Ultrastructural features of immunoreactive somatostatin neurons in the rat caudate nucleus. *J. Neurosci.* 2:1267–1274.

DiFiglia, M., Aronin, N., and Martin, J. B. (1982) Light and electron microscopic localization of immunoreactive leu-enkephalin in the monkey basal ganglia. *J. Neurosci.* 2:303–320.

Dimova, R., Vuillet, J., Nieoullon, A., and Kerkerian-Le Goff, L. (1993) Ultrastructural features of the choline acetyltransferase–containing neurons and relationships with nigral dopaminergic and cortical afferent pathways in the rat striatum. *Neuroscience* 53:1059–1071.

Donoghue, J. P., and Herkenham, M. (1986) Neostriatal projections from individual cortical fields conform to histochemically distinct striatal compartments in the rat. *Brain Res.* 365:397–403.

Doucet, G., Descarries, L., and Garcia, S. (1986) Quantification of the dopamine innervation in adult rat neostriatum. *Neuroscience* 19:427–445.

Dubé, L., Smith, A. D., and Bolam, J. P. (1988) Identification of synaptic terminals of thalamic or cortical origin in contact with distinct medium-size spiny neurons in the rat neostriatum. *J. Comp. Neurol.* 267:455–471.

Dunwiddie, T., and Lynch G. (1978) Long-term potentiation and depression of synaptic responses in the rat hippocampus: Localization and frequency dependency. *J. Physiol.* 276:353–367.

Earle, M. L., and Davies, J. A. (1991) The effect of methamphetamine on the release of glutamate from striatal slices. *J. Neural Transm. Gen. Sect.* 86:217–222.

Féger, J., Deniau, J. M., de Champlain, J., and Feltz, P. (1979) A survey of electrophysiology and pharmacology of neostriatal input-output relations. In I. Divac and R. G. E. Öberg (eds.), *The Neostriatum.* Oxford: Pergamon Press, pp. 71–102.

Ferkany, J. W., and Coyle, J. T. (1983) Kainic acid selectively stimulates the release of endogenous excitatory acidic amino acids. *J. Pharmacol. Exp. Ther.* 225:399–406.

Flaherty, A. W., and Graybiel, A. M. (1993) Output architecture of the primate putamen. *J. Neurosci.* 13:3222–3237.

Freund, T. F., Powell, J. F., and Smith, A. D. (1984) Tyrosine hydroxylase-immunoreactive boutons in synaptic contact with identified striatonigral neurons, with particular reference to dendritic spines. *Neuroscience* 13:1189–1215.

Frotscher, M., Rinne, U., Hassler, R., and Wagner, A. (1981) Termination of cortical afferents on identified neurons in the caudate nucleus of the cat. *Exp. Brain Res.* 41:329–337.

Fuxe, K. (1965) Evidence for the existence of monoamines in the central nervous system. IV. Distribution of monoamine nerve terminals in the central nervous system. *Acta Physiol. Scand. Suppl.* 247:39–85.

Fuxe, K., and Agnati, L. F. (1991) *Volume Transmission in the Brain: Novel Mechanisms for Neural Transmission.* New York: Raven Press.

Galli, T., Godeheu, G., Artaud, F., Desde, J. M., Pittaluga, A., Barbeito, L., Glowinski, J., and Cheramy, A. (1991) Specific role of *N*-acetyl-aspartyl-glutamate in the in vivo regulation of dopamine release from dendrites and nerve terminals of nigrostriatal dopaminergic neurons in the cat. *Neuroscience* 42:19–28.

Garcia-Munoz, M., and Chavez-Noriega, L. (1986) Striatal dopaminergic terminal field excitability. In G. N. Woodruff, J. A. Poat, and P. J. Roberts (eds.), *Dopamine Systems and Their Regulation* Weinheim, Germany, VCH Publishers, pp. 469–471.

Garcia-Munoz, M., Arbuthnott, G. W., and Rutherford, A. (1989) The afterhyperpolarization in striatal cells. In A. R. Crossman and M. A. Sambrook (eds.), *Neural Mechanisms in Disorders of Movement.* London: John Libbey, pp. 75–79.

Garcia-Munoz, M., Young, S. J., and Groves, P. M. (1991a) Terminal excitability of the corticostriatal pathway. I. Regulation by dopamine receptor stimulation. *Brain Res.* 551:195–206.

Garcia-Munoz, M., Young, S. J., and Groves, P. M. (1991b) Terminal excitability of the corticostriatal pathway. II. Regulation by glutamate receptor stimulation. *Brain Res.* 551:207–215.

Garcia-Munoz, M., Young, S. J., and Groves, P. M. (1992) Presynaptic long-term changes in excitability of the corticostriatal pathway. *Neuro Report* 3:357–360.

Garcia-Munoz, M., Young, S. J., Patino, P., and Groves, P. M. (1994) Long-lasting changes in excitability of corticostriatal terminals following tetanic stimulation. In G. Percheron, J. S. McKenzie, and J. Féger (eds.), *Basal Ganglia IV. New Data and Concepts on the Structure and Function of the Basal Ganglia.* New York: Plenum Press, pp. 245–254.

Gerfen, C. R. (1984) The neostriatal mosaic: Compartmentalization of corticostriatal input and striatonigral output systems. *Nature* 311:461–464.

Gerfen, C. R. (1989) The neostriatal mosaic: Striatal patch-matrix organization is related to cortical lamination. *Science* 246:385–388.

Gerfen, C. R. (1991) Substance P (neurokinin-1) receptor mRNA is selectively expressed in cholinergic neurons in the striatum and basal forebrain. *Brain Res.* 556:165–170.

Gerfen, C. R. (1992) The neostriatal mosaic: Multiple levels of compartmental organization in the basal ganglia. *Annu. Rev. Neurosci.* 15:285–320.

Gerfen, C. R., and Sawchenko, P. E. (1985) A method for anterograde axonal tracing of chemically specified circuits in the central nervous system: combined *Phaseolus vulgaris*–leucoagglutinin (PHA-L) tract tracing and immunohistochemistry. *Brain Res.* 343:144–150.

Gerfen, C. R., Baimbridge, K. G., and Miller, J. J. (1985) The neostriatal mosaic: Compartmental distribution of calcium binding protein and parvalbumin in the basal ganglia of the rat and monkey. *Proc. Natl. Acad. Sci. U. S. A.* 82:8780–8784.

Gerfen, C. R., Baimbridge, K. G., and Thibault, J. (1987a) The neostriatal mosaic: III. Biochemical and developmental dissociation of patch-matrix mesostriatal systems. *J. Neurosci.* 7:3935–3944.

Gerfen, C. R., Herkenham, M., and Thibault, J. (1987b) The neostriatal mosaic: II. Patch- and matrix-directed mesostriatal dopaminergic and non-dopaminergic systems. *J. Neurosci.* 7:3915–3934.

Gerfen, C. R. & Young III, W. S. (1988) Distribution of striatonigral and striatopallidal peptidergic neurons in both patch and matrix compartments: An *in situ* hybridization histochemistry and fluorescent retrograde tracing study. *Brain Res.* 460:161–167.

Giordano, M., and Prado-Alcalá, R. A. (1986). Retrograde amnesia induced by post-trial injection into caudate-putamen. Protective effect of the negative reinforcer. *Pharmacol. Biochem. Behav.* 24:905–909.

Graybiel, A. M. (1984) Correspondence between the dopamine islands and striosomes of the mammalian striatum. *Neuroscience* 13:1157–1187.

Graybiel, A. M. (1990) Neurotransmitters and neuromodulators in the basal ganglia. *Trends Neurosci.* 13:244–254.

Graybiel, A. M., and Ragsdale, C. W. (1978) Histochemically distinct compartments in the striatum of human, monkey and cat demonstrated by acetylthiocholinesterase staining. *Proc. Natl. Acad. Sci. U. S. A.*, 75:5723–5726.

Graybiel, A. M., and Hickey, T. L. (1982) Chemospecificity of ontogenetic units in the striatum: demonstration by combining [^{3}H]thymidine neuronography and histochemical staining. *Proc. Natl. Acad. Sci. U. S. A*, 79:198–202.

Graybiel, A. M., and Chesselet, M.-F. (1984) Compartmental distribution of striatal cell bodies expressing [Met] enkephalin-like immunoreactivity. *Proc. Natl. Acad. Sci. U. S. A.* 81:7980–7984.

Graybiel, A. M., Ragsdale, C. W., Yoneoka, E. S., and Elde, R. P. (1981) An immunohistochemical study of enkephalins and other neuropeptides in the striatum of the cat with evidence that the opiate peptides are arranged to form mosaic patterns in register with the striosomal compartments visible by acetylcholinesterase staining. *Neuroscience* 6:377–397.

Grofová, I. (1975) The identification of striatal and pallidal neurons projecting to substantia nigra. An experimental study by means of retrograde axonal transport of horseradish peroxidase. *Brain Res.* 91:286–291.

Groves, P. M. (1980) Synaptic endings and their postsynaptic targets in neostriatum: synaptic specializations revealed from analysis of serial sections. *Proc. Natl. Acad. Sci. U. S. A.* 77:6926–6929.

Groves, P. M. (1983) A theory of the functional organization of the neostriatum and the neostriatal control of voluntary movement. *Brain Res. Rev.* 5:109–132.

Groves, P. M., Wilson, C. J., Young, S. J., and Rebec, G. V. (1975) Self-inhibition by dopaminergic neurons. *Science* 190:522–529.

Groves, P. M., Fenster, G. A., Tepper, J. M., Nakamura, S., and Young, S. J. (1981) Changes in dopaminergic terminal excitability induced by amphetamine and haloperidol. *Brain Res.* 221:425–431.

Groves, P. M., Martone, M., Young, S. J., and Armstrong, D. M. (1988) Three-dimensional pattern of enkephalin-like immunoreactivity in the caudate nucleus of the cat. *J. Neurosci.* 8:892–900.

Groves, P. M., Linder, J. C., and Young, S. J. (1994). 5-Hydroxydopamine-labeled dopaminergic axons: Three dimensional reconstructions of axons, synapses, and postsynaptic targets in rat neostriatum. *Neuroscience* 58:593–604.

Guthrie, P. B., Segal, M., and Kater, S. B. (1991) Independent regulation of calcium revealed by imaging dendritic spines. *Nature* 354:76–80.

Haracz, J. L., Tschanz, J. T., Greenberg, J., and Rebec, G. V. (1989) Amphetamine-induced excitations predominate in single neostriatal neurons showing motor-related activity. *Brain Res.* 489:365–368.

Hattori, T., Fibiger, H. C., McGeer, P. L. and Maler, L. (1973) Analysis of the fine structure of the dopaminergic nigrostriatal projection by electron microscopic autoradiography. *Exp. Neurol.* 4:599–611.

Hattori, T., Takada, M., Moriizumi, T., and van der Kooy, D. (1991) Single dopaminergic nigrostriatal neurons from two chemically distinct synaptic types: Possible transmitter segregation within neurons. *J. Comp. Neurol.* 309:391–401.

Herkenham, M., and Pert, C. B. (1981) Mosaic distribution of opiate receptors, parafascicular projections and acetylcholinesterase in rat striatum. *Nature* 291:415–418.

Herrera-Marschitz, M., Goiny, M., Utsumi, H., Ferre, S., Guix, T., and Ungerstedt, U. (1990) Regulation of cortical and striatal dopamine and acetylcholine release by glutamate mechanisms assayed in vivo with microdialysis: In situ stimulation with kainate-, quisqualate- and NMDA-receptor agonists. In G. Lubec and G. A. Rosenthal (eds.), *Amino Acids: Chemistry, Biology and Medicine.* Leiden: ESCOM Science Publishers B.V., pp. 599–604.

Hirsch, J. C., and Crepel, F. (1992) Postsynaptic calcium is necessary for the induction of LTP and LTD of monosynaptic EPSPs in prefrontal neurons: An *in vitro* study in the rat. *Hippocampus* 10:173–175.

Ito, M. (1989). Long-term depression. *Ann. Rev. Neurosci.* 12:85–102.

Izzo, P. N., Graybiel, A. M., and Bolam, J. P. (1987) Characterization of substance P- and [met]enkephalin-immunoreactive neurons in the caudate nucleus of cat and ferret by a single section Golgi procedure. *Neuroscience* 20:577–587.

Jacobs, B. L., Steinfels, G. F., and Strecker, R. E. (1984) Dopaminergic unit activity in freely moving animals: A review. In E. Usdin, A. Carlsson, A. Dahlström, and J. Engel (eds.), *Catecholamines: Neuropharmacology and Central Nervous System—Theoretical Aspects.* New York: Alan R. Liss, pp. 393–399.

Jiménez-Castellanos, J., and Graybiel, A. M. (1987) Subdivisions of the dopamine-containing A8-A9-A10 complex identified by their differential mesostriatal innervation of striosomes and extrastriosomal matrix. *Neuroscience* 23:223–242.

Jiménez-Castellanos, J., and Graybiel, A. M. (1989). Compartmental origins of striatal efferent projections in the cat. *Neuroscience* 32:297–321.

Johnson, J. G., Gerfen, C. R., Haber, S. N., and van der Kooy, D. (1990) Mechanisms of striatal pattern formation: Conservation of mammalian compartmentalization. *Dev. Brain Res.* 57:93–102.

Kalsner, S., and Westfall, T. C. (1990) Presynaptic receptors and the question of autoregulation of neurotransmitter release. *Ann. N. Y. Acad. Sci.* 604:652–655.

Kawagoe, K. T., Garris, P. A., Wiedemann, D. J., and Wightman, R. M. (1992) Regulation of transient dopamine concentration gradients in the microenvironment surrounding nerve terminals in the rat striatum. *Neuroscience* 51:55–64.

Kawaguchi, Y., Wilson, C. J., and Emson, P. C. (1989) Intracellular recording of identified neostriatal patch and matrix spiny cells in a slice preparation preserving cortical inputs. *J. Neurophysiol.* 62:1052–1068.

Kemp, J. M., and Powell, T. P. S. (1971) The termination of fibres from the cerebral cortex and thalamus upon dendritic spines in the caudate nucleus: A study with the Golgi method. *Philos. Trans. R. Soc. Lond. [Biol]* 262:429–439.

Kimura, H., McGeer, P. I., Peng, J. H., and McGeer, E. G. (1981) The central cholinergic system studied by choline acetyltransferase immunohistochemistry in the cat. *J. Comp. Neurol.* 200:151–201.

Kimura, M., Aosaki, T., and Ishida, A. (1993) Neurophysiological aspects of the differential roles of the putamen and caudate nucleus in voluntary movement. In H. Narabayashi, T. Nagatsu, N. Yanagisawa, and Y. Mizuno (eds.), *Parkinson's Disease. From Basic Research to Treatment.* New York: Raven Press, pp. 62–70.

Kita, T., Kita, H., and Kitai, S. T. (1985) Local stimulation induced GABAergic response in rat striatal slice preparations: Intracellular recordings on QX-314 injected neurons. *Brain Res.* 360:304–310.

Kita, H., Kosaka, T., and Heizmann, C. W. (1990) Parvalbumin-immunoreactive neurons in the rat neostriatum: A light and electron microscopic study. *Brain Res.* 536:1–15.

Kitai, S. T., and Surmeier, D. J. (1993) Cholinergic and dopaminergic modulation of potassium conductances in neostriatal neurons. In H. Narabayashi, T. Nagatsu, N. Yanagisawa, and Y. Mizuno (eds.), *Parkinson's Disease. From Basic Research to Treatment.* New York: Raven Press, pp. 40–52.

Koch, C., and Zador, A. (1993) The function of dendritic spines: Devices subserving biochemical rather than electrical compartmentalization. *J. Neurosci.* 13:413–422.

Koch, C., Zador, A., and Brown, T. H. (1992) Dendritic spines: Convergence of theory and experiment. *Science* 256:973–974.

Kornhuber, J., and Kornhuber, M. E. (1983) Axo-axonic synapses in the rat striatum. *Eur. J. Neurol.* 22:433–436.

Kowall, N. W., Quigley, B. J., Krause, J. E., Lu, F., Kosofsky, B. E., and Ferrante, R. J. (1993) Substance P and substance P receptor histochemistry in human neurodegenerative diseases. *Regul. Pep.* 46:174–185.

Lapper, S. R., and Bolam, J. P. (1992) Input from the frontal cortex and the parafascicular nucleus to cholinergic interneurons in the dorsal striatum of the rat. *Neuroscience* 51:533–545.

Laursen, A. M. (1963) Corpus striatum. *Acta Physiol. Scand. Suppl.* 59(211):11–106.

Leviel, V., Gobert, A., and Guibert, B. (1990) The glutamate-mediated release of dopamine in the rat striatum: further characterization of the dual excitatory-inhibitory function. *Neuroscience* 39:305–312.

Levy, W. B., and Steward, O. (1983) Temporal contiguity requirements for long-term associative potentiation/depression in the hippocampus. *Neuroscience* 8:791–797.

Lighthall, J. W., and Kitai, S. T. (1983) A short duration GABAergic inhibition in identified neostriatal medium spiny neurons: In vitro slice study. *Brain Res. Bull.* 11:103–110.

Malach, R., and Graybiel, A. M. (1986) Mosaic architecture of the somatic sensory-recipient sector of the cat's striatum. *J. Neurosci.* 6:3436–3458.

Manley, M. S., Young, S. J., Martone, M. E., and Groves, P. M. (1991) Three-dimensional reconstruction of the neuropeptide network in the human caudate nucleus. *Soc. Neurosci. Abstr.* 17:962.

Marsden, C. D. (1989) Slowness of movement in Parkinson's disease. *Mov. Disord.* 4:S26–S37.

Martin, D., Bustos, G. A., Bowe, M. A., Bray, S. D., and Nadler, J. V. (1991a) Autoreceptor regulation of glutamate and aspartate release from slices of the hippocampal CA1 area. *J. Neurochem.* 56:1647–1655.

Martin, L. J., Hadfield, M. G., Dellovade, T. L., and Price, D. L. (1991b) The striatal mosaic in primates: Patterns of neuropeptide immunoreactivity differentiate the ventral striatum from the dorsal striatum. *Neuroscience* 43:397–417.

Martone, M., Butters, N., Payne, M., Becker, J. T., and Sax, D. S. (1984) Dissociations between skill learning and verbal recognition in amnesia and dementia. *Arch. Neurol.* 41:965–970.

Martone, M. E., Young, S. J., Armstrong, D. M., and Groves, P. M. (1991) Organization of cholinergic perikarya in the caudate nucleus of the cat with respect to heterogeneities in enkephalin and substance P staining. In G. Bernardi, M. B. Carpenter, G. D. Di Chiara, M. Morelli, and P. Stanzione (eds.), *The Basal Ganglia III.* New York: Plenum Press, pp. 39–48.

Martone, M. E., Armstrong, D. M., Young, S. J., and Groves, P. M. (1992) Ultrastructural examination of enkephalin and substance P input to cholinergic neurons within the rat neostriatum. *Brain Res.* 594:253–262.

Martone, M. E., Armstrong, D. M., Young, S. J., and Groves, P. M. (1993) Cholinergic neurons are distributed preferentially in areas rich in substance P-like immunoreactivity in the caudate nucleus of the adult cat. *Neuroscience* 56:567–579.

Maura, G., Carbone, R., and Raiteri, M. (1989) Aspartate-releasing nerve terminals in rat striatum possess D-2 dopamine receptors mediating inhibition of release. *J. Pharmacol. Exp. Ther.* 251:1142–1146.

McGeer, E. G., McGeer, P. L., and Singh, K. (1978) Kainate-induced degeneration of neostriatal neurons: Dependency upon cortico-striatal tract. *Brain Res.* 139:381–383.

Mehta, A. K., and Ticku, M. K. (1990) Role of *N*-methyl-D-aspartate (NMDA) receptors in experimental catalepsy in rats. *Life Sci.* 46:37–42.

Mercuri, N. B., Calabresi, P., Stefani, A., Stratta, F., and Bernardi, G. (1991) GABA depolarizes neurons in the rat striatum: An in vivo study. *Synapse* 8:38–40.

Misgeld, U., Wagner, A., and Ohno, T. (1982) Depolarizing IPSPs and depolarization by GABA of rat neostriatum cells in vitro. *Exp. Brain Res.* 45:108–114.

Moghaddam, B., Gruen, R. J., Roth, G. H., Bunney, B. S., and Adams, R. N. (1990) Effect of L-glutamate on the release of striatal dopamine: In vivo dialysis and electrochemical studies. *Brain Res.* 518:55–60.

Moriizumi, T., and Hattori, T. (1992) Ultrastructural morphology of projections from the medial geniculate nucleus and its adjacent region to the basal ganglia. *Brain Res. Bull.* 29:193–198.

Morris, R. G. M., Kandel, E. R., and Squire, L. R. (1988) The neuroscience of learning and memory: Cells, neural circuits and behaviour. *Trends Neurosci.* 11:125–127.

Müller, W., and Conner, J. A. (1991) Dendritic spines as individual neuronal compartments for synaptic Ca^{2+} responses. *Nature* 354:73–76.

Müller, W., and Connor, J. A. (1992) Ca^{2+} signalling in postsynaptic dendrites and spines of mammalian neurons in brain slice. *J. Physiol. (Paris)* 86:57–66.

Naudon, L., Dourmap, N., Leroux-Nicollet, I., and Costentin, J. (1992) Kainic acid lesion of striatum increases dopamine release but reduces 3-methoxytyramine level. *Brain Res.* 572:247–249.

Öberg, R. G. E., and Divac, I. (1979) "Cognitive" functions of the neostriatum. In I. Divac and R. G. E. Öberg (eds.), *The Neostriatum.* Oxford: Pergamon Press, pp. 291–313.

O'Brien, T. F., Faissner, A., Schachner, M., and Steindler, D. A. (1992) Afferent boundary interactions in the developing neostriatal mosaic. *Dev. Brain Res.* 65:259–267.

O'Keefe, J., and Nadel, L. (1978). *The Hippocampus as a Cognitive Map.* New York: Oxford University Press.

Olson, L., Seiger, A., and Fuxe, K. (1972) Heterogeneity of striatal and limbic dopamine innervation: Highly fluorescent islands in developing and adult rats. *Brain Res.* 44:283–288.

Packard, M. G., and McGaugh, J. L. (1992) Double dissociation of fornix and caudate nucleus lesions on acquisition of two water maze tasks: Further evidence for multiple memory systems. *Behav. Neurosci.* 106:439–446.

Park, M. R., Lighthall, J. W., and Kitai, S. T. (1980) Recurrent inhibition in rat neostriatum. *Brain Res.* 194:359–369.

Parthasarathy, H. B., Schall, J. D., and Graybiel, A. M. (1992) Distributed but convergent ordering of corticostriatal projections: Analysis of the frontal eye field and the supplementary eye field in the macaque monkey. *J. Neurosci.* 12:4468–4488.

Pasik, T., and Pasik, P. (1982) Serotoninergic afferents in the monkey neostriatum. *Acta Biol. Hung.* 33:277–288.

Pasik, P., Pasik, T., and DiFiglia, M. (1979) The internal organization of the neostriatum in mammals. In I. Divac and R. G. E. Öberg (eds.), *The Neostriatum.* Oxford: Pergamon Press, pp. 5–36.

Pasik, P., Pasik, T., Holstein, G. R., and Saavedra, J. P. (1984) Serotoninergic innervation of the monkey basal ganglia: An immunocytochemical, light and electron microscopy study. In J. S. McKenzie, R. E. Kemm, and L. N. Wilcock (eds.), *The Basal Ganglia.* New York: Plenum Press, pp. 115–129.

Penney, J. B., and Young, A. B. (1983) Speculations on the functional anatomy of basal ganglia disorders. *Ann. Rev. Neurosci* 6:73–94.

Penny, G. R., Afsharpour, S., and Kitai, S. T. (1986) The glutamate decarboxylase–, leucine enkephalin–, methionine enkephalin– and substance P–immunoreactive neurons in the neostriatum of the rat and cat: Evidence for partial population overlap. *Neuroscience* 17:1011–1045.

Penny, G. R., Wilson, C. J., and Kitai, S. T. (1988) Relationship of the axonal and dendritic geometry of spiny projection neurons to the compartmental organization of the neostriatum. *J. Comp. Neurol.* 269:275–289.

Phelps, P. E., Houser, C. R., and Vaughn, J. E. (1985) Immunocytochemical localization of choline acetyltransferase within the rat neostriatum: A correlated light and electron microscopic study of cholinergic neurons and synapses. *J. Comp. Neurol.* 238:286–307.

Pickel, V. M., Beckley, S. C., Joh, T. H., and Reis, D. J. (1981) Ultrastructural immunocytochemical localization of tyrosine hydroxylase in the neostriatum. *Brain Res.* 225:373–385.

Prado-Alcalá, R. A., Grinberg, J. Z., Arditti, L., Garcia-Munoz, M., Prieto, G. H., and Brust-Carmona, H. (1975) Learning deficits produced by chronic and reversible lesions of the corpus striatum in rats. *Physiol. Behav.* 15:283–287.

Ragsdale, C. W., and Graybiel, A. M. (1981) The fronto-striatal projections in the cat and monkey and its relationship to inhomogeneities established by acetylcholinesterase. *Brain Res.* 208:259–266.

Ragsdale, C. W., and Graybiel, A. M. (1990) A simple ordering of neocortical areas established by the compartmental organization of their striatal projections. *Proc. Natl. Acad. Sci. U. S. A.* 87: 6196–6199.

Ribak, C. E., Vaughn, J. E., and Roberts, E. (1980) GABAergic nerve terminals decrease in the substantia nigra following hemitransections of the striatonigral and pallidonigral pathways. *Brain Res.* 192:413–420.

Rolls, E. T., Thorpe, S. J., and Maddison, S. P. (1983). Responses of striatal neurons in the behaving monkey. 1. Head of the caudate nucleus. *Behav. Brain Res.* 7:179–210.

Rolls, E. T., Thorpe, S. J., Boytim, M., Szabo, I., and Perritt, D. I. (1984) Responses of striatal neurons in the behaving monkey. 3. Effects of iontophoretically applied dopamine on normal responsiveness. *Neuroscience* 12:1201–1212.

Rowlands, G. J., and Roberts, P. J. (1980) Activation of dopamine receptors inhibits calcium-dependent glutamate release from cortico-striatal terminals *in vitro. Eur. J. Pharmacol.* 62:241–242.

Rutherford, A., Garcia-Munoz, M., and Arbuthnott, G. W. (1988) An afterhyperpolarization recorded in striatal cells *in vitro:* Effect of dopamine administration. *Exp. Brain Res.* 71:399–405.

Ryan, L. J., Young, S. J., Segal, D. S., and Groves, P. M. (1989) Antidromically identified striatonigral projection neurons in the chronically implanted behaving rat: relations of cell firing to amphetamine-induced behaviors. *Behav. Neurosci.* 103:3–14.

Sadikot, A. F., Parent, A., Smith, Y., and Bolam, J. P. (1992) Efferent connections of the centromedian and parafascicular thalamic nuclei in the squirrel monkey: A light and electron microscopic study of the thalamostriatal projection in relation to striatal heterogeneity. *J. Comp. Neurol.* 320:228–242.

Scatton, B., Worms, P., Lloyd, K. G., and Bartholini, G. (1982) Cortical modulation of striatal function. *Brain Res.* 232:331–343.

Schmidt, W. J., and Bury, D. (1988) Behavioral effects of *N*-methyl-D-aspartate in the anteromedial striatum of the rat. *Life Sci.* 43:545–549.

Schultz, W., Apicella, P., and Ljungberg, T. (1993) Responses of monkey dopamine neurons to reward and conditioned stimuli during successive steps of learning a delayed response task. *J. Neurosci.* 13:900–913.

Selemon, L. D., and Goldman-Rakic, P. S. (1985) Longitudinal topography and interdigitation of corticostriatal projections in the rhesus monkey. *J. Neurosci.* 5:776–794.

Shepherd, G. M. (1990) *The Synaptic Organization of the Brain.* New York: Oxford University Press.

Siggins, G. R. (1978) Electrophysiological role of dopamine in striatum: excitatory or inhibitory? In M. A. Lipton, A. DiMascio, and K. F. Killam (eds), *Psychopharmacology: A Generation of Progress.* New York: Raven Press, pp. 143–157.

Smiley, J. F., and Goldman-Rakic, P. S. (1993) Heterogeneous targets of dopamine synapses in monkey prefrontal cortex demonstrated by serial section electron microscopy: A laminar analysis using the silver-enhanced diaminobenzidine sulfide (SEDS) immunolabeling technique. *Cereb. Cortex* 3:223–238.

Soghomonian, J.-J., Descarries, L., and Watkins, K. C. (1989) Serotonin innervation in adult rat neostriatum. II. Ultrastructural features: A radioautographic and immunocytochemical study. *Brain Res.* 481:67–86.

Somogyi, P., Bolam, J. P., and Smith, A. D. (1981) Monosynaptic cortical input and local axon collaterals of identified striatonigral neurons. A light and electron microscopic study using the Golgi-peroxidase transport-degeneration procedure. *J. Comp. Neurol.* 195:567–584.

Steindler, D. A., O'Brien, T. F., and Cooper, N. G. F. (1988) Glycoconjugate boundaries during early postnatal development of the neostriatal mosaic. *J. Comp. Neurol.* 267:357–369.

Stevens, J. R. (1989) The search for an anatomic basis for schizophrenia: review and update. In E. Muller (ed.), *Neurology and Psychiatry: A Meeting of the Minds.* Basel: Karger, pp. 64–87.

Surmeier, D. J., Eberwine, J., Wilson, C. J., Cao, Y., Stefani, A., and Kitai, S. T. (1992) Dopamine receptor subtypes colocalize in rat striatonigral neurons. *Proc. Natl. Acad. Sci. U. S. A.* 89:10178–10182.

Takagi, H., Somogyi, P., Somogyi, J., and Smith, A. D. (1983) Fine structural studies on a type of somatostatin-immunoreactive neuron and its synaptic connections in the rat neostriatum: A correlated light and electron microscopic study. *J. Comp. Neurol.* 214:1–16.

Takagi, H., Mizuta, H., Matsuda, T., Inagaki, S., Tateishi, K., and Hamaoka, T. (1984) The occurrence of cholecystokinin-like immunoreactive neurons in the rat neostriatum: Light and electron microscopic analysis. *Brain Res.* 309:346–349.

Takeuchi, H., Young, S. J., and Groves, P. M. (1982) Dopaminergic terminal excitability following arrival of the nerve impulse: the influence of amphetamine and haloperidol. *Brain Res.* 245:47–56.

Tennyson, V. M., Heikkila, R., Mytilineou, C., Côté, L., and Cohen, G. (1974) 5-Hydroxydopamine "tagged" neuronal boutons in rabbit striatum: Interrelationship between vesicles and axonal membrane. *Brain Res.* 82:341–348.

Tepper, J. M., and Groves, P. M. (1990) In vivo electrophysiology of central nervous system terminal autoreceptors. *Ann. N. Y. Acad. Sci.* 604:470–487.

Tepper, J. M., Young, S. J., and Groves, P. M. (1984) Autoreceptor-mediated changes in dopaminergic terminal excitability: Effects of increases in impulse flow. *Brain Res.* 309:309–316.

Tepper, J. M., Groves, P. M., and Young, S. J. (1985) The neuropharmacology of the autoinhibition of monoamine release. *Trends Pharmacol. Sci.* 6:251–256.

Triarhou, L. C., Norton, J., and Ghetti, B. (1988) Synaptic connectivity of tyrosine hydroxylase immunoreactive nerve terminals in the striatum of normal, heterozygous and homozygous weaver mutant mice. *J. Neurocytol.* 17:221–232.

Trulson, M. E., and Jacobs, B. L. (1979) Effects of *d*-amphetamine on striatal unit activity and behavior in freely moving cats. *Neuropharmacology* 18:735–738.

Tyler, E. C., Lovinger, D. M., and Merritt, A. (1992) Short- and long-term synaptic depression in neostriatum. *Soc. Neurosci. Abstr.* 18:1351.

van der Kooy, D., and Fishell, G. (1987) Neuronal birthdate underlies the development of striatal compartments. *Brain Res.* 401:155–161.

Voorn, P., Jorritsma-Byham, B., Van Dijk, C., and Buijs, R. M. (1986) The dopaminergic innervation of the ventral striatum in the rat: A light- and electron-microscopical study with antibodies against dopamine. *J. Comp. Neurol.* 251:84–99.

Vuillet, J., Kerkerian, L., Salin, P., and Nieoullon, A. (1989) Ultrastructural features of NPY-containing neurons in the rat striatum. *Brain Res.* 477:241–251.

Walsh, J. P. (1991) Long-term potentiation (LTP) of excitatory synaptic input to medium spiny neurons of the rat striatum. *Soc. Neurosci. Abstr.* 17:852.

Walsh, J. P. (1993) Depression of excitatory synaptic input in rat striatal neurons. *Brain Res.* 608:123–128.

Wang, J. K. T. (1991) Presynaptic glutamate receptors modulate dopamine release from striatal synaptosomes. *J. Neurochem.* 57:819–822.

West, M. O., Carelli, R. M., Pomerantz, M., Cohen, S. M., Gardner, J. P., Chapin, J. K., and Woodward, D. J. (1990) A region in the dorsolateral striatum of the rat exhibiting single-unit correlations with specific locomotor limb movements. *J. Neurophysiol.* 64:1233–1246.

Westerink, B. H. C., Santiago, M., and De Vries, J. B. (1992) The release of dopamine from nerve terminals and dendrites of nigrostriatal neurons induced by excitatory amino acids in the conscious rat. *Naunyn-Schmiedebergs Arch. Pharmacol.* 345:523–529.

Whishaw, I. Q., Mittleman, G., Bunch, S. T., and Dunnet, S. B. (1987) Impairments in the acquisition, retention, and selection of spatial navigation strategies after medial caudate-putamen lesions in rats. *Behav. Brain Res.* 24:125–138.

Wickens, J. R., Alexander, M. E., and Miller, R. (1991) Two dynamic modes of striatal function under dopaminergic-cholinergic control: Simulation and analysis of a model. *Synapse* 8:1–12.

Williams, G. V., and Millar, J. (1990) Concentration-dependent actions of stimulated dopamine release on neuronal activity in rat striatum. *Neuroscience* 39:1–16.

Wilson, C. J. (1984) Passive cable properties of dendritic spines and spiny neurons. *J. Neurosci.* 4:281–297.

Wilson, C. J. (1990) Basal ganglia. In G. Shepherd (ed.), *The Synaptic Organization of the Brain.* New York: Oxford University Press, pp. 279–316.

Wilson, C. J. (1992) Dendritic morphology, inward rectification, and the functional properties of neostriatal neurons. In T. McKenna, J. Davis, and S. F. Zornetzer (eds.), *Single Neuron Computation.* San Diego: Academic Press, pp. 141–171.

Wilson, C. J., and Groves, P. M. (1980). Fine structure and synaptic connections of the common spiny neuron of the rat neostriatum: A study employing intracellular injection of horseradish peroxidase. *J. Comp. Neurol.* 194:599–615.

Wilson, C. J., Groves, P. M., Kitai, S. T., and Linder, J. C. (1983). Three-dimensional structure of dendritic spines in the rat neostriatum. *J. Neurosci.* 3:383–398.

Wilson, C. J., Kita, H., and Kawaguchi, Y. (1989) GABAergic interneurons, rather than spiny cell axon collaterals, are responsible for the IPSP responses to afferent stimulation in neostriatal spiny neurons. *Soc. Neurosci. Abstr.* 15:907.

Yamamoto, B. K., and Davy, S. (1992) Dopaminergic modulation of glutamate release in striatum as measured by microdialysis. *J. Neurochem.* 58:1736–1742.

Yetarian, E. H., and Van Hoesen, G. W. (1978) Cortico-striate projections in the rhesus monkey: the organization of certain cortico-caudate connections. *Brain Res.* 139:43–63.

Young, S. J., Royer, S. M., Groves, P. M., and Kinnamon, J. C. (1987) Three-dimensional reconstructions from serial micrographs using the IBM PC. *J. Electron Microsc. Tech.* 6:207–217.

Zigmond, M. J., Berger, T. W., Grace, A. A., and Stricker, E. M. (1989) Compensatory responses to nigrostriatal bundle injury: Studies with 6-hydroxydopamine in an animal model of Parkinsonism. *Mol. Chem. Neuropathol.* 10:185–200.

Zigmond, M. J., Hastings, T. G., and Abercrombie, E. D. (1992) Neurochemical responses to 6-hydroxydopamine and L-dopa therapy: Implications for Parkinson's disease. *Ann. N. Y. Acad. Sci.* 648:71–86.

Editors' Commentary on Part I

The preceding chapters provide complementary perspectives on the fundamental information processing functions of the basal ganglia. In chapter 1, a simple conceptual model of basal ganglia information processing was presented, based on the salient anatomy and physiology of the system, to provide a framework for the interpretation of the more comprehensive anatomy, physiology, and computational models described in subsequent chapters. Functionally, the proposed conceptual model performs two main types of information processing tasks related to the planning and execution of behaviors. The first is a *pattern recognition* task that involves discriminating firing patterns present in the highly convergent input to individual striatal spiny neurons from many different cerebral cortical neurons. It was hypothesized that this multidimensional input provides a sample of the many sensory and internal events and states prevalent at a given moment, and that the pattern recognized by a spiny neuron, after being trained by reinforcement learning, would represent a complex context of behavioral significance to the organism. The second main type of information processing task postulated is the *registration* of detected contexts into units of working memory. Houk suggested that this occurs as a disinhibitory process that initiates positive feedback in corticothalamic circuits of the frontal lobes (or as an inhibitory process that terminates preexisting positive feedback). By registering (or canceling) the detected contexts in working memory, they become converted into sustained signals that can be used subsequently to plan and coordinate complex behaviors over time. In this commentary, we review some of the key points presented in chapters 2 through 4 within the framework of this *pattern recognition/registration model*

LINKING THE MODEL TO BEHAVIOR

Anatomically, the striatum appears to be a bridge between diverse sensory-derived, memory-derived, and motivational input signals and an equally diverse set of goal-setting, planning, and motor control outputs. One question we can ask then is, given their privileged access to cortical signals, in what behaviorally relevent ways are striatal spiny cells able to process this information? In chapter 2, Schultz and colleagues presented evidence linking striatal activity to recognition of specific behavioral events from "intention to outcome." Perhaps most compelling were the responses of three types of striatal neurons that were time-locked to separate events in the animal's expected future. In effect, these spiny neurons seem to be able to recognize current contextual patterns that have, in the past, preceded certain salient events such as the delivery of a food reward. These neurons clearly appear to be engaging in contextual pattern recognition. While some of the associations leading to this expectation signal may be formed at the cortical level, it seems likely that the recognition of the entire context happens at the level of the spiny neuron since the neuronal architecture of these cells is so well matched to the problem of pattern classification, as will be discussed momentarily.

Schultz raises the question as to whether striatal neurons are better related to the recognition of events and contexts or to motor planning and the control of behavior. He finds data supporting either conclusion. The likely resolution of this dilemma is that both conclusions are appropriate and in fact quite complementary. The instances of correlations with events and states help to define how striatal neurons respond to their diverse inputs, whereas the instances of correlations with overt behavior instead help define how the outputs of the neurons may participate in controlling the animal's reactions and behavior. It is reasonable that the experimenter cannot always decipher the context that caused a cell to fire, or the influence that the cell had on the animal's behavior. Thus, some cells would appear better related to context detection and others to behavioral control, and in select cases the relations to both might be evident.

SPINY NEURON INPUTS

The pattern-classifying ability of the pattern recognition/registration model relies importantly on a rich convergence of diverse input modalities at the level of individual striatal spiny neurons. Indeed, in chapter 3, Wilson indicates that each spiny neuron receives inputs from approximately 10,000 *different* afferent axons. This ratio would strongly support pattern classification if the individual inputs were sufficiently rich in information content, i.e., statistically independent of each other. Some of the anatomical findings that are reviewed in the next set of chapters will help us to assess the actual potential for diversity in the input vector to a striatial neuon.

The excitation-based explanation for the dual-state firing pattern of spiny neurons developed in chapter 3 places important restrictions on the temporal and spatial patterning of cortical inputs required for crossing the threshold between these two states. Wilson predicted that only relatively coherent inputs spread over a large portion of the dendritic tree would be appropriate for boosting a spiny neuron into its "up" state. This coincidence requirement points to the need for a balance between richness, or diversity, of inputs and some degrees of commonality. Indeed, an input vector that is too rich might not have enough coherence to toggle the neuron's states with any degree of regularity.

SHARP THRESHOLDS AND COMPETITIVE INTERACTIONS

The discriminatory power of the pattern recognition/registration model relies on a sharply defined threshold between the firing and quiescent states of the spiny neuron. While the firing patterns of spiny neurons clearly have sharp thresholds between up and down states, Wilson actually described a potential division of the up state into firing and nonfiring modes. Although this further division fits nicely with the observation that spiny neurons are largely silent, it is not clear what effect it would have on the conceptual model.

Competitive interactions have often been assumed in hypotheses of basal ganglia function, motivated both by psychological argument and by anatomical evidence suggesting an extensive network of GABAergic collaterals within the striatum. Thus, the absence of intracellular evidence for collateral inhibition reported in chapter 3 is surprising. However, the final word is probably not in yet, due to current limitations in our knowledge of cellular mechanisms. For example, we have little idea about the functions of the several peptides that are discussed in chapter 4, and their modulatory actions may turn out to be a necessary condition for the expression of collateral inhibition. Although independence between neighboring spiny neurons is consistent with the single-layer pattern recognition model of the striatum discussed in chapter 1, the presence of competitive interactions between adjacent cells could be a very useful feature, as will be highlighted in several subsequent chapters in this volume.

DOPAMINERGIC INNERVATION

The pattern recognition/registration model also depends importantly on a reinforcing influence of dopaminergic fibers on the synaptic weights between corticostriatal and spiny neurons. In chapter 4, Grove and colleagues provide a microanatomical review that clearly demonstrates the extent and character of this innervation. Dopaminergic innervation appears diffuse when viewed at a microscopic level—thus a given fiber can be expected to branch to many spine zones on at least several cells within a given striatal compartment. However, the innervation appears more topographically specific when viewed more globally, and probably cannot be considered a global reinforcing input. The mechanism of dopamine influence is far from simple, with evidence for both presynaptic and postsynaptic effects, as well as LTP and LTD of synaptic efficiencies. These issues will be explored further in part III of this book.

II Motor Functions and Working Memories

5 Adaptive Neural Networks in the Basal Ganglia

Ann M. Graybiel and Minoru Kimura

The question of what the basal ganglia do is often posed by asking whether the basal ganglia have a primary function in and of themselves in the control of action, or whether their function is mainly modulatory. That is, should we look for operations explicitly directed toward, for example, calculation of movement direction or force, or do the basal ganglia add, instead, conditional signals to the motor-programming mechanism whose computations of direction and force occur elsewhere?

In this chapter, we take the position that the basal ganglia may indeed carry out higher-level computations related to the details of motor programs, but that the activity of the basal ganglia may also have unique characteristics permitting ongoing and long-term modifiability of motor and nonmotor programs. This is not to take the view that the basal ganglia are unique in exhibiting neuroplasticity, for this is probably a universal phenomenon of nerve cells and neuronal networks. Rather, we suggest that the basal ganglia may be in a privileged position to receive information motivating change in behavior as well as information related to the action programs to be developed (figure 5.1).

To make these points we raise three sets of findings: first, evidence that the organization of cortical inputs to the striatopallidal system provides possible substrates for conditional processing of incoming sensorimotor data; second, evidence that specializations of the basal ganglia may favor reinforcement-related inputs to influence the sensorimotor processing; and third, evidence that for at least one population of striatal neurons (the tonically active neurons, or TANs), acquisition of novel responsiveness does occur during the course of behavioral conditioning in monkeys, and dopamine is necessary for the expression of these new responses.

TEMPLATES FOR MODIFIABLE PROGRAMMING

We start with the idea that the basal ganglia may figure prominently in adaptive motor control, and that their architecture may be set up for

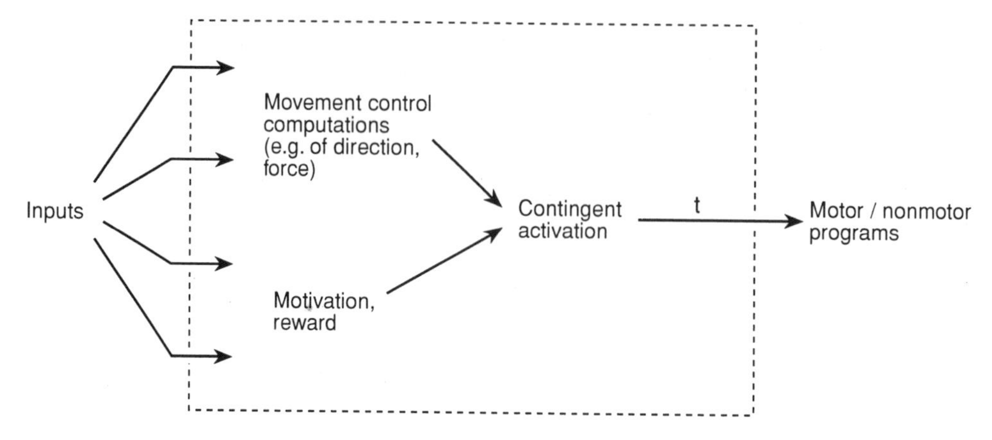

Figure 5.1 Hypothesis that basal ganglia (*box*) are involved in contingent activation of motor and nonmotor programs destined for further processing by thalamocortical and brainstem targets prior to implementation in actions. t, time.

such a role. The hypothesis derives from electrophysiologically guided tracer studies in cat and monkey suggesting that systematic remapping occurs in the corticostriatal projection (Malach and Graybiel, 1986; Parthasarathy et al., 1992; Flaherty and Graybiel, 1991, 1993b, 1994). The evidence for the somatic sensory and motor cortex is as follows:

1. Any one small part of the body map in SI cortex projects to multiple, partly interconnected zones in the sensorimotor striatum. Such zones have been called matrisomes to indicate that they are patchy zones in the matrix compartment of the striatum, and to distinguish them from the patchy zones that form the chemically distinct striosomes (Graybiel, 1984, 1990; Graybiel et al., 1991; Flaherty and Graybiel, 1991, 1993a,b, 1994). Different body part representations are sent to different sets of input matrisomes, though closely related body part projections may overlap. *Thus the information sent from SI to striatum is topographically specified but is broadly distributed: the body map is broken up.*

2. When inputs from homologous body part representations in SI cortical areas (say, area 3b and area 1 foot regions) are injected with transportable tracers in the squirrel monkey, the two sets of input patches labeled largely overlap (figure 5.2). Projections from nonhomologous sites in SI cortex largely to not overlap (Flaherty and Graybiel, 1991). *Thus the particular distributions of corticostriatal connections from homologous sites in different SI cortical areas are coordinated according to internal principles of striatal architecture.*

3. Inputs from homologous body part representations in SI and MI cortex also overlap in the striatum of the squirrel monkey, whereas those from homologous sites in SI and contralateral MI tend not to overlap (Flaherty and Graybiel, 1993b). *Thus cortical areas with different*

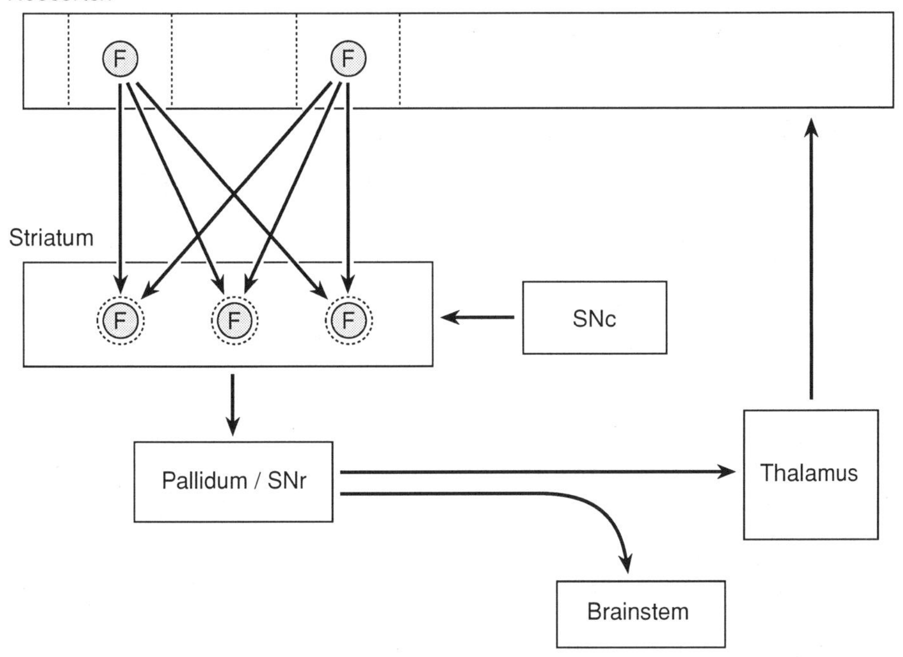

Figure 5.2 Divergence and convergence characterize corticostriatal projections from the somatic sensory (SI) cortex. In the example shown, an area of SI foot cortex sends divergent projections to a set of patches ("input matrisomes") in the striatum. Homologous sites in a different SI area send connections to overlapping sets of these striatal input patches. Dashes around matrisomes indicate that they are not sharply bounded. F, foot; SNc, substantia nigra pars compacta, SNr, substantia nigra pars reticulata. (Based on Flaherty and Graybiel, 1991, 1993b, 1994.)

types of body map also exhibit overlap of corticostriatal inputs, and this is not a trivial consequence of a connectivity rule that has all potentially related sites overlap

There is further evidence that these characteristics of the projections from somatic sensory and motor cortex to the striatum are not unique:

1. Overlap has been shown also for corticostriatal inputs from the frontal eye field and supplementary eye field in the macaque monkey (Parthasarathy et al., 1992). These inputs, like those from SI and MI, are distributed as sets of partly interconnected input patches (matrisomes). The eye field matrisomes lie adjacent to striatal sites with inputs from nearby, but not eye field, cortex. *This suggests that there are rules for corticostriatal input whereby inputs to the striatum related to a similar function can have coordinated terminal sites, and whereby adjacency as well as overlap is regulated.*

2. In macaque monkey, there are patchy inputs from premotor and prefrontal cortex to the matrix of the caudate nucleus and ventral

putamen (Eblen and Graybiel, 1992, 1993). *Thus input matrisomes are present throughout the striatum, including the nonsensorimotor striatum.*

3. A highly restricted set of cortical inputs reaches striosomes. In macaque monkey, the only strong striosome-selective inputs found so far derive from cortex in the caudal part of the medial prefrontal cortex and from the caudal orbitofrontal neocortex closest to the limbic cortex proper (Eblen and Graybiel, 1992, 1993). *Thus particular corticostriatal inputs from cortex close to the limbic system favor the chemically distinct striosomal system.*

DISTRIBUTED BASAL GANGLIA THROUGHPUTS

What are the advantages of such distributed modular circuits? In the case of matrisomes, we want to put forward two views:

Hypothesis 1

Matrisomes may allow similar versions of a given input to be distributed nearly simultaneously to multiple sites so that different computations can be performed concurrently before the resulting information related to a particular representation or parameter is sent to output sites. To give a concrete example, if different computations are needed for multiple aspects of movement planning about a given joint (e.g., muscle selection, movement direction, joint configuration, or coordinate regulation of different joints), there would be the possibility of doing different local computations at each of the different patches before the results are passed on to the cortex or elsewhere. The computations performed would depend on the differing neighbor relations of the different patches. Even though the different matrisomes related to a particular input could serve as individual processors, their activity could be temporally coordinated to their shared inputs.

It is not just the afferent connections of the striatum that have a clumpy organization; the projection neurons do too. In the squirrel monkey, retrograde tracer injections placed in either segment of the globus pallidus label clusters and bands of striatal medium-sized neurons ("output matrisomes") in the extrastriosomal matrix (Giménez-Amaya and Graybiel, 1990, 1991; Flaherty and Graybiel, 1993a, 1994). *These findings suggest that projection neurons in the striatum are locally ordered according to their output targets in the pallidum.*

Few facts are yet known about this clustering of output neurons, but they suggest interesting characteristics for the throughput of the basal ganglia (figure 5.3).

1. Striatal neurons projecting to each segment of the globus pallidus have a clumpy arrangement. Individual neurons project to one segment or to the other, but some neurons projecting to the external segment of

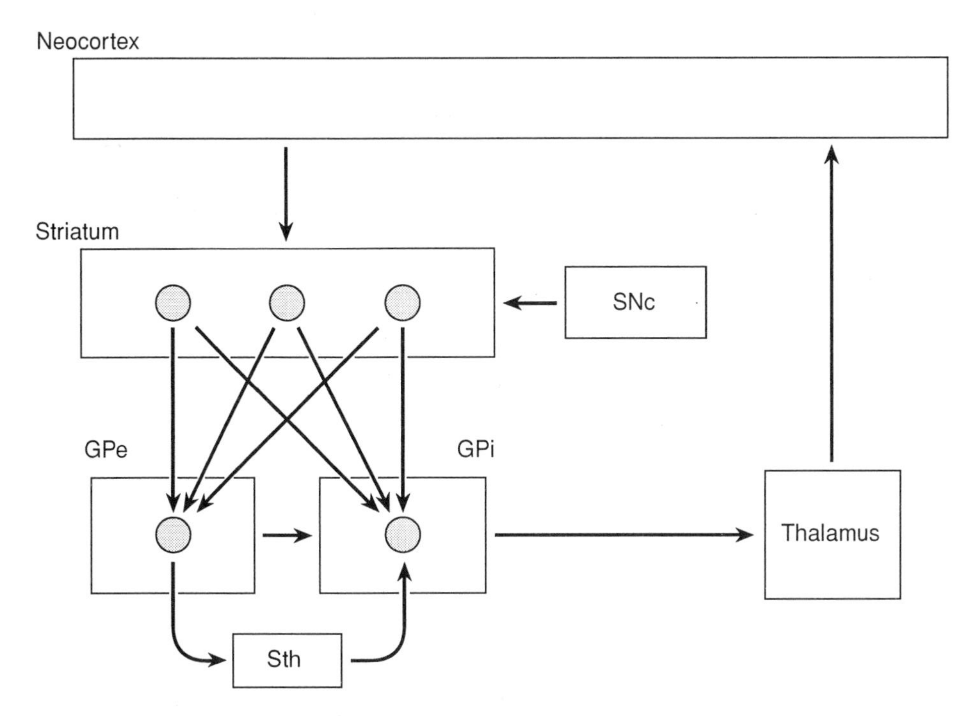

Figure 5.3 Modular divergence and convergence in the striatopallidal pathway. Distributed clusters of projection neurons in the striatum ("output matrisomes") can project to small sites in either segment of the globus pallidus. Dual tracer experiments suggest that the same clusters can project both to the external segment (GPe) and to the internal segment (GPi), although single projection neurons within the clusters project to only one segment. Sth, subthalamic nucleus; SNc, substantia nigra pars compacta. (Based on Flaherty and Graybiel, 1993b, 1994.)

the globus pallidus (GPe) are intermingled with those projecting to the internal segment (GPi); i.e., GPe-projecting matrisomes and GPi-projecting matrisomes overlap (Flaherty and Graybiel, 1993a). *This arrangement suggests that there may be a modular "template" in the striatum for redistributing information to the pallidum, and that there is provision for sending matched signals to GPe and to GPi.*

2. Nearby sites in the pallidum can receive information from different sets of nearby output matrisomes. (Flaherty and Graybiel, 1993a; see also Hoover and Strick, 1993). *This means that there can be distinct channeling of information through the striatopallidal system in addition to modular convergence.*

3. When striatal inputs and outputs have been conjointly labeled in single hemispheres in the squirrel monkey (Flaherty and Graybiel, 1994), instances have been found in which the input patches match the output patches with striking consistency (figure 5.4). *Thus some cortical inputs to the basal ganglia may be dispersed within the striatum only to be brought together again at the next stage in basal ganglia processing.*

These experiments are technically demanding, and though cortical sites could be identified physiologically (by recording and stimulation), the pallidal sites could not be in the acute preparations used. We thus do not know whether, for example, inputs related to the foot representation in S3 and MI, after diverging in the striatum, converge onto a "foot representation" in the pallidum or onto some other functional unit. The fact nonetheless remains that some cortex can distribute inputs to a set of clumps in the striatum that in turn project to small sites in GPe or GPi, or both.

Thus matrisomes may set up redistribution systems for throughputs, including, in some cases, systems bringing the distributed information back together in the pallidum either in the original mapping framework or in new frameworks, or both.

Hypothesis 2

Clustering of striatal afferents into small modules in the striatum could allow spatial and temporal interactions between afferent inputs in the modules so that their effects can be locally coordinated in the time domain. This could, in turn, promote firing of target projection neurons

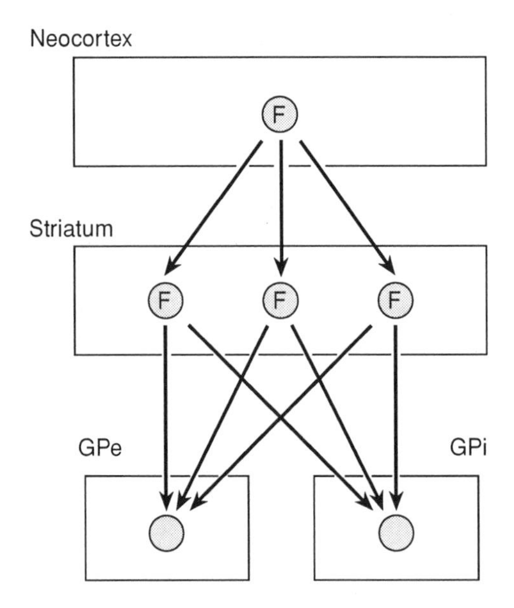

Figure 5.4 Reconvergence in the corticostriatopallidal pathway. Dual anterograde and retrograde tracer experiments suggest that information from somatic sensory cortex can, at least in some instances, diverge in the striatum only to reconverge in a small site in the pallidum. The example shown is based on an experiment in which a small anterograde tracer injection in the area SI foot region labeled multiple patches, and a small retrograde tracer injection in the pallidum of the same hemisphere labeled neurons in a largely overlapping set of patches. F, foot; GP, globus pallidus. (From Flaherty and Graybiel, 1994.)

(and interneurons) and also local activity-dependent modifiability in the input-output functions of the striatum.

According to Wilson (1993) and his colleagues, the projection neurons of the striatum are two-state neurons, either being hyperpolarized (down state) or depolarized (up state). They have to be brought to a threshold state of depolarization before being able to fire. But most afferents to the striatum have been shown to be of the en passant type—passing from one cell's dendritic tree to another much as do parallel fibers of the cerebellum (Fox et al., 1971, 1972). It is highly unlikely that a few afferent inputs to the spines of a projection neuron would fire the neuron. Therefore, it is unlikely that a particular cortical afferent fiber could, by itself, strongly affect a target projection neuron. However, in an input matrisome, there should be many such afferent fibers firing in a coordinated or even synchronous fashion, and this might lead to neural firing by the projection neurons.

The firing of projection neurons would thus be a form of contingent activity, dependent (a) on which afferents projecting to a given input patch are active, (b) on whether the active afferents fire in a temporally coordinated way, (c) on what the afferent connectivity of the neighboring regions of the striatum is—which could enhance or diminish the activity in any particular patch—, and (d) on what the status is of the interneurons near and in the patch.

This proposal about the compartmental architecture of the striatum suggests that the compartments could serve to maximize conditional responsiveness of the projection neurons. These neurons, known for their lack of spontaneous activity, would be brought to threshold when they receive coordinated afferent activation and would then fire in coordinated local groups. Interestingly, given the very large dendrites of pallidal and nigral neurons to which these output cells project, one benefit of local coordination of striatal output neuron activity could be to allow differential summing at their synapses in the target structures. The spatiotemporal coordination of striatal afferents and efferents should also strongly favor synaptic plasticity, for example by hebbian mechanisms involving conventional or unconventional extracellular signals.

PRIVILEGED ACCESS OF BASAL GANGLIA TO INPUTS FROM LIMBIC SYSTEM

Compared to other motor system structures such as the cerebellum, red nucleus, and motor cortex, the basal ganglia receive a large input from the limbic system. This input takes several forms, but it is striking that the amygdala, the hippocampal formation, and the brainstem midline limbic continuum (including the limbic midbrain area and midline thalamus) all project to one or more parts of the striatum, as do the neocortical areas most tightly linked to these limbic structures.

These limbic-related inputs have preferred distributions in the ventral striatum, but some also reach into the caudate nucleus and putamen, including some limbic-related inputs which seek out the neurochemically specialized striosomal system. *These facts suggest that a specialized function of the basal ganglia may be to allow incentive-related signals from the limbic system to help determine whether a movement (or nonmotor function) should be carried out.*

Striosomes, like matrisomes, form distributed systems. They are spread throughout the caudate nucleus and at least part (perhaps all) of the putamen. Accordingly, limbic information reaching sets of related striosomes could be simultaneously distributed to widely separate sites in the striatum. *Thus, ongoing information processing in different striatal districts—with widely different inputs—could be influenced coordinately by signals related to the emotional significance of the information.*

The connections between striosomes and the surrounding matrix appear to be strongly constrained. Many projection neurons have one or more dendrites that avoid crossing striosomal borders (Bolam et al.,

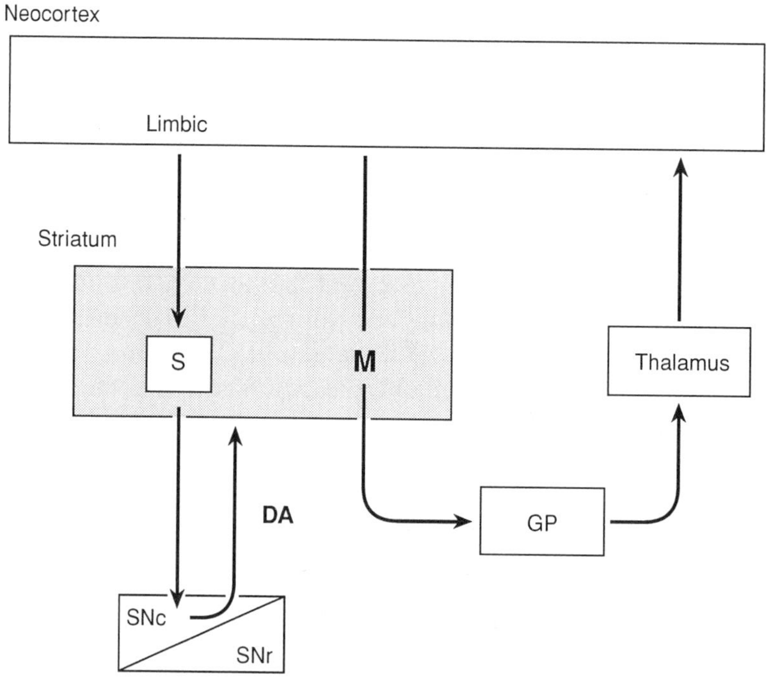

Figure 5.5 Striosomes may represent one source of motivation-related input to the dopaminergic nigrostriatal system. Experiments suggest that striosomes receive related inputs from cortex (shown) and elsewhere (not shown) and project to or near to the substantia nigra pars compacta, whose dopamine-containing cells project to the striatum. S, striosome; M, matrix; SNc, substantia nigra pars compacta; SNr, substantia nigra pars reticulata; DA, dopamine; GP, globus pallidus.

1988; Kawaguchi et al., 1989, 1990; Walker et al., 1993). Not all striatal neurons avoid dendritic crossings, however, and one class of cells whose dendrites freely cross are the cholinergic interneurons (Kawaguchi, 1992; Wilson et al., 1990; Aosaki et al., 1994c). *This suggests that the influence of the limbic inputs to striosomes could be conditionally accessed by neurons of the matrix as well, depending on the state of intrinsic network activity in the striatum.*

Striosomes preferentially target sites of the substantia nigra that are close to, and may indeed include, the dopamine-containing pars compacta (Gerfen, 1984; Jiménez-Castellanos and Graybiel, 1989). These dopamine-containing pars compacta neurons are critical for the normal operation of the striatum. As discussed below, there is now evidence that in behaving monkeys many of these neurons respond preferentially to incentive-related stimuli (Schultz, 1986; Schultz and Romo, 1990; Apicella et al. 1991, Schultz et al., 1993). *If striosomes indeed target dopamine-containing midbrain neurons, then they could form part of a mechanism capable of influencing reward learning as represented in the nigrostriatal system* (figure 5.5).

NEUROPLASTICITY IN BASAL GANGLIA CIRCUITS DURING SENSORIMOTOR CONDITIONING

The above discussion has highlighted potential substrates for temporal coordination and for neuroplasticity in the basal ganglia. We have recently obtained physiological evidence in the behaving primate that strongly supports the concept that timing and adaptive plasticity are both critical features of striatal functions (Aosaki et al., 1992, 1993, 1994a,b). We exposed macaque monkeys to a classical conditioning task in which conditioning stimuli came to signal the forthcoming presentation of rewards, and a behavioral response to the conditioning stimuli was established.

Clicks and light flashes were used as conditioning stimuli, juice was delivered as reward, and the monkey learned to lick for juice when a click or light flash occurred. Recordings from TANs made before, during, and after the training showed that before conditioning, only a small percentage of the TANs (11–21%) responded to the clicks, but that a large percentage of them (51–74%) came to respond to the clicks following conditioning. Similar findings were made with light (light-emitting diode) stimuli as conditioning stimuli.

These acquired neuronal responses were greatly diminished when the dopaminergic fibers in the striatum were destroyed by unilateral intrastriatal infusion of the neurotoxin MPTP (Aosaki et al., 1994c). The responses were not reacquired despite prolonged retraining after the MPTP injection. There were no such changes in the percentage of responsive TANs on the contralateral, uninjected side. *On the basis of these findings we conclude that (1) plastic changes can occur in the associative network of the striatum, and (2) that dopamine acting in the striatum is necessary for the expression of these learned responses.*

Analysis of TAN distributions in the striatum (Aosaki et al., 1994a) suggests that roughly half of these neurons lie at the borders of striosomes. None were found in striosomes. This distribution is significantly different from what would be expected by chance, based on the proportional areas of striosomes and matrix. *Thus, the conditional activity of TANs may be expressed in relation to differential activity in striosomes and matrix.*

The TANs are thought to be striatal interneurons (Kimura et al., 1990) and, judging from their tonic firing and prolonged afterhyperpolarization, characteristic of cholinergic neurons in the rodent striatum (Wilson et al., 1990; Kawaguchi, 1992), the TANs may be cholinergic interneurons of the primate striatum. Cholinergic neurons, as noted above, have dendrites that can cross striosome-matrix boundaries. Other candidate interneurons include somatostatin-containing and parvalbumen-containing populations. The firing characteristics of these cells, at least as judged from slice experiments in rodents (Kawaguchi, 1992), do not resemble those of TANs, however. Thus, a reasonable hypothesis is that the TANs undergoing changes during conditioning are cholinergic interneurons, and that they are able to collect reinforcement-related information from striosomes, along with other inputs, and to influence projection neurons in the matrix. Our experiments suggest that the sensory inputs to such cells can acquire increased synaptic strength through behavioral conditioning, and that this plasticity is gated by dopaminergic nigrostriatal inputs.

The general view suggested by these findings is that at least a subset of the activity patterns of neurons in the primate striatum share the following characteristics: they are conditional, they are achieved through learning, depend on cholinergic-dopaminergic interactions, and are influenced by activity in striosomes. It is interesting that in in vitro experiments, neural plasticity in the form of long-term depression (LTD) is also dependent on cholinergic and dopaminergic receptors (Calabresi et al., 1992).

TIMING AS A CRITICAL CONTROL VARIABLE IN BASAL GANGLIA NETWORKS

One unexpected finding in studying the activity of TANs during conditioning is that TANs recorded in widely separated sites in the striatum responded in a temporally coordinated way to the conditioning stimulus once learning was achieved (Aosaki et al., 1994a,c). This suggests that, after learning, a stimulus that evokes a conditioned behavioral response can trigger concurrent TAN responses across a large part of the striatum, the input machinery of the basal ganglia. What comes to mind is a "now go" signal to the motor system. Just such a signal may be what is missing in the parkinsonian patient, who has lost the permissive gating on dopamine of this signaling system.

The response of TANs to conditioning stimuli following learning includes a sharp reduction of TAN firing that is time-locked to the stimulus presentation (Aosaki et al., 1994a). *We have proposed elsewhere that the time-locked silencing response and subsequent onset of firing could be a latch-on or reset signal that serves to bind distributed striatal circuits in the time domain* (Graybiel et al., 1994) (figure 5.6).

EFFECTS OF BASAL GANGLIA PROCESSING ON ACTION PLANS

The output of the basal ganglia is directed mainly toward the brainstem and toward the frontal lobes. Evolutionarily, one could assume that the brainstem connections were well developed before those with the frontal neocortex. In the best-studied brainstem output system, the connection from the nigral pars reticulata to the superior colliculus, Hikosaka and Wurtz (1983a–d) and Chevalier and Deniau (1990) have shown that signals from the pars reticulata release the activity of neurons in the superior colliculus that participate in the initiation of saccadic eye movements. Memory-related as well as sensory inputs trigger changes in the pars reticulata neurons in the monkey. In these instances

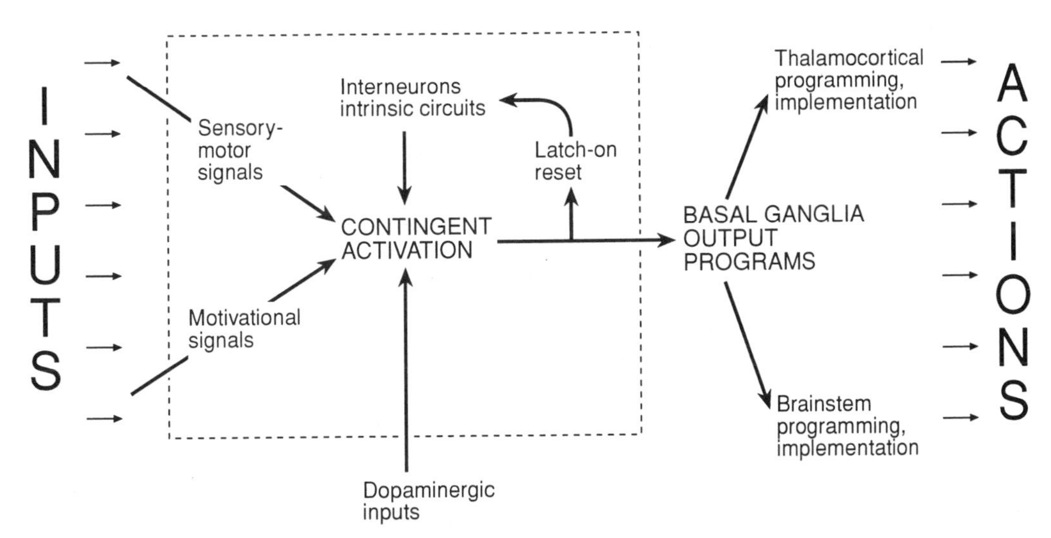

Figure 5.6 Hypothesis is that the striatum (*box*) serves as part of an adaptive motor control system. Both extrinsic signals from the cerebral hemispheres and brainstem and intrinsic signals generated by interneurons and intrinsic circuits contribute to contingent activation of projection neurons and modifiable reset and latch-on functions promoting adaptive adjustments in striatal activity. Basal ganglia output programs contribute to programming and implementation of actions through projections to thalamocortical and brainstem processing networks.

the basal ganglia help to provide a timing cue for the saccade—the release of presaccadic activity—and they help to provide this signal depending on particular internal (mnemonic) and external (sensorimotor) conditions.

There is as yet no explicit information about the comparable effects of basal ganglia signals on the neocortex. There is also no a priori reason to assume that the functions of the ascending basal ganglia system would be parallel to, or elaborate versions of, the downstream functions. We propose, however, that this may be so: that the basal ganglia send to the thalamus and cortex signals about whether to move and when to move. The evidence reviewed suggests that some of the "when" and "whether" conditionals are dependent on inputs to the basal ganglia, and that these importantly involve limbic-related inputs via the striosomes and the substantia nigra complex (see figure 5.6). Other conditional activity appears to reflect ongoing changes in the intrinsic associative circuits of the striatum and may depend on the modular architecture of the striatum. Adaptive changes in behavior thus may critically involve short-term and long-term changes in neural processing in the basal ganglia.

ACKNOWLEDGMENTS

Preparation of this manuscript was supported by NIH Javits Award R01 NS25529 and a Human Frontier Science Program Award.

REFERENCES

Aosaki, T., Ishida, A., Watanabe, K., Imai, H., Graybiel, A. M., and Kimura, M. (1992) Effects of dopaminergic agents on the tonically active neurons of the striatum in hemiparkinsonian monkeys. *Soc. Neurosci. Abstr.* 18:693.

Aosaki, T., Tsubokawa, H., Watanabe, K., Graybiel, A. M., and Kimura, M. (1993) Tonically active neurons in the primate striatum acquire responses to sensory stimuli during behavioral conditioning. *Soc Neurosci Abstr* 19:1585.

Aosaki, T., Graybiel, A. M., and Kimura, M. (1994a) Temporal and spatial characteristics of the tonically active neurons of the striatum (submitted for publication).

Aosaki, T., Kimura, M., and Graybiel, A. M. (1994b) Nigrostriatal dopamine system affects acquired neural responses in striatum of behaving monkeys. *Science,* in press.

Aosaki, T., Tsubokawa, H., Watanabe, K., Graybiel, A. M., and Kimura, M. (1994c) Responses of tonically active neurons in the primate's striatum undergo systematic changes during behavioral sensory-motor conditionings. *J Neurosci,* in press.

Apicella, P., Scarnati, E., and Schultz, W. (1991) Tonically discharging neurons of monkey striatum respond to preparatory and rewarding stimuli. *Exp. Brain Res.* 84:672–675.

Bolam, J. P., Izzo, P. N., and Graybiel, A. M. (1988) Cellular substrate of the histochemically-defined striosome/matrix system of the caudate nucleus: A combined Golgi and immunocytochemical study in cat and ferret. *Neuroscience* 24:853–875.

Calabresi, P., Maj, R., Pisani, A., Mercuri, N. B., and Bernardi, G. (1992) Long-term synaptic depression in the striatum: Physiological and pharmacological characterization. *J. Neurosci.* 12:4224–4233.

Chevalier, G., and Deniau, J. M. (1990) Disinhibition as a basic process in the expression of striatal functions. *Trends Neurosci.* 13:277–280.

Eblen, F., and Graybiel, A. M. (1992) Striosome/matrix affiliations of prefronto-striatal projections in the monkey. *Soc. Neurosci. Abstr.* 18:309.

Eblen F., and Graybiel, A. M. (1993) Highly restricted inputs to striosomes from prefrontal cortex in the macaque monkey. *Soc. Neurosci. Abstr.* 19:1435.

Flaherty, A. W., and Graybiel, A. M. (1991) Corticostriatal transformations in the primate somatosensory system. Projections from physiologically mapped body-part representations. *J. Neurophysiol.* 66:1249–1263.

Flaherty, A. W., and Graybiel, A. M. (1993a) Output architecture of the primate putamen. *J. Neurosci.* 13:3222–3237.

Flaherty, A. W., and Graybiel, A. M. (1993b) Two input systems for body representation in the primate striatal matrix: Experimental evidence in the squirrel monkey. *J. Neurosci.* 13:1120–1137.

Flaherty, A. W., and Graybiel, A. M. (1994) Input-output organization of the primate sensorimotor striatum. *J. Neurosci.* 14:599–610.

Fox, C. A., Andrade, A., Hillman, D. E., and Schwyn, R. C. (1971/1972) The spiny neurons in the primate striatum: A Golgi and electron microscopic study. *J. Hirnforsch.* 13:181–201.

Gerfen, C. R. (1984) The neostriatal mosaic: Compartmentalization of corticostriatal input and striatonigral output systems. *Nature* 311:461–464.

Giménez-Amaya, J.-M., and Graybiel, A. M. (1990) Compartmental origins of the striatopallidal projection in the primate. *Neuroscience* 34:111–126.

Giménez-Amaya, J.-M., and Graybiel, A. M. (1991) Modular organization of projection neurons in the matrix compartment of the primate striatum. *J. Neurosci.* 11:779–791.

Graybiel, A. M. (1984) Neurochemically specified subsystems in the basal ganglia. *Ciba Found. Symp.* 107:114–143.

Graybiel, A. M. (1990) Neurotransmitters and neuromodulators in the basal ganglia. *Trends Neurosci.* 13:244–254.

Graybiel, A. M., Flaherty, A. W., and Giménez-Amaya, J.-M. (1991) Striosomes and matrisomes. In G. Bernardi, M. B. Carpenter, G. di Chiara, M. Morelli, and P. Stanzione (eds), *The Basal Ganglia III.* New York: Plenum Press, pp. 3–12.

Graybiel, A. M., Aosaki, T., Flaherty, A. W., and Kimura, M. (1994) The basal ganglia and adaptive motor control. Submitted for publication.

Hikosaka, O., and Wurtz, R. H. (1983a) Visual and oculomotor functions of monkey *Macaca mulatta* substantia nigra pars reticulata. 4. Relation of substantia nigra to superior colliculus. *J. Neurophysiol.* 49:1285–1301.

Hikosaka, O., and Wurtz, R. H. (1983b) Visual and oculomotor functions of monkey *Macaca mulatta* substantia nigra pars reticulata. 2. Visual responses related to fixation of gaze. *J. Neurophysiol.* 49:1254–1267.

Hikosaka, O., and Wurtz, R. H. (1983c) Visual and oculomotor functions of monkey *Macaca mulatta* substantia nigra pars reticulata. 3. Memory contingent visual and saccade responses. *J. Neurophysiol.* 49:1268–1284.

Hikosaka, O., and Wurtz, R. H. (1983d) Visual and oculomotor functions of monkey *Macaca mulatta* substantia nigra pars reticulata. 1. Relation of visual and auditory responses to saccades. *J. Neurophysiol.* 49:1230–1253.

Hoover, J. E., and Strick, P. L. (1993) Multiple output channels in the basal ganglia. *Science* 259:819–821.

Jiménez-Castellanos, J., and Graybiel, A. M. (1989) Evidence that histochemically distinct zones of the primate substantia nigra pars compacta are related to patterned distributions of nigrostriatal projection neurons and striatonigral fibers. *Exp. Brain Res.* 74:227–238.

Kawaguchi, Y. (1992) Large aspiny cells in the matrix of the rat neostriatum in vitro: Physiological identification, relation to the compartments and excitatory postsynaptic currents. *J. Neurophysiol.* 67:1669–1682.

Kawaguchi, Y., Wilson, C. J., and Emson, P. C. (1989) Intracellular recording of identified neostriatal patch and matrix spiny cells in a slice preparation preserving cortical inputs. *J. Neurophysiol.* 62:1052–1068.

Kawaguchi, Y., Wilson, C. J., and Emson, P. C (1990) Projection subtypes of rat neostriatal matrix cells revealed by intracellular injection of biocytin. *J. Neurosci.* 10:3421–3438.

Kimura, M., Kato, M., and Shimazaki, H. (1990) Physiological properties of projection neurons in the monkey striatum to the globus pallidus. *Exp. Brain. Res.* 82:672–676.

Malach, R., and Graybiel, A. M. (1986) Mosaic architecture of the somatic sensory-recipient sector of the cat's striatum. *J. Neurosci.* 6:3436–3458.

Parthasarathy, H. B., Schall, J. D., and Graybiel, A. M. (1992) Distributed but convergent ordering of striatal projections: The frontal eye field and the supplementary eye field in the monkey. *J. Neurosci.* 12:4468–4488.

Schultz, W. (1986) Responses of midbrain dopamine neurons to behavioral trigger stimuli in the monkey. *J. Neurophysiol.* 56:1439–1461.

Schultz, W., and Romo, R. (1990) Dopamine neurons of the monkey midbrain: Contingencies of responses to stimuli eliciting immediate behavioral reactions. *J. Neurophysiol.* 63:607–624.

Schultz, W., Apicella, P., and Ljungberg, T. (1993) Responses of monkey dopamine neurons to reward and conditioned stimuli during successive steps of learning a delayed response task. *J. Neurosci.* 13:900–913.

Walker, R. H., Arbuthnott, G. W., Baughman, R. W., and Graybiel, A. M. (1993) Dendritic domains of medium spiny neurons in the primate striatum: relationships to striosomal borders. *J. Comp. Neurol.* 337:614–628.

Wilson, C. J. (1993) The generation of natural firing patterns in neostriatal neurons. *Prog. Brain Res.* 99:277–297.

Wilson, C. J., Chang, H. T., and Kitai, S. T. (1990) Firing patterns and synaptic potentials of identified giant aspiny interneurons in the rat neostriatum. *J. Neurosci.* 10:508–519.

6 Macro-organization of the Circuits Connecting the Basal Ganglia with the Cortical Motor Areas

Peter L. Strick, Richard P. Dum, and Nathalie Picard

In this chapter, we briefly present some of our recent anatomical results on the macro-organization of the circuits that link the basal ganglia with the skeletomotor areas in the frontal lobe. Our results suggest that each cortical motor area projects *most densely* to a topographically distinct region of the caudate and putamen. This anatomical arrangement creates multiple "input channels" in the striatum. Similarly, each of these cortical areas appears to be influenced by projections from a topographically distinct region of the internal segment of the globus pallidus (GPi). This arrangement creates multiple "output channels" in GPi. Connections within the basal ganglia tend to connect the input channel related to a particular cortical area with the output channel that innervates the same cortical area. These observations lead us to propose that the underlying structural framework for basal ganglia interactions with the skeletomotor areas of the cerebral cortex is multiple "closed loops."

Before proceeding, it is important to note some of the recent changes in our concepts about the anatomy of the cortical motor areas. It is now clear that the frontal lobe contains multiple motor areas involved in the generation and control of limb movement (for references and review, see Dum and Strick, 1991a,b, 1993). The primary motor cortex receives input from at least six premotor areas in the frontal lobe. Two of these areas are found on the lateral surface of the hemisphere (the dorsal premotor area [PMd] and the ventral premotor area [PMv]) and four of the premotor areas are found on the medial wall of the hemisphere (the supplementary motor area [SMA] and three cingulate motor areas buried in the banks of the cingulate sulcus). Each of the premotor areas projects not only to the primary motor cortex but also directly to the spinal cord (e.g., Dum and Strick, 1991b; He et al., 1993). In fact, the number of corticospinal neurons in the premotor areas equals or exceeds the number in the primary motor cortex. Consequently, multiple parallel pathways from the frontal lobe can contribute to the generation of motor output.

ORGANIZATION OF CORTICAL INPUT TO THE BASAL GANGLIA

We have begun to examine how each of the cortical motor areas projects upon the striatum. Our initial studies focused on projections from the SMA, PMv, and primary motor cortex because the location of arm representation in these areas is relatively well defined and these areas are known to be densely interconnected (for reviews, see Dum and Strick, 1991a,b; He et al., 1993). We were particularly interested in testing the concept that cortical areas which are densely interconnected project to overlapping regions of the striatum (Yeterian and Van Hoesen, 1978).

In several experiments, we employed a double-labeling strategy and injected different tracers into the arm representations of two or more motor areas in the same animal. For example, in one animal the arm area of the PMv was injected with tritiated amino acids, and the arm area of the SMA was injected with wheat germ agglutinin conjugated to horseradish peroxidase (WGA-HRP). At the cortical level, efferents from each injection site terminated in the region of the other injection site (not shown). This indicates that tracer was injected into corresponding portions of the body map in each premotor area. Within the striatum, the two premotor areas projected to comparable rostrocaudal levels of the putamen (figure 6.1). There was some overlap in the anterograde transport from each injection site. However, the arm area of the SMA terminated most densely in the dorsal part of the putamen, whereas the arm area of the PMv terminated most densely in the ventral part of the putamen. Each premotor area appeared to send its densest corticostriatal projection to a separate region of the putamen.

These initial results indicated that a more quantitative approach was required to characterize patterns of corticostriatal termination. There-

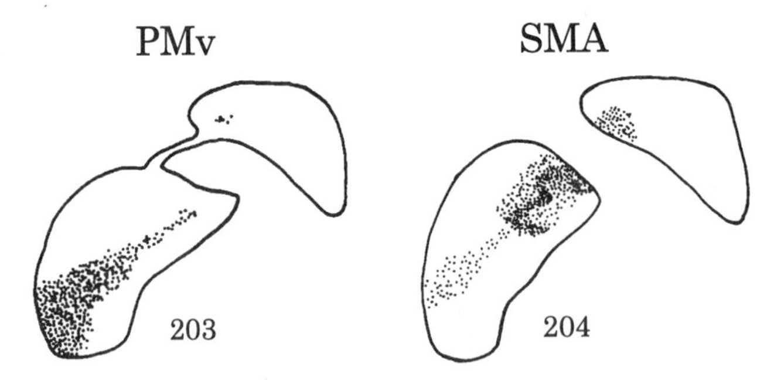

Figure 6.1 Corticostriatal projections from the arm areas of the ventral premotor area (PMv) and the supplementary motor area (SMA). The arm area of the PMv was injected with tritiated amino acids. In the same animal, the arm area of the SMA was injected with WGA-HRP. The *stippling* indicates the anterograde transport from each injection site found on adjacent sections through the putamen and caudate. Note that there is little overlap in the regions of the putamen which receive dense input from each premotor area.

fore, we have begun to examine patterns of termination using computer-based image capture and gradient density analysis. Our first study using this approach examined corticostriatal terminations in the putamen after a large injection of WGA-HRP into the arm area of the primary motor cortex (figure 6.2). In general, we found that the arm area of the primary motor cortex projected to the same rostrocaudal level of the putamen as the arm areas of the SMA and PMv. Nevertheless, the region of the putamen that received dense input from the primary motor cortex appeared to be largely ventral to the region that received dense input from the SMA and dorsal to the region that received dense input from the PMv (cf. figures 6.1 and 6.2). The arm area of the primary motor cortex did have some terminations in the dorsal region of the putamen which is the site of dense terminations from the SMA. However, gradient density analysis showed that, in general, the motor cortex terminations in this region reached only the lowest density levels. Consequently, our initial results suggest that efferents from each motor area terminate most densely in a separate region of the putamen. We propose that this anatomical arrangement creates distinct "input channels" in the striatum.

At this point, one might ask whether the focal nature of corticostriatal terminations might lead to similar patterns of functional activity. We have begun to probe this issue by examining patterns of metabolic activation in the putamen during movement using the 2-deoxyglucose (2DG) technique (Sokoloff et al., 1977). We conditioned a monkey to perform licking movements for a juice reward. During the task, the animal was seated in a primate chair which restrained his head and supported his trunk and both arms. When the monkey performed the task without making any extraneous movements, it was injected with a bolus of [^{14}C] 2-deoxy-D-glucose. We then followed the appropriate procedures to demonstrate 2DG uptake in activated regions of brain tissue. 2DG labeling in the putamen was examined using gradient density analysis (figure 6.3). The region of the putamen that was activated during orofacial movements was located just ventral and adjacent to the region which, in other animals, received dense input from the arm area of the primary motor cortex (cf. figures 6.2 and 6.3). This observation is consistent with the known somatotopic organization of the putamen (e.g., Crutcher and DeLong, 1984; Alexander and DeLong, 1985). Other than the topographic shift, the pattern of metabolic activation in the putamen during orofacial movements is in many respects comparable to the pattern of termination in the putamen of efferents from the arm area of the primary motor cortex. This result suggests that the pattern of metabolic activation in the putamen may actually reflect the anatomical arrangement of corticostriatal connections.

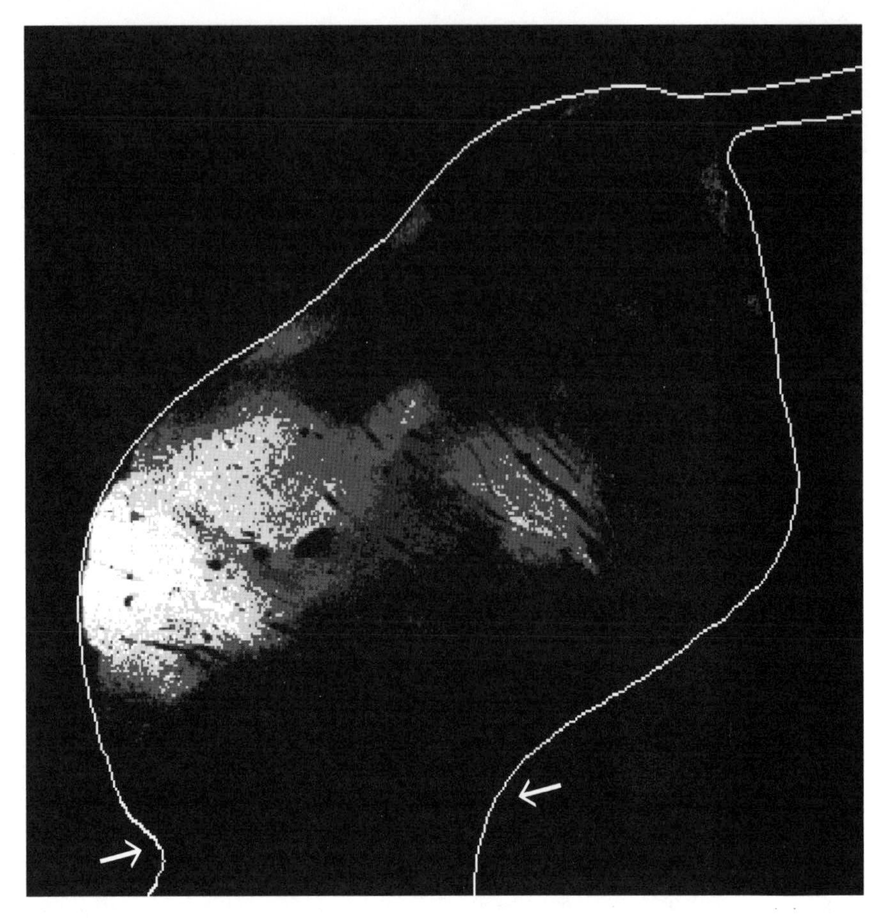

Figure 6.2 Primary motor cortex terminations in the putamen. The arm area of the primary motor cortex was injected with WGA-HRP. Anterograde labeling in the putamen was examined using gradient density analysis. Note that, in comparison with figure 6.1, the regions of the putamen that receive a dense projection from the arm area of the primary motor cortex are largely separate from those that receive dense projections from either the arm area of the ventral premotor area or the supplementary motor area.

ORGANIZATION OF BASAL GANGLIA OUTPUT TO THE CEREBRAL CORTEX

We have employed a newly developed technique, transneuronal transport of herpes simplex virus, to examine multiple stages in the organization of basal ganglia connections with the motor areas of cerebral cortex. Virus transport provides a unique method for labeling a chain of synaptically linked neurons (for references and review, see Strick and Card, 1992). This technique enables us to trace the cortical connections of basal ganglia neurons through two or more synapses.

We have worked most extensively with two strains of herpes simplex virus type 1 (HSV-1), the H129 and McIntyre-B strains. These strains

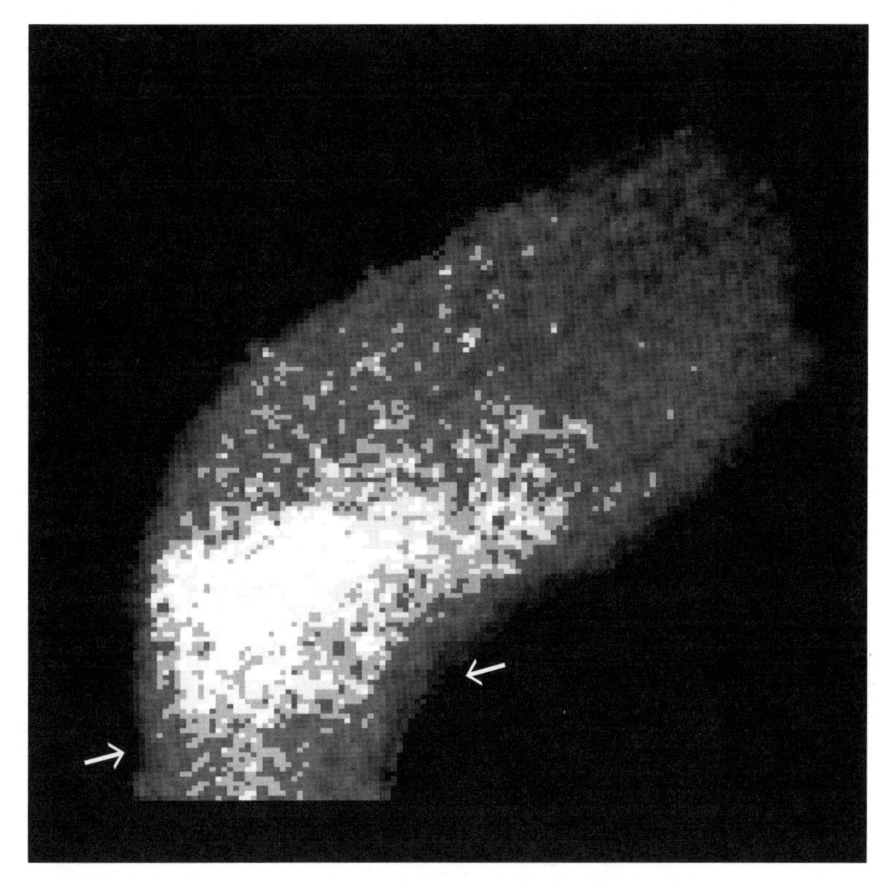

Figure 6.3 Metabolic activation in the putamen associated with orofacial movements. A monkey was conditioned to lick for fluid reward. Then the animal was injected with ^{14}C-labeled 2-deoxyglucose during task performance. Labeling in the putamen was examined using gradient density analysis. The arrows in this figure are placed in the same relative location as those in figure 6.2. Note that the metabolic activation in the putamen during orofacial movements is located just ventral to the region which receives dense input from the arm area of the primary motor cortex.

are transported transneuronally in different directions (Zemanick et al., 1991). The H129 strain is transported transneuronally in the *anterograde* direction (figures 6.4 and 6.5). For example, 3 days after injection of the H129 strain into the arm area of primary motor cortex, virus was transported in the anterograde direction from first-order neurons at the injection site to second-order neurons in the putamen. The labeled neurons in the putamen were found in regions that receive input from the arm area of the primary motor cortex (cf. figures 6.2 and 6.5). Five days after cortical injection of the H129 strain, there was a second stage of anterograde transneuronal transport from second-order neurons in the putamen to third-order neurons in the external (GPe) and internal segments (GPi) of the globus pallidus (see figures 6.4 and 6.5). The ratio

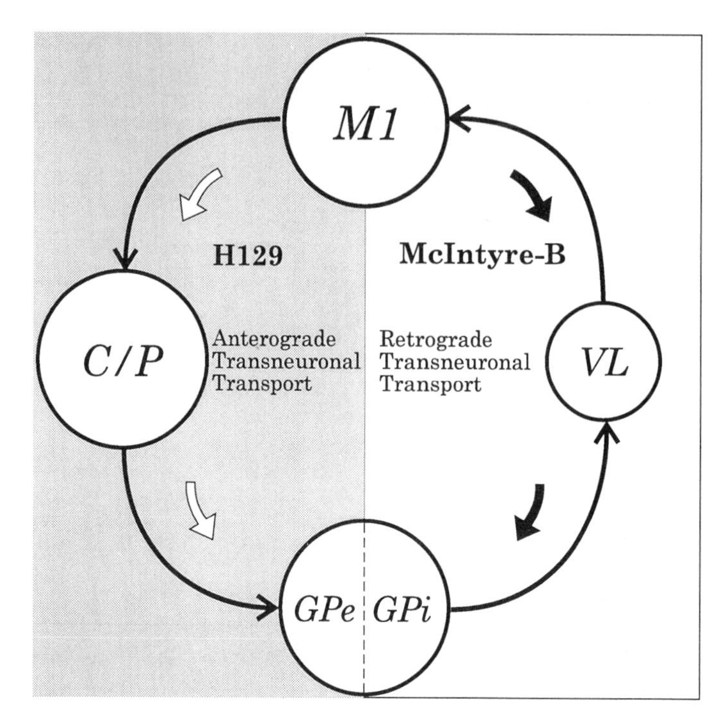

Figure 6.4 Patterns of herpes simplex virus type 1 (HSV-1) transneuronal transport in basal ganglia circuits. Different strains of HSV-1 are transported transneuronally in different directions. The H129 strain is transported transneuronally in the anterograde direction. After injections of this strain into the primary motor cortex (M1), virus moves from the injection site to label second-order neurons in the caudate and putamen (C/P) and then third-order neurons in the external (GPe) and internal segments (GPi) of the globus pallidus. In contrast, the McIntyre-B strain is transported transneuronally in the retrograde direction. After injections of the McIntyre-B strain into M1, virus moves from the injection site to label first-order neurons in the ventrolateral thalamus (VL), and then second-order neurons in GPi.

of labeled neurons in GPe to GPi was 10:1. The labeled neurons in the globus pallidus were confined largely to ventral regions of the caudal third of GPe. This location overlaps the region of GPe where electrophysiological studies have found neurons with activity changes related to arm movements (e.g., DeLong et al., 1985; Anderson and Horak, 1985; Hamada et al., 1990; Mink and Thach, 1991). Taken together, these results suggest that the pathway from the arm area of the primary motor cortex through the putamen to GPe targets a highly localized region of GPe.

In contrast to the H129 strain, the McIntyre-B strain of HSV-1 is transported transneuronally in the *retrograde* direction (Zemanick et al., 1991; Hoover and Strick, 1993) (see figures 6.4 and 6.5). For example, 3 days after injections of McIntyre-B strain into the arm area of the primary motor cortex, densely labeled neurons were found in subdivisions

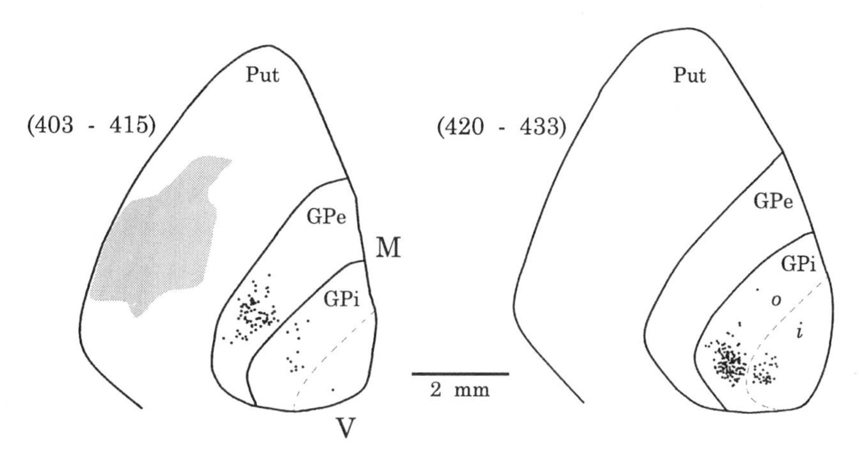

Figure 6.5 Transneuronal transport of herpes simplex virus type 1 (HSV-1) to external (GPe) and internal (GPi) segments of globus pallidus. The arm area of the primary motor cortex was injected with different strains of HSV-1. (*Left*) Strain H129, animal Z6. (*Right*) Strain McIntyre-B, animal Z10. The two diagrams are outlines of cross sections through the putamen (Put) and GPe and GPi. A dashed line indicates the border between the outer (*o*) and inner (*i*) portions of GPi. Six sections, 100 to 150 μm apart, were overlapped to produce each diagram. The shading in the diagram of Z6 indicates the location of large numbers of labeled neurons in the putamen. Note that this region is comparable to the region labeled by anterograde transport after injections of a conventional tracer into the arm area of the primary motor cortex (see figure 6.2). The small dots in Z6 and Z10 indicate the location of labeled neurons in GPe and GPi. Note that most of the labeled neurons were found in GPe after anterograde transneuronal transport of the H129 strain. In contrast, all of the labeled neurons were found in GPi after retrograde transneuronal transport of the McIntyre-B strain. M, medial; V, ventral. (Adapted from Zemanick et al., 1991).

of the ventrolateral thalamus that are known to innervate the primary motor cortex (e.g., ventralis posterior lateralis pars oralis [VPLo] and ventralis lateralis pars oralis, [VLo]) (e.g., Holsapple et al., 1991). Five days after the cortical injections of the McIntyre-B strain, virus was transported transneuronally in the retrograde direction from first-order neurons in the ventrolateral thalamus to second-order neurons in GPi (see figures 6.4 and 6.5). Virus antigen densely filled the labeled GPi cells, clearly marking their somata and primary dendrites. The labeled primary dendrites coursed obliquely through GPi, but were most concentrated in the regions of GPi that contained labeled cell bodies.

The labeled neurons in GPi were confined almost exclusively to midrostrocaudal levels of the nucleus and formed two distinct clusters, one in the outer portion of GPi and the other in its inner portion (figures 6.5 and 6.6). The ratio of labeled neurons in the outer portion to the inner portion was 4:1. The location of labeled neurons in GPi was comparable to the region where electrophysiological studies have found GPi neurons that were inhibited by stimulation of the primary motor cortex (Yoshida et al., 1993). This region of GPi also contains neurons with activity related to arm movements (e.g., DeLong et al., 1985; Anderson and Horak, 1985; Hamada et al., 1990; Mink and Thach, 1991; Yoshida et al., 1993). These results support the proposal that the arm area of the primary motor cortex is the target of output from the basal ganglia (for references and discussion, see Holsapple et al., 1991). Furthermore, these observations indicate that the projection from GPi through the ventrolateral thalamus to the arm area of the primary motor cortex originates from a specific region within GPi.

In recent studies, we have used retrograde transneuronal transport of the McIntyre-B strain to examine projections from GPi to the arm representations of two premotor areas, the SMA and PMv (Hoover and

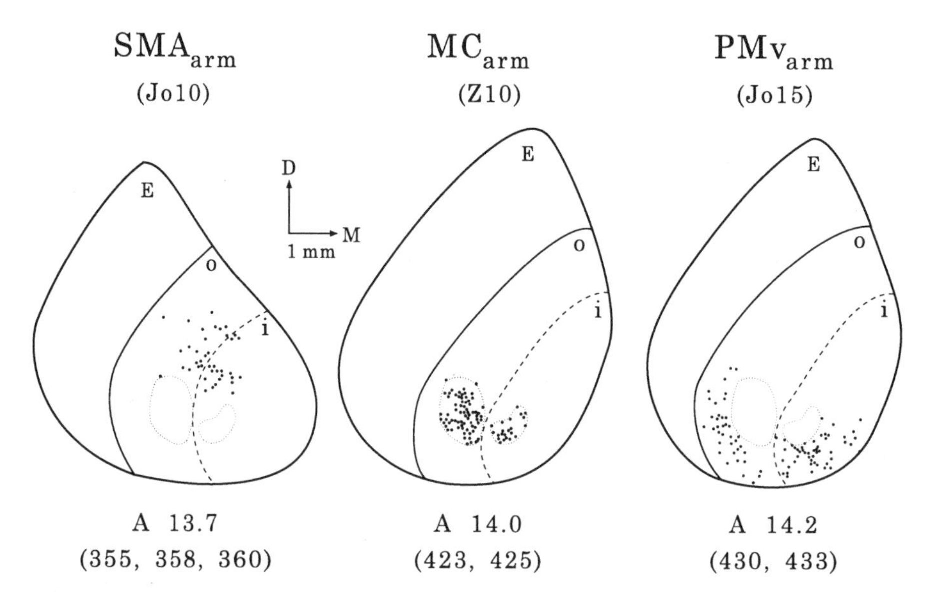

Figure 6.6 Origin of pallidal projections to three cortical motor areas. This figure illustrates representative coronal sections through the globus pallidus of animals that received injections of the McIntyre-B strain of herpes simplex virus type 1 into the arm representations of the supplementary motor area (SMA) (animal Jo10, *left*), primary motor cortex (MC) (animal Z10, *middle*), or ventral premotor area (PMv) (animal Jo15, *right*). The solid dots indicate the positions of neurons labeled by retrograde transneuronal transport of virus. Labeled neurons are charted from two or three adjacent sections (section numbers at bottom in parentheses). For comparison, the fine dotted lines in the diagrams on the right and left indicate the region of GPi-containing neurons labeled from the primary motor cortex in animal Z10. The thick solid line indicates the outline of the globus pallidus (GP). The thin solid line indicates the border between GPe and GPi. The dashed line indicates the border between the inner and outer portions of the internal segment of GP (GPi). E, external segment of GP; i, inner portion of GPi; o, outer portion of GPI; D, dorsal; M, medial. (Reproduced from Hoover and Strick, 1993.)

Strick, 1993). Like virus injections into the primary motor cortex, virus injections into the arm representation of the SMA or PMv labeled many neurons in GPi (see figure 6.6). An average of 1300 GPi neurons were labeled by transneuronal transport in each experiment. These labeled neurons were found in both the inner and outer portions of GPi.

The portion of GPi containing a high density of labeled neurons was consistently limited to a 1- to 2-mm region in the middle of its anteroposterior extent. Within this region, the dorsoventral location of labeled pallidal neurons varied depending on the location of the cortical injection site. Virus injections into the arm representation of the SMA labeled neurons in a dorsal region of GPi (see figure 6.6, animal *Jo10*). In contrast, injections into the arm representation of the PMv labeled neurons mainly in ventrolateral portions of GPi (see figure 6.6, animal *Jo15*). Neurons labeled after injections into the arm area of the primary motor cortex were located between the two groups of neurons labeled by the SMA and PMv injections (see figure 6.6, animal *Z10*). These observations indicate that pallidothalamocortical projections target premotor and primary motor areas of the cerebral cortex. Furthermore, our results suggest that each of these cortical areas receives a projection from a topographically distinct set of GPi neurons. We propose that this anatomical arrangement creates distinct "output channels" in GPi (figure 6.7).

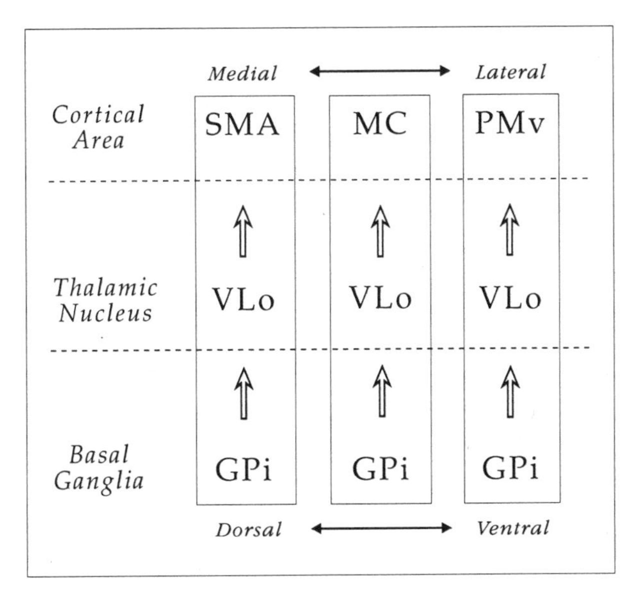

Figure 6.7 Output channels in internal segment of globus pallidus (GPi). The arm representations in the supplementary motor area (SMA), primary motor cortex (MC), and ventral premotor area (PMv) are each the target via the pars oralis of the ventral lateral thalamus (VLo), of projections from distinct regions of GPi. (reproduced from Hoover and Strick, 1993.)

SUMMARY AND CONCLUSIONS

We believe that the anatomical connections briefly outlined in this chapter reveal some of the basic structural framework for interactions between the basal ganglia and the cerebral cortex (figure 6.8). Our observations suggest that the primary focus of inputs from different cortical motor areas (e.g., the SMA, primary motor cortex, and PMv) is to topographically separate regions of putamen. This arrangement creates distinct input channels in the "motor" portion of the striatum. Each of these input channels may be somatotopically organized.

There is considerable evidence that the input channels related to the skeletomotor areas of cerebral cortex are topographically separate from those associated with other cortical areas, such as the prefrontal cortex (for references and review, see Alexander et al., 1986). However, there is also convincing evidence that a given input channel may receive projections from more than one cortical area (e.g., Yeterian and Van Hoesen, 1978; Selemon and Goldman-Rakic, 1985; Flaherty and Graybiel, 1991, 1993; Parthasarathy et al., 1992). For example, Flaherty and Graybiel (1991, 1993) have shown that inputs from the representations of homologous body parts in the primary somatosensory and primary motor cortex overlap in small zones in the striatum. On the other hand, homologous sites in the ipsilateral primary somatosensory cortex and *contralateral* primary motor cortex do not project to the same small zones (even though they may project to the same input channel). Thus, on the local level, the rules for overlap of corticostriatal inputs appear to be complex. Our prediction is that, at the macro level, a given input channel receives its densest projection from one or perhaps two closely related cortical areas. These projections would be considered the "defining input" for the channel. Each input channel in the striatum may also receive projections from other cortical areas, but these projections would be less dense than those of the defining inputs. At this point, these predictions should be tested by performing a complete analysis of projections from skeletomotor and related areas of the cerebral cortex onto the sensorimotor portion of the putamen. Clearly, the patterns of convergence of different cortical systems in the striatum remain to be defined.

Our results using retrograde transneuronal transport of HSV-1 suggest that topographically separate regions of GPi project to different cortical areas. Further support for this suggestion comes from a recent physiological study which found that pallidal neurons responding to inputs from different cortical areas were localized within separate regions of GPi (Yoshida et al., 1993). We propose that the clustering of output neurons which project to a common cortical area creates distinct output channels in the motor portion of GPi.

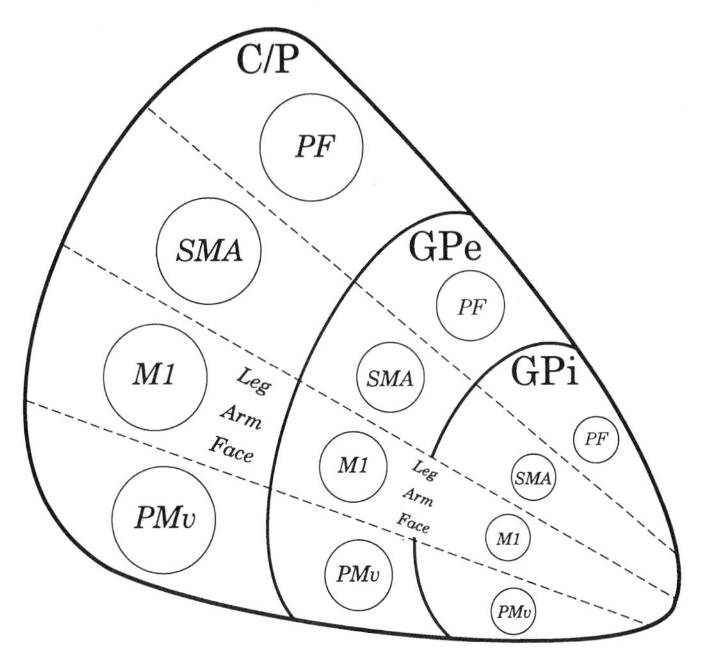

Figure 6.8 Organization of basal ganglia loops with the motor areas of cerebral cortex. This diagram schematically represents our proposal about the input-output organization of circuits which link the caudate and putamen (C/P) and external (GPe) and internal (GPi) segments of globus pallidus with the cortical motor areas. See text for details. PF, prefrontal cortex; SMA, supplementary motor area; M1, primary motor cortex, PMv, ventral premotor area.

Recently, we have used single-neuron recording techniques in awake trained primates to examine some of the response properties of neurons in different output channels of GPi (Mushiake and Strick, 1993; Strick et al., 1993). We found that pallidal neurons that were preferentially active during the performance of remembered sequences of movement were located dorsally in GPi, potentially in the output channel that innervates the SMA. This output channel may be especially involved in the guidance of movements based on internal cues or determining the serial order of motor behavior or both. On the other hand, those pallidal neurons that were preferentially active during the performance of sequences of movements guided by external cues were located ventrally in GPi, potentially in the output channel that innervates the PMv. This output may be particularly concerned with guiding movement based on external cues. Thus, our results suggest that individual output channels are preferentially involved in different aspects of motor behavior (see also Nambu et al., 1990).

Like the input channels to the striatum, the output channels related to the skeletomotor areas of cerebral cortex also appear to be topographically separate from those associated with other cortical areas such as the prefrontal cortex (for references and review, see Alexander et al., 1986). Preliminary data from anatomical experiments in our laboratory which used retrograde transneuronal transport of HSV-1 from area 46 in the frontal lobe supports this view (Middleton and Strick, unpublished observations). However, the number of distinct output channels in GPi, the cortical areas innervated by each output channel, and whether this arrangement is characteristic of all pallidal projections to cerebral cortex remain to be determined.

Finally, anatomical studies have provided considerable evidence that interconnections between the striatum, GPe, and GPi are largely along the radial dimension (e.g., Hazrati and Parent, 1992; for further references and review, see Parent and Hazrati, 1993). As a consequence, these pathways would tend to connect the input channel related to a particular cortical area with the output channel innervating the same cortical area. This anatomical arrangement would create a closed loop which we propose is the underlying structural framework for basal ganglia interactions with the cerebral cortex. If our proposal is correct, many such closed loops, each with a similar internal organization, interconnect the basal ganglia with different regions of the cerebral cortex. A challenge for future studies is to determine what computational operation is performed in these loops that can be commonly applied to motor and nonmotor areas of the cerebral cortex.

ACKNOWLEDGMENTS

This work was supported by funds to P.L.S. from the VA Medical Research Service and by U.S. Public Health Service grants NS2957 and NS24328 and a fellowship to N.P. from the Medical Research Council of Canada.

REFERENCES

Alexander, G. E., and DeLong, M. R. (1985) Microstimulation of the primate neostriatum. II. Somatotopic organization of striatal microexcitable zones and their relation to neuronal response properties. *J. Neurophysiol.* 53:1417–1430.

Alexander, G. E., DeLong, M. R., and Strick, P. L. (1986) Parallel organization of functionally segregated circuits linking basal ganglia and cortex. *Annu. Rev. Neurosci.* 9:357–381.

Anderson, M. E., and Horak, F. B. (1985) Influence of the globus pallidus on arm movements in monkeys. III. Timing of movement-related information. *J. Neurophysiol.* 54:433–448.

Crutcher, M. D., and DeLong, M. R. (1984) Single cell studies of the primate putamen. I. Functional organization. *Exp. Brain Res.* 53:233–243.

DeLong, M. R., Crutcher, M. D., and Georgopoulos, A. P. (1985) Primate globus pallidus and subthalamic nucleus: Functional organization. *J. Neurophysiol.* 53:530–543.

Dum, R. P., and Strick, P. L. (1991a) Premotor areas: Nodal points for parallel efferent systems involved in the central control of movement. In D. R. Humphrey and H.-J. Freund (eds.), *Motor Control: Concepts and Issues*. London: John Wiley & Sons Ltd., pp. 383–397.

Dum, R. P., and Strick, P. L. (1991b) The origin of corticospinal projections from the premotor areas in the frontal lobe. *J. Neurosci.* 11:667–689.

Dum, R. P., and Strick, P. L. (1993) The cingulate motor areas. In B. A. Vogt and M. Gabriel (eds.), *The Neurobiology of Cingulate Cortex and Limbic Thalamus*. Boston: Birkhauser, pp. 415–441.

Flaherty, A. W., and Graybiel, A. M. (1991) Corticostriatal transformations in the primate somatosensory system. Projections from physiologically mapped body-part representations. *J. Neurophysiol.* 66:1249–1263.

Flaherty, A. W., and Graybiel, A. M. (1993) Two input systems for body representation in the primate striatal matrix: Experimental evidence in the squirrel monkey. *J. Neurosci.* 13:1120–1137.

Hamada, I., DeLong, M. R., and Mano, N. (1990) Activity of identified wrist-related pallidal neurons during step and ramp wrist movements in the monkey. *J. Neurophysiol.* 64:1892–1906.

Hazrati, L.-N., and Parent, A. (1992) The striatopallidal projection displays a high degree of anatomical specificity in the primate. *Brain Res.* 592:213–227.

He, S.-Q., Dum, R. P., and Strick, P. L. (1993) Topographic organization of corticospinal projections from the frontal lobe: Motor areas on the lateral surface of the hemisphere. *J. Neurosci.* 13:952–980.

Holsapple, J. W., Preston, J. B., and Strick, P. L. (1991) The origin of thalamic inputs to the 'hand' representation in the primary motor cortex. *J. Neurosci.* 11:2644–2654.

Hoover, J. E., and Strick, P. L. (1993) Multiple output channels in the basal ganglia. *Science* 259:819–821.

Mink, J. W., and Thach, W. T. (1991) Basal ganglia motor control: I. Nonexclusive relation of pallidal discharge to five movement modes. *J. Neurophysiol.* 65:273–300.

Mushiake, H., and Strick, P. L. (1993) Activity of pallidal neurons during sequential movements. *Soc. Neurosci. Abstr.* 19:1584.

Nambu, A., Yoshida, S., and Jinnai, K. (1990) Discharge patterns of pallidal neurons with input from various cortical areas during movement in the monkey. *Brain Res.* 519:183–191.

Parent, A., and Hazrati, L.-N. (1993) Anatomical aspects of information processing in primate basal ganglia. *Trends Neurosci.* 16:111–116.

Parthasarathy, H. B., Schall, J. D., and Graybiel, A. M. (1992) Distributed but convergent ordering of striatal projections: The frontal eye field and the supplementary eye field in the monkey. *J. Neurosci.* 12:4468–4488.

Selemon, L. D., and Goldman-Rakic, P. S. (1985) Longitudinal topography and interdigitation of corticostriatal projections in the rhesus monkey. *J. Neurosci.* 5:776–794.

Sokoloff, L., Reivich, M., Kennedy, C., DesRosiers, M. H., Patlak, C. S., Pettigrew, K. S., Sakurada, O., and Shinohara, M. (1977) The [^{14}C]-deoxyglucose method for the measurement of local cerebral glucose utilization: Theory, procedure, and normal values in the conscious and anesthetized albino rat. *J. Neurochem.* 28:879–916.

Strick, P. L., and Card, J. P. (1992) Transneuronal mapping of neural circuits with alpha herpesviruses. In J. P. Bolam (ed.), *Experimental Neuroanatomy: A Practical Approach,* Oxford: Oxford University Press, pp. 81–101.

Strick, P. L., Hoover, J. E., and Mushiake, H. (1993) Evidence for "output channels" in the basal ganglia and cerebellum. In N. Mano, I. Hamada, and M. R. DeLong (eds.), *Role of the Cerebellum and Basal Ganglia in Voluntary Movement.* Amsterdam: Elsevier, pp. 171–180.

Yeterian, E. H., and Van Hoesen, G. W. (1978) Cortico-striate projections in the rhesus monkey. The organization of certain cortico-caudate connections. *Brain Res.* 139:43–63.

Yoshida, S., Nambu, A., and Jinnai, K. (1993) The distribution of the globus pallidus neurons with input from various cortical areas in the monkeys. *Brain. Res.* 611:170–174.

Zemanick, M. C., Strick, P. L., and Dix, R. D. (1991) Transneuronal transport of herpes simplex virus type 1 in the primate motor system: Transport direction is strain dependent. *Proc. Natl. Acad. Sci. U. S. A.* 88:8048–8051.

7 Toward a Circuit Model of Working Memory and the Guidance of Voluntary Motor Action

Patricia S. Goldman-Rakic

The prefrontal or granular frontal cortex, as a component of the frontal lobe, has traditionally been considered functionally more related to central motor structures than to sensory systems. Nevertheless, compared with the primary motor cortex and premotor areas, especially those with direct projections to the spinal cord, the prefrontal role in motor control is still quite obscure. Three issues seem paramount. First, what is the distinctive contribution of prefrontal compared with premotor or motor or any other cortical area to behavior? Second, how do the functions of prefrontal cortex become integrated with those of other areas involved in motor control? A related issue here is the nature, degree, and purpose of parallel pathways described between frontal lobe and other cortical areas and the neostriatum. Third, what are the cellular mechanisms underlying prefrontal function? These are interrelated subjects and progress into one aspect is likely to have ramifications for the other aspects of the cortical control of action. In this chapter, I present some ideas developed from work in this and other laboratories.

Recently, it has become clear that multiple areas in the dorsolateral prefrontal cortex, in particular areas 8, 46, 12, and 45, play an essential role in what has been termed memory-guided performance. I have proposed that a working memory process is the fundamental specialization of prefrontal cortex and the mechanism for directing responses by *internal* representations, which can be considered the basis for memory-guided responding (Goldman-Rakic, 1987). Further, this process distinguishes the prefrontal contribution to behavior from those systems of the brain and cortex that guide behavior by associative processes, by sensory guidance, or by prepotent reflexive mechanisms. These latter processes are considered to be the province of posterior association regions, including the hippocampal formation—regions of the cerebrum which have been accorded a major role in a large fraction of implicit and associatively learned behaviors and are considered the storage sites for the facts, events, instructions, concepts, rules, and habits that are the products of long-term conditioning and practice. Our contention is that these products of learning and past experience are accessed by prefrontal neurons which process them and amalgamate them with the ongoing stream of information that is currently being experienced. With respect to motor action, we should not fail to heed that the utterances of language, which are directed by purely *representational* processes and not by external stimuli, are guided by an on-line processor in one or more areas of the prefrontal cortex. There is ample evidence that the brain respects the distinction between associative, sensory-guided behavior and memory-guided, i.e., representationally guided responding. Since this theory and its neural underpinnings has been reviewed in numerous chapters and reviews (e.g., Goldman-Rakic, 1987, 1988, 1990; Goldman-Rakic et al., 1993), this chapter focuses on and proposes some possible models of interaction between prefrontal circuits and components of basal ganglia circuitry.

PHYSIOLOGICAL FEATURES OF PREFRONTAL NEURONS REVEALED BY DELAYED-RESPONSE TASKS

By now it is well known that prefrontal neurons increase their discharge rate during the delay period of delayed-response trials, as originally shown by Fuster and Alexander (1971) and Kubota and Niki (1971). More recent studies have explored the role of the dorsolateral prefrontal cortex in motor behavior with an oculomotor version of the delayed-response task (Boch and Goldberg, 1989; Barone and Joseph, 1989, Funahashi et al., 1989, 1990, 1991) In our own studies prefrontal neurons have been shown to have "memory fields," that is, to increase their firing when a particular target, and only that target, disappears from view and has to be recalled several seconds later (Funahashi et al., 1989). Figure 7.1 displays an example of a prefrontal neuron with a *memory field*, i.e., the same neuron preferentially codes the same location trial after trial for one of four target locations. In the example shown, the neuron has its "best direction" for the 90-degree target, though many neurons generally have memory fields contralateral to the hemisphere in which the neurons are recorded. Further, consistent with the concept of a memory field, delay activity in prefrontal neurons is tuned for the distance of a stimulus from the fovea as well as for its direction (O Scalaidhe and Goldman-Rakic, 1993). Importantly, the prefrontal neurons in these studies are activated solely during the time that the information is held "on line." As soon as a motor action based on that information is initiated, the neuron's activity returns to baseline (Funahashi et al., 1991). In this way, memories for location of objects are mapped transiently in the prefrontal cortex and new information is updated continually. Finally, in instances where the mnemonic activity of a neuron is not maintained throughout the delay period, the animal generally makes an error (Funashashi et al., 1989). These and other results provide strong evidence at a cellular level for a role of prefrontal

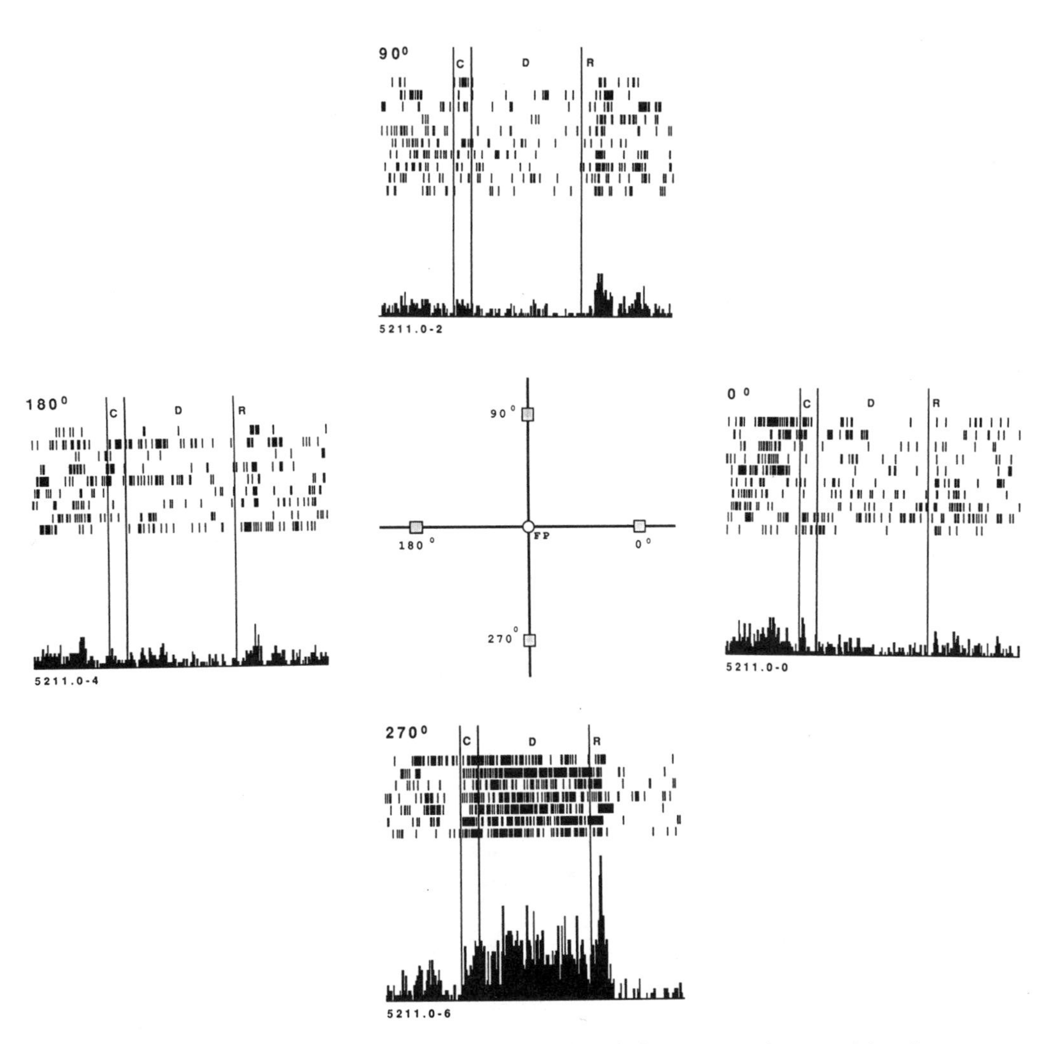

Figure 7.1 Neuronal activity from a single cell during an oculomotor delayed-response task with four target locations. Neuronal discharge is increased during the delay period only for one of the four target locations. This preferential activation is considered a "memory field" and in this case the neuron's best direction is the 270-degree direction.

neurons in representational processes, i.e., maintenance of information in the *absence* of the stimulus that was initially present.

Prefrontal neurons with memory fields have complex features related to the real-time events in a delayed-response task. They may register a response to the presention of a visual stimulus, followed by tonic activation during the delay period (delay-period activity) and also express a presaccadic activation (Funahashi et al., 1991). The presaccadic responses like the delay-period and cue-period activity, are directionally specific, generally favoring contralateral or, as shown in figure 7.1, vertical visual fields. Thus the same cell that encodes a target direction in working memory may also issue (or be issued) a motor command.

At the same time, activity of such a neuron may be suppressed for the opponent direction where instead a *postsaccadic* response can often be seen, possibly serving as a signal that a response has been executed (see figure 7.1). About 200 to 400 ms may elapse between the peak of presaccadic activation and the peak of the postsaccadic responses in the population of neurons studied by Funahashi et al. (1991). Since the occurrence of the postsaccadic response coincides with the return of neuronal activity to baseline firing rate, we have posited that this signal is a feedback signal possibly in the form of a corollary discharge from motor centers (Goldman-Rakic et al., 1990; Funahishi et al., 1991). As suggested below, this feedback signal from oculomotor structures could arrive via the thalamocortical pathway.

INHIBITION IN MEMORY CIRCUITS

Inhibition also plays an important role in prefrontal circuitry. First, the inhibition is important in the construction of a specific memory field in several ways. First and foremost, many neurons have opponent memory fields, i.e., some are inhibited in the direction opposite that of their best directions (Funahashi et al., 1989). Thus, the same neuron is activated under some conditions (prospective memory for the opponent) and is inhibited under other conditions (memory for its nonpreferred, opponent, target). This pattern of activity also has to be accounted for. Some prefrontal neurons do not exhibit enhanced activity at all but rather are inhibited preferentially for a given target direction (figure 7.2). Finally, a small fraction of prefrontal neurons do not have memory fields but are activated in the delay period equally or nearly so for all directions, which we termed *omnidirectional* (Funahashi et al., 1989). Similar omnidirectional neurosonal profiles are a more consistent finding in recordings from the posterior cingulate regions (Carlson et al., 1993) with which the prefrontal cortex is connected (Selemon and Goldman-Rakic, 1989) and it is possible that this pattern in prefrontal neurons is a reflection of prefrontal-posterior cingulate circuitry. The unique firing patterns occurring during the same behavior task in a different but interconnected area reinforce the view that prefrontal cortex is perhaps uniquely concerned with the informational content of a memorandum and not the nonspecific anticipation associated with responses in any or all directions.

Another source of inhibition is that necessary to sharply turn off activation (or inhibition) related to the delay, and finally there must be a tonic source of inhibition when the neuron is not firing at its maximal rate as in the intertrial interval, before the cue or after the response. Recent studies by Frazer Wilson and Seamus O Scalaidhe in this laboratory have begun to identify the contribution of inhibitory interneurons in the carving of memory fields. Based on physiological studies of

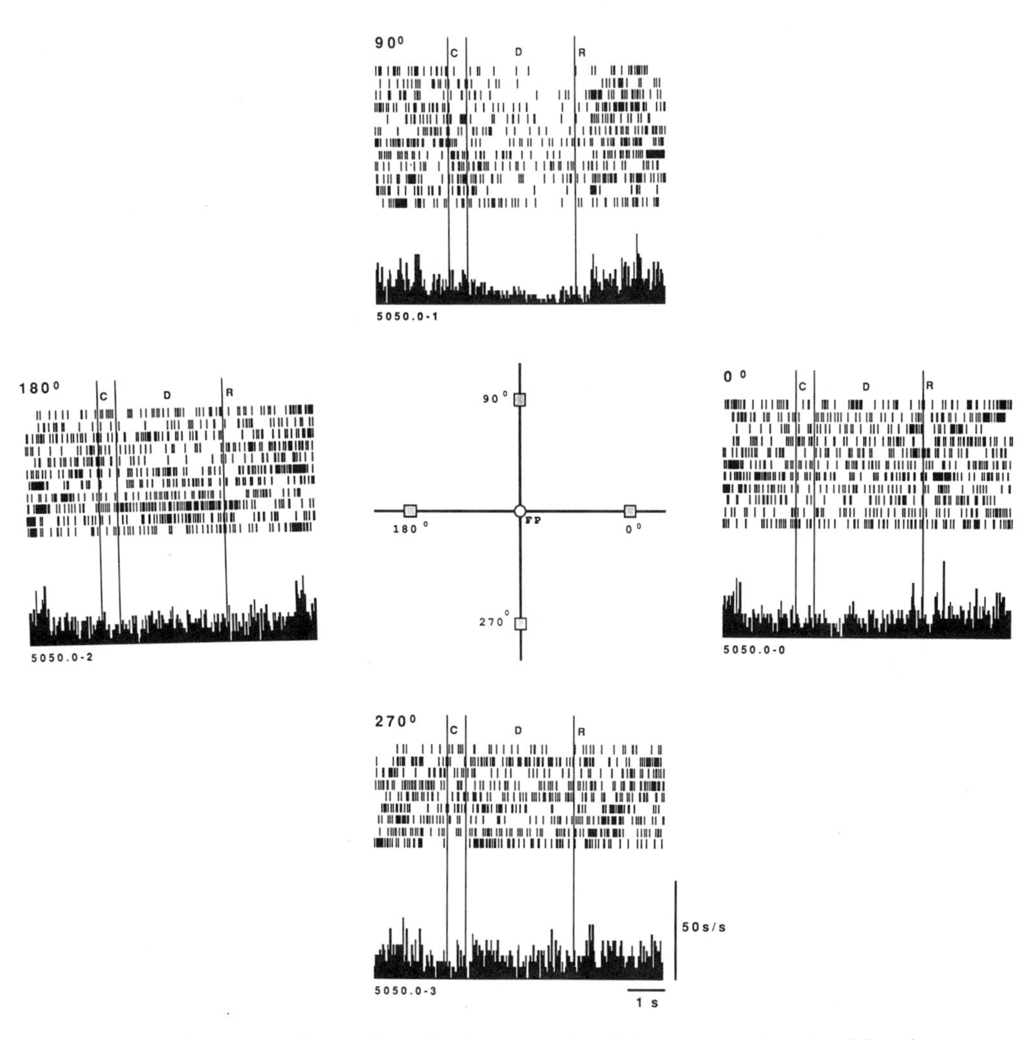

Figure 7.2 Neuronal activity from a single cell during an oculomotor delayed-response task with four target locations. Neuronal discharge is decreased during the delay period only for one of the four target locations. This preferential activation is considered a "memory field" and in this case the neuron's best direction is the 90-degree direction.

identified cells in slice preparations (Schwartzkroin and Kunkel, 1985; McCormick et al., 1985), we hypothesized that thin, fast-spiking neurons represented interneurons, presumably GABAergic whereas slow, thick-spiking neurons were indicative of pyramidal firing. Using the waveform of prefrontal neurons as evidence of their neuronal classification as interneuron or pyramidal neuron, Wilson et al. (1994) have shown that pairs of neurons (a putative pyramidal and a GABAergic interneuron) so identified have inverse activation patterns and exhibited phased relationships in the timing of their responses. Interestingly, the fast-spiking putative interneurons had receptive fields that were comparable in size and selectivity to those of the putative pyramidal cells.

and the majority of them responded to stimuli with increases in firing rate, while their paired putative pyramidal cell was decreased in firing (as shown in figure 7.3). These findings demonstrate that projection and interneurons can be differentiated in vivo and are functionally symbiotic in association cortex as they are in sensory areas. The various physiological properties described above provide a framework for analysis of complex functions carried out by cortical networks and cellular elements. The next section considers what these elements might be.

SENSORY CODING IN PREFRONTAL CORTEX: ROLE OF LOCAL AND LONG-TRACT CORTICOCORTICAL CONNECTIONS

It is reasonable to suggest that directionally specific delay-period activity must originate as a sensory signal (as in our laboratory experiments) or as a memory or idea from long-term stores. The most likely circuitry available to provide such information is that of the posterior parietal cortex, and, perhaps additionally, input arriving via the polysensory area of the superior temporal sulcus (Selemon and Goldman-Rakic, 1988). Indeed, both the parietal and prefrontal areas that are anatomically interconnected show enhanced metabolic activity when monkeys perform delayed-response tasks (Friedman and Goldman-Rakic, 1994). Further, recent studies in my laboratory have recorded neuronal activity in the posterior parietal of monkeys performing the same oculomotor task that we employed in studies of prefrontal physiology. Our evidence shows that parietal neurons have functional properties in relation to task parameters similar to those of prefrontal neurons and that both prefrontal and parietal neurons are functionally interactive (Chafee and Goldman-Rakic, in preparation). Furthermore, both parietal and prefrontal neurons tested in delayed-response tasks register the occurrence of the sensory cue, and, in both areas the "best direction" of the cue-period activity of a prefrontal neuron is highly correlated with that of its delay-period activity (Funahashi et al., 1990; Chafee and Goldman-Rakic, in preparation). The prefrontal targets of parietal or other corticocortical projections are either columns of pyramidal cells in the superficial layers of the cortex or the distal dendrites in these layers of neurons in lower strata, e.g., layer V. The physiological evidence described above indicates that there must be a system of the mutual feedforward antagonism between cells with different best directions that could explain opponent memory fields.

DIRECTIONAL SIGNALING.: ROLE OF THE CORTICOSTRIATONIGRAL/PALLIDAL PROJECTIONS

From the considerations discussed above, the pyramidal cell population that receives information input could be a heterogeneous group. At one

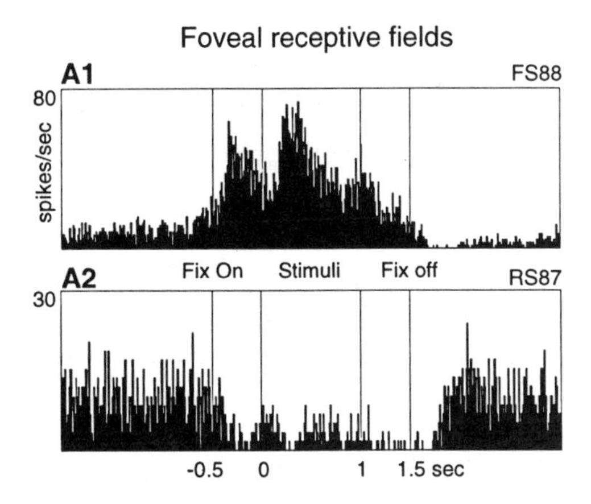

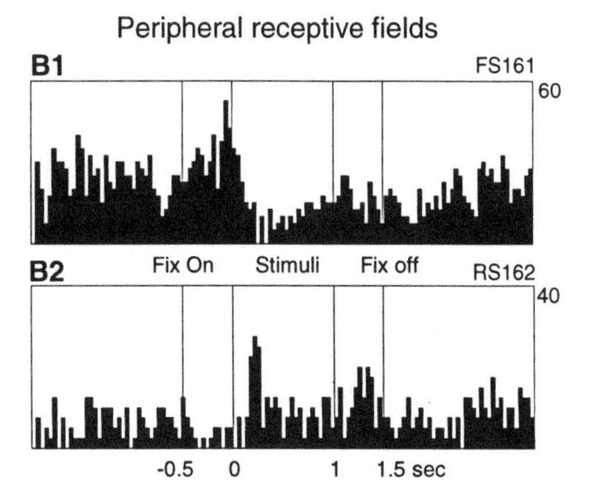

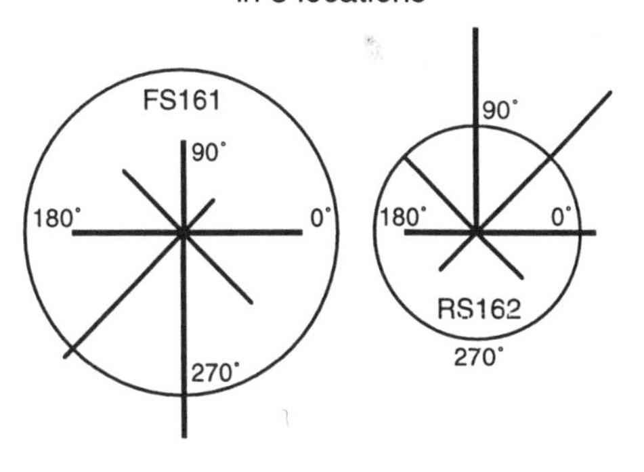

Figure 7.3 *A1* and *A2,* Inverted responses of an FS/RS pair (50 μm apart) during the Pic task. Each tick represents the occurrence of an action potential; each row of ticks represents a single trial. Bin width = 12 ms. *B1* and *B2,* Inverted responses of an FS/RS pair (200 μm apart) in the RF task. These neurons responded maximally to stimuli 13 degrees above (RS162), or 9 degrees right and 9 degrees above the fixation point (FS161). Bin width = 40 ms. *B3,* Vector plots illustrating the overall tuning of the pair-graded increases in the regular-spiking cell firing correspond to graded decreases in the fast-spiking cell, and vice versa. Each vector represents response magnitude plotted relative to a stimulus location. Firing rates are normalized so that the maximum vector length equals 100%. The circles represent spontaneous firing rates. FS, fast spike; RS, regular spike; Pic, picture task; RFtask, receptive field task.

level, some of them appear to be capable of directing a skeletal response; others, an oculomotor response; some may project to the premotor areas, and others to subcortical structures, etc. It is possible, further, that each particular cluster of prefrontal neurons could ultimately output to neurons controlling specific muscle groups, whether they be oculomotor or skeletal. The major source of the descending efferent pathways in layer V. In their respective targets, the terminals of these prefrontal projection neurons exhibit not only somatotopic organization (Flaherty and Graybiel, 1991) but modularity (Goldman and Nauta, 1976) and their projections are highly segregated, as described below.

The modular character of the prefrontal corticostriatal (Goldman and Nauta, 1977a) and prefrontocollicular (Goldman and Nauta, 1976) pathways has been appreciated since the early studies employing axonal transport as a method for tracing pathways (Goldman and Nauta, 1976). In addition, the same injections labeled corticocortical pathways and showed them to take the form of spatially periodic columns in their various cortical targets (Goldman and Nauta, 1977b). These findings indicated that cortical efferent pathways were organized in a manner similar, for example, to sensory afferent systems in visual cortex, and they provided the groundwork for understanding the prefrontal cortex in modular terms and ultimately in terms of a set of parallel processors.

Selemon and Goldman-Rakic (1985) took these findings one step further. Using a double-label strategy to study the possible convergence of projections from the principal sulcus and from a number of other cortical areas to the neostriatum, they discovered that each cortical area apparently projected upon a different set of neostriatal cells. Rather than convergence of cortical inputs in their common target, our data revealed that the projections from anatomically and functionally allied cortical areas remained segregated and sometimes interdigitated in the common areas of the caudate nucleus to which they projected (Selemon and Goldman-Rakic, 1985). Although suspected on the basis of physiological recordings, these data provided the strong anatomical evidence for parallelism for corticostriatal pathways *by area.* However, it is possible that parallelism extends to directional mapping such that a column of prefrontal neurons may project to a cluster of striatal neurons with like directional preferences.

Before discussing the evidence for this possibility, it is necessary to mention that the degree of compartmentalization and parallelism in corticostriatal terminations remains at issue. Recently, Parthasrathy and Graybiel (1992) provided evidence that rather than remaining segregated, projections from functionally related oculomotor cortical areas converge in the neostriatum. At prime face, this finding conflicts with the Selemon and Goldman-Rakic (1985) study described above, which showed that functionally related prefrontal and and parietal zones exhibited little convergence. The differences between these studies could be related to the functional organization of the system under study (higher association cortices vs. somatotopically related areas) or possibly to methodological factors (e.g., means of aligning the findings on adjacent sections). Explanation is lacking but needed for these apparent inconsistencies as they suggest different principles of corticostriatal organization—topographic vs. functional association governing corticostriatal projections.

Pathway tracing studies conducted on nonhuman primates have also elucidated a number of more direct pathways from prefrontal association cortex to cortical motor centers. The newest data concern transcortical routes to premotor regions containing either forelimb or hindlimb representations. Thus, as illustrated in figure 7.4, projections have been documented from the principal sulcus to: (1) the rostral supplementary motor area (Barbas and Pandya, 1989; McGuire et al., 1991; Bates and Goldman-Rakic, 1993); (2) the recently described cingulate premotor areas (Bates and Goldman-Rakic, 1993); and (3) the lateral premotor cortex (Barbas and Mesulam, 1985)—areas that contain forelimb representations and that, in turn, project either to the primary motor cortex or spinal cord.

PRE- AND POSTSACCADIC NEURONAL ACTIVITY: POSSIBLE ROLE OF THALAMIC INNERVATION.

It is well known that corticostriatal fibers terminate on the medium spiny neurons of the neostriatum which, in turn, project upon the substantia nigra *or* globus pallidus. Two types of study have been carried out to examine the organization of these important pathways employing the powerful strategy of double labeling. In one study, carried out in my laboratory, fast blue (FB) and diamidino yellow (DY) were injected into the substanbtia nigra and globus pallidus, respectively, in the same monkey and labeled cells in the neostriatum were charted with an X-Y plotting system. The findings of these two studies establish that (1) the putamen and caudate each project to both the globus pallidus and substantia nigra; (2) the neostriatal projection neurons to the globus pallidus and substantia nigra are organized in cellular clusters (figure 7.5; see also Selemon and Goldman-Rakic, 1990); and (3) these cell clusters for the most part originate in the matrix compartment.

The other double-label study was carried out by Parent and Hazrati (1993). They injected the anterograde tracers—*Phaseolus vulgaris* leukoagglutinin (PHA-L) and biocytin—into adjacent portions of the putamen in squirrel monkey and the consequent labeling was examined in the globus pallidus and substantia nigra (Parent and Hazrati, 1993). In line with previous results, including those mentioned above, the putamen was found to project both to the globus pallidus and substantia nigra, to terminate in distinct arborizations, and to maintain segregated terminal fields in those structures.

Taken together these two studies provide evidence, in primates, of a remarkable degree of parcellation, segregation, or parallelism in the corticostriatonigral (pallidal) circuitry. Given the high degree of parallel structuring in the corticostriatal terminal fields as described originally by Selemon and Goldman Rakic (1990), it is almost obvious, though still unproven, that the corticostriatal system must be matched to the striatonigral and striatopallidal efferents, on the one hand, and to the corticostriatal inputs (from cortical columns of neurons with specified

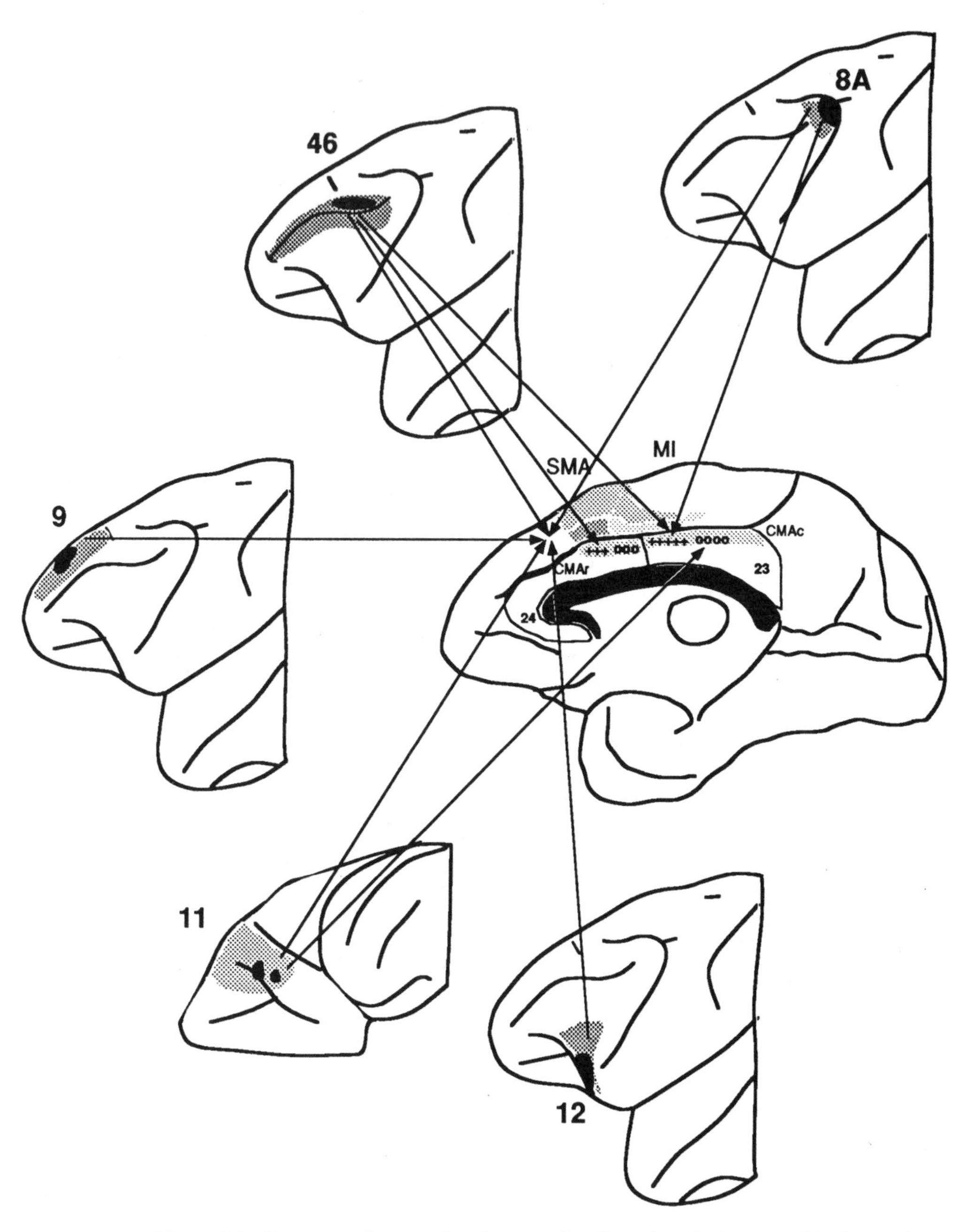

Figure 7.4 Summary of connections between 2prefrontal cortical areas and both cingulate and rostral supplementary motor areas (SMA). Forelimb (plus signs) and hindlimb (circles) representations are noted for the rostral and caudal cingulate motor areas (CMAr and CMAc, respectively). Each input shows the injection site in black, surrounded by stippling to indicate the extent of the relevant cytoarchtectonic subdivisions. MI, primary motor cortex. (From Bates and Goldman-Rakic, 1993.)

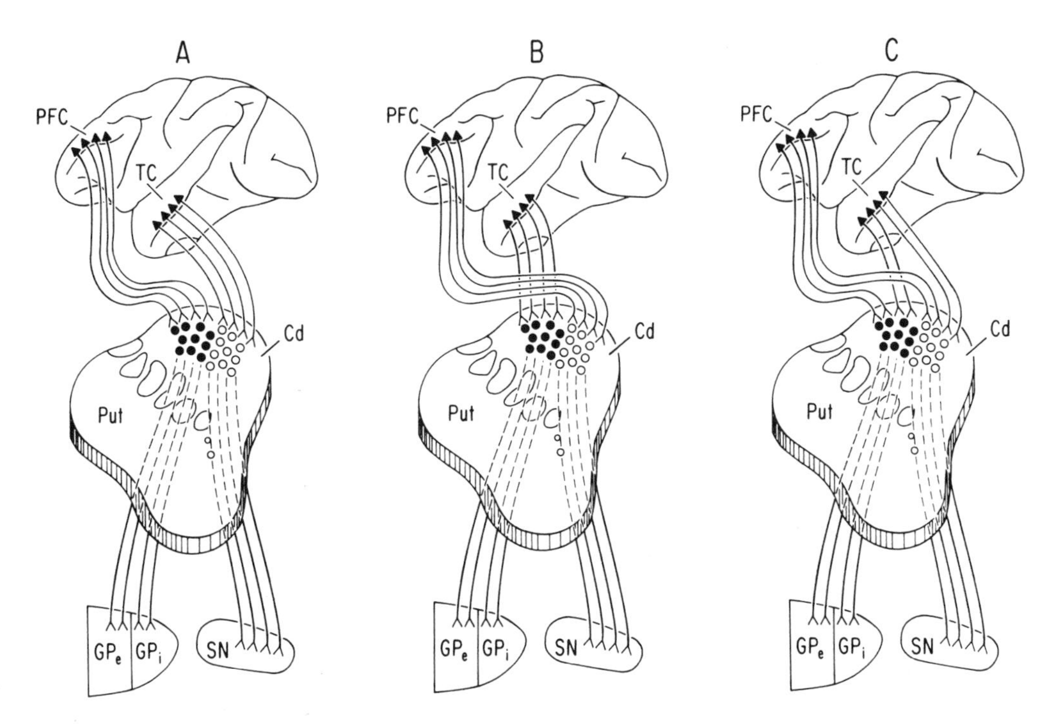

Figure 7.5 Diagram depicting theoretically plausible modular arrangements of the corticostriatonigral/pallidal projection systems. PFC, prefrontal cortex; TC, temporal cortex; Cd, caudate; Put, putamen; GPe, external, and GPi, internal segments of globus pallidus; SN, substantia nigra. (From Selemon and Goldman-Rakic, 1989.)

directional preferences), on the other. However, how the hookup is made has not been worked out and figure 7.5, taken from Selemon and Goldman-Rakic (1989), models a number of potential arrangements, all of which preserve the modularity of the loop circuitry.

Both the substantia nigra and globus pallidus project via the thalamus back to the prefrontal cortex, parcelling feedforward projections onto clusters of cells in the thalamus (Ilinsky et al., 1985). We examined this issue some time ago by injecting the substantia nigra with tritiated amino acids to anterogradely label the nigrothalamic projections (Ilinsky et al. 1985). Our findings showed that the substantia nigra had more widespread projections to the thalamus than previously indicated; terminal fields were found in the magnocellular, parvocellular, and paralamellar subdivisions of the mediodorsal nucleus as well as in the ventral anterior nucleus. Most striking, however, was the patchiness of the nigrothalamic arborization, which could only be described as forming distinct clusters in each of its terminations, not unlike those described by Goldman and Nauta (1976) in the corticopetal projections, and not unlike those described by Parent and Hazrati (1993) in the striatopetal projections. The segregated pathways through the paralamellar, parvocellular, and magnocellular portions of the mediodorsal nucleus are indicative of further segregation in the loop circuitry. The compartmental nature of the descending part of the corticostriatal loop circuitry appears thus to be maintained in its ascending trajectory. Again, the linkage between the striatonigral and the nigrothalamic projection systems has yet to be made in specific terms, but the pattern is clearly one of compartmentalization in the feedforward projections of the paleostriatum to the thalamus.

The thalamocortical systems of connections should perhaps be considered separately from the corticostriatothalamocortical circuitry because it may serve a dual purpose—both as a feedback and feedforward pathway to the cortex. As a segment of the corticostriatal loop circuitry, its specific functions are unclear. Generally, this pathway is considered a feedforward excitatory path critical to initiation of a motor response. If so, it could be the source of the presaccadic burst in prefrontal neurons. However, the presaccadic burst could also reflect an important feedback signal, information that a signal has been deployed to the superior colliculus to initiate an eye movement. The superior colliculus, in turn, projects to the mediodorsal nucleus and could, through that pathway, issue the signal that is registered at the cortex as a postsaccadic response. These pathways are also compartmentalized and could convey specific topographic information. For example, following injections of horseradish peroxidase (HRP) into various cytoarchitectonic areas of prefrontal cortex, Goldman-Rakic and Porrino (1985) observed that the thalamocortical projections arose from distinct cellular groupings within the mediodorsal thalamus and other thalamic nuclei rather than from a homogeneous array of corticofugal projection neurons (figure 7.6). Compartmentalization of the feedback circuits provides the type of circuit organization that would be ideal for a corollary discharge function.

Any theory of basal ganglia function will ultimately have to attend to the role played by the corticothalamic projections. This is often a neglected feature of the thalamus as a system, even though by some accounts the corticothalamic innervation is as dense or denser than the thalamocortical connections. A corticothalamocortical reverberating circuitry may provide some of the amplification necessary to sustain delay-period activation. Anatomical studies of this system in prefrontal cortex have shown that the bulk of corticothalamic neurons reside in layer VI, although a subset also lie in the upper strata of layer V (e.g., see Giguere and Goldman-Rakic, 1988). The layer VI cells project not only back to the thalamus but also to layer IV where they could amplify the thalamocortical projections and any other inputs to that layer. According to this notion, cells and dendrites in layer IV and lower layer III of the prefrontal cortex could be studded with excitatory synapses from both the thalamus and from distant sensory associational cortical areas.

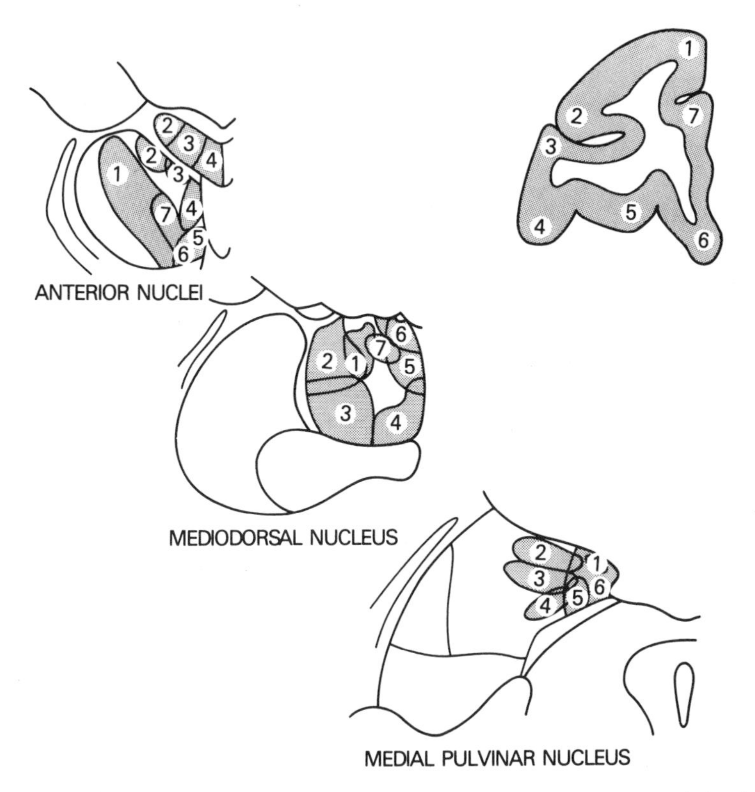

Figure 7.6 Summary diagram illustrating the subnuclear origins of thalamic projections to various subdivisions (indicated by number) of the primate prefrontal cortex. The topography of thalamocortical projections is shown for the anterior nuclei, the mediodorsal nucleus, and the medial pulvinar nucleus. (From Goldman-Rakic and Porrino, 1985.)

SUMMARY: A CIRCUIT MODEL FOR WORKING MEMORY

A major theme running through the literature on the cortical control of motor action is the differentiation of two response-related neuronal activity patterns—one that is tied to the preparation and organization of the movements themselves and one that is referenced to the target direction or goal of a *movement,* independent of the movement itself (see minireview in Goldman-Rakic et al., 1992). Both types of coding are carried out by prefrontal neurons and are dissociable one from another (Niki and Watanabe, 1976; Funahashi et al., 1993). Thus, there are at least two types of pyramidal neuron with respect to motor programming and an inference can be made that they are connected by local circuitry. According to the view being developed here, when behavior is memory-guided, a preparatory signal for a specified direction of eye movement originates first and foremost as a representation held temporarily in prefrontal memory circuits as *directional* delay-period activity and then is transmitted locally to other prefrontal neurons, and to distant cortical *and* subcortical motor centers. In the case of eye movements, e.g., a neuron in the caudal principal sulcus, containing the preparatory signal for an eye movement to the 90-degree location, may project to the frontal eye field, the medial supplementary eye field, the caudate nucleus, and the superior colliculus. In the case of a skeletal movement, the outflow would be to the basal ganglia, and to the premotor areas where it serves as a relay to primary motor cortex, or as a particular corticospinal outflow and hence a particular response of a limb. According to this view, the preparatory directional delay-period activation, observed in remote structures like the parietal (Gnadt and Andersen 1988; Chafee et al., in preparation), premotor (Tanji et al., 1980; Tanji and Kurata, 1985; Alexander and Crutcher, 1990); motor (Lurito et al., 1991), frontal eye field (Funahashi et al., 1989), and subcortical structures (Hikosaka and Wurtz 1983; Hikosaka et al. 1989; Brotchie et al., 1991) may depend on the integrity of prefrontal cortex *under conditions of representational processing*. The latter point bears emphasis because the aforementioned regions can control motor action under many other conditions and particularly when responses are sensory-guided. According to the modular working memory hypothesis of prefrontal cortex (Goldman-Rakic, 1987), the prefrontal component is essential only when the movement is guided by representational mechanisms. Delay-period activation in premotor neurons can presumably be activated independently by external sensory cues presumably driven by parietal or temporal lobe imputs to these areas which can guide behavior directed by external cues.

Anatomical and physiological data support the notion that the motor component of working memory functions (spatial and nonspatial) are carried out by multiple dedicated and parallel networks of corticosubcortical structures that make up the prefrontostriatothalamocortical loop circuitry reviewed above. Whenever the memory field of a prefrontal neuron in layer V is activated, presumably by corticocortical information flow, this directional information is presumably conveyed to the basal ganglia via the corticostriatal pathway. At the end of the delay, a phasic response heralds the initiation of a motor response by an average of 73 ms. At the same time, a signal can be recorded in the basal ganglia prior to a response with a median latency of 105 ms before the response (Hikosaka et al., 1989; Alexander and Crutcher, 1990). Neuronal activity in the substantia nigra is simultaneously depressed for about 100 ms by activation of the striatonigral projections (Hikosaka and Wurtz, 1983, 1985a,b). Finally, a contralateral eye movement follows the initial presaccadic burst in concert with phasic responses in the superior colliculus and thalamus (as described in Chevalier and Deniau, 1990). A speculative suggestion is that the disinhibition of thalamic activity results in a feedback postsaccadic excitation of the layer V neuron that is recorded in the prefrontal cortex an average of 130 ms following the execution of the oculomotor response.

A challenging question is the mechanism maintaining the tonic activation or delay-period activity. Is it an example of a reverberating, or "reentrant" circuit and if so, which extrinsic connections are involved? Based on retrograde tracing data (Goldman-Rakic and Porrino, 1985), several classes of thalamic neuron can be identified. It is possible that one of these classes is the generator of reverberatory delay-period activity while the other class is the feedback pathways for the phasic pre- and postsaccadic signals. The delay-period activity could also be sustained by intrinsic or local circuit reciprocal connections between nearby pyramidal cells. Recent studies of intrinsic circuits (Levitt et al., 1993; Kritzer and Goldman-Rakic, 1993) are providing data on the local connections that may help to resolve this issue.

According to this model, corticopetal cells in layer V can influence (disinhibit) downstream neurons (basal ganglia; tectum) that will release an eye movement with particular direction and amplitude; cells in the same layer that output to the globus pallidus and premotor areas can exert the same type of motor control over forelimb movements. The important point is that the prefrontal neurons come into play only when these responses are genuinely memory-guided; when they are sensory-guided, the prefrontal cortex is dispensable and premotor circuits are sufficient. Motor control in this scheme is redundant in one strong sense; many areas can command a particular movement; they are differentially essential, however, depending on the conditions or level of guidance of the movement, whether by a single stimulus, a conditional stimulus or, at the highest level, by the memory of a stimulus, i.e., by a representation or concept. Only in the last situation do prefrontal cortical areas lodge the motor command.

REFERENCES

Alexander, G. E., and Crutcher, M. D. (1990) Neural representations of the target (goal) of visually guided arm movements in three motor areas of the monkey. *J. Neurophysiol.* 64:164–178.

Barbas, H., and Mesulam, M.-M. (1985) Cortical afferent input to the principalis region of the rhesus monkey. *Neuroscience* 5:619–637.

Barbas, H., and Pandya, D. N. (1989) Architecture and intrinsic connections of the prefrontal cortex in the rhesus monkey. *J. Comp. Neurol.* 286:353–375.

Barone, P., and Joseph, J.-P. (1989) Prefrontal cortex and spatial sequencing in macaque monkeys. *Exp. Brain Res.* 78:447–464.

Bates, J. F., and Goldman-Rakic, P. S. (1993) Prefrontal connections of medial motor areas in the rhesus monkey. *J. Comp. Neurol.* 335:1–18.

Boch, R. A., and Goldberg, M. E. (1989) Participation of prefrontal neurons in the preparation of visually guided eye movements in the rhesus monkey. *J. Neurophysiol.* 61:1064–1084.

Brotchie, P., Iansek, R., and Horne, M. K. (1991) Motor function of the monkey globus pallidus. *Brain* 114:1685–1702.

Carlson, S., Mikami, A., and Goldman-Rakic, P. S. (1993) Omnidirectional delay activity in the monkey posterior cingulate cortex during the performance of an oculomotor delayed response task. *Soc. Neurosci. Abstr.* 19:800.

Chevalier, G., and Deniau, J. M. (1990) Disinhibition as a basic process in the expression of striatal functions. *Trends Neurosci.* 13:277–280.

Flaherty, A. W., and Graybiel, A. M. (1991) Corticostriatal transformations in the primate somatosensory system. Projections from physiologically mapped body-party representations. *J. Neurophysiol.* 66:1249–1263.

Friedman, H. R., and Goldman-Rakic, P. S. (1994) Coactivation of prefrontal cortex and inferior parietal cortex in working memory tasks revealed by 2DG functional mapping in the rhesus monkey. *J. Neurosci.*, 14:2775–2788.

Funahashi, S., Bruce, C. J., and Goldman-Rakic, P. S. (1989) Mnemonic coding of visual space in the monkey's dorsolateral prefrontal cortex. *J. Neurophysiol.* 61:331–349.

Funahashi, S., Bruce, C. J., and Goldman-Rakic, P. S. (1990) Visuospatial coding in primate prefrontal neurons revealed by oculomotor paradigms. *J. Neurophysiol.* 63:814–831.

Funahashi, S., Bruce, C. J., and Goldman-Rakic, P. S. (1991) Neuronal activity related to saccadic eye movements in the monkey's dorsolateral prefrontal cortex. *J. Neurophysiol.* 65:1464–1483.

Funahashi, S., Chafee, M. V., and Goldman-Rakic, P. S. (1993) Prefrontal neuronal activity in rhesus monkeys performing a delayed anti-saccade task. *Nature* 365:753–756.

Fuster, J. M., and Alexander, G. E. (1971) Neuron activity related to short-term memory. *Science* 173:652–654.

Giguere, M., and Goldman-Rakic, P. S. (1988) Mediodorsal nucleus: Areal, laminar and tangential distribution of afferents and efferents in the frontal lobe of rhesus monkeys. *J. Comp. Neurol.* 277:195–213.

Gnadt, J. W., and Andersen, R. A. (1988) Memory related motor planning activity in posterior parietal cortex of macaque. *Exp. Brain Res.* 70:216–220.

Goldman, P. S., and Nauta, W. J. H. (1976) Autoradiographic demonstration of a projection from prefrontal association cortex to the superior colliculus in the rhesus monkey. *Brain Res.* 116:145–149.

Goldman, P. S., and Nauta, W. J. H. (1977a) An intricately patterned prefrontocaudate projection in the rhesus monkey. *J. Comp. Neurol.* 171:369–386.

Goldman, P. S., and Nauta, W. J. H. (1977b) Columnar distribution of corticocortical fibers in the frontal association, limbic, and motor cortex of the developing rhesus monkey. *Brain Res.* 122:393–413.

Goldman-Rakic, P. S. (1987) Circuitry of primate prefrontal cortex and regulation of behavior by representational memory. In V. B. Mountcastle, F. Plum, and S. R. Geiger (eds.), *Handbook of Physiology*, Vol. I, sec. 1. New York: Oxford University Press, pp. 373–417.

Goldman-Rakic, P. S. (1988) Topography of cognition: Parallel distributed networks in primate association cortex. *Annu. Rev. Neurosci.* 11:137–156.

Goldman-Rakic, P. S. (1990) Cellular and circuit basis of working memory in prefrontal cortex of nonhuman primates. *Prog. Brain Res.* 85:325–336.

Goldman-Rakic, P. S., and L. Porrino (1985) The primate mediodorsal (MD) nucleus and its projection to the frontal lobe. *J. Comp. Neurol.* 242:535–560.

Goldman-Rakic, P. S., Funahashi, S., and Bruce, C. J. (1990) Neocortical memory circuits. *Q. J. Quant. Biol.* 55:1025–1038.

Goldman-Rakic, P. S., Bates, J. F., and Chafee, M. V. (1992) The prefrontal cortex and internally generated motor acts. *Curr. Opin. Neurobiol.* 2:830–835.

Goldman-Rakic, P. S., Chafee, M., and Friedman, H. (1993) Allocation of function in distributed circuits. In T. Ono, L. R. Squire, M. E. Raichle, D. I. Perrett, and M. Fukuda (eds.), *Brain Mechanisms of Perception and Memory: From Neuron to Behavior.* New York: Oxford University Press, pp. 445–456.

Hikosaka, O., and Wurtz, R. H. (1983) Visual and oculomotor functions of monkey substantia nigra pars reticulata. III. Memory-contingent visual and saccade responses. *J. Neurophysiol.* 49:1268–1284.

Hikosaka, O., and Wurtz, R. H. (1985a) Modification of saccadic eye movements by GABA-related substances. I. Effect of muscimol and bicuculline in monkey superior colliculus. *J. Neurophysiol.* 53:266–291.

Hikosaka, O., and Wurtz, R. H. (1985b) Modification of saccadic eye movements by GABA-related substances. I. Effect of muscimol in monkey substantia nigra pars reticulata. *J. Neurophysiol.* 53:292–308.

Hikosaka, O., Sakamoto, M., and Usui, S. (1989) Functional properties of monkey caudate neurons III. Activities related to expectation of target and reward. *J. Neurophysiol.* 61:814–832.

Ilinsky, I., Jouandet, M., and Goldman-Rakic, P. S. (1985) Organization of the nigrothalamocortical system in the rhesus monkey. *J. Comp. Neurol.* 236:315–330.

Kritzer, M. F., and Goldman-Rakic, P. S. (1993) Intrinsic circuits of human prefrontal cortex. *Soc. Neurosci. Abstr.* 19:1444.

Kubota, K., and Niki, H. (1971) Prefrontal cortical unit activity and delayed cortical unit activity and delayed alternation performance in monkeys. *J. Neurophysiol.* 34:337–347.

Levitt, J. B., Lewis, D. A., Yoshioka, T., and Lund, J. S. (1993) Topography of pyramidal neuron intrinsic connections in macaque monkey prefrontal cortex (areas 9 and 46). *J. Comp. Neurol.* 338:360–376.

Lurito, J. T., Georgakopoulos, T., and Georgopoulos, A. P. (1991) Cognitive spatial-motor processes. *Exp. Brain Res.* 87:562–580.

McCormick, D. A., Connors, B. W., Lighthall, J. W., and Prince, D. A. (1985) Comparative electrophysiology of pyramidal and sparsely spiny stellate neurons of the neocortex. *J. Neurophysiol.* 54:782–806.

McGuire, P. K., Bates, J. F., and Goldman-Rakic, P. S. (1991) Interhemispheric integration: I. Symmetry and convergence of the corticocortical connections of the left and the right principal sulcus (PS) and the left and the right supplementary motor area (SMA) in the rhesus monkey. *Cereb. Cortex* 1:390–407.

Niki, H., and Watanabe, M. (1976) Prefrontal unit activity and delayed response: Relation to cue location versus direction of response. *Brain Res.* 105:78–88.

O Scalaidhe, S. P., and Goldman-Rakic, P. S. (1993) Memory fields in the prefrontal cortex of the macaque. *Soc. Neurosci. Abstr.* 19:800.

Parent, A., and Hazrati, L.-N. (1993) Anatomical aspects of information processing in primate basal ganglia. *Trends Neurosci.* 16:111–116.

Parthasarathy, H. B., Schall, J. D., and Graybiel, A. M. (1992) Distributed but convergent ordering of corticostriatal projections: Analysis of the frontal eyefield and the supplementary eyefield in the macaque monkey. *J. Neurosci.* 11:4468–4488.

Schwartzkroin, P. A., and Kunkel, D. D. (1985) Morphology of identified CA1 regions of guinea pig hippocampus. *J. Comp. Neurol.* 232:205–218.

Selemon, L. D., and Goldman-Rakic, P. S. (1985) Longitudinal topography and interdigitation of corticostriatal projections in the rhesus monkey. *J. Neurosci.* 5:776–794.

Selemon, L. D., and Goldman-Rakic, P. S. (1988) Common cortical and subcortical target areas of the dorsolateral prefrontal and posterior parietal cortices in the rhesus monkey: A double label study of distributed neural networks. *J. Neurosci.* 8:4049–4068.

Selemon, L. D., and Goldman-Rakic, P. S. (1990) Topographic intermingling of striatonigral and striatopallidal neurons in the rhesus monkey. *J. Comp. Neurol.* 297:359–376.

Tanji, J., and Kurata, K. (1985) Contrasting neuronal activity in supplementary and precentral motor cortex of monkeys. I. Responses to instruction determining motor responses to forthcoming signals of different modalities. *J. Neurophysiol.* 53:129–141.

Tanji, J., Tanuguchi, K., and Saga, T. (1980) Supplementary motor area: Neuronal response to motor instructions. *J. Neurophysiol.* 43:60–68.

Wilson, F. A. W., O Scalaidhe, S. P., and Goldman-Rakic, P. S. (1994) Functional synergism between putative GABAergic and pyramidal neurons in prefrontal cortex. *Proc. Natl. Acad. Sci. U. S. A.*, 91:4009–4013.

8 Modeling the Roles of Basal Ganglia in Timing and Sequencing Saccadic Eye Movements

Michael A. Arbib and Peter F. Dominey

In chapter 7, Goldman-Rakic has elegantly summarized experimental data on the interactions between the basal ganglia and the "working memory" systems of prefrontal cortex. The first part of this chapter presents a model of these interactions in the control of saccadic eye movements (Dominey and Arbib, 1992), while the second part models the possible role of corticostriatal plasticity in learning how to embed saccades in more complex or conditional patterns of behavior (Dominey et al., submitted for publication).

BASAL GANGLIA AND THE CONTROL OF WORKING MEMORY

The substrate for our model (figure 8.1) is the brainstem saccade generator, which we model with a variant of Scudder's (1988) elaboration of Robinson's control-theoretic models (van Gisbergen et al., 1981), which showed how the topographic code of target position in superior colliculus might be transformed into the appropriate time course of burst neuron activity to move the eyes to the target. What we wish to explore here is the way in which cortical systems and the basal ganglia can play down upon this collicular-plus-brainstem system to yield an increasingly rich pattern of oculomotor behaviors. In this section, we discuss the role of working memory in delay saccades.[1]

Superior colliculus (SC) guides the brainstem to move the eye to sample the visual input in different directions. The output of the retina travels through various way stations to the posterior parietal cortex (PP). Dynamic remapping of visual information is controlled by corollary discharge to PP, with PP projecting to the quasivisual (QV) cells in SC, yielding a model of the double saccade paradigm (see Dominey and Arbib, 1992, for details). PP then sends target information (whether relayed from current visual input, or remapped) to the frontal eye fields (FEF). FEF projects to the brainstem (thus allowing generation of eye movements even after lesion of SC) but here we emphasize the FEF projection to SC, focusing on two types of cells, the Fon ("fixation on")

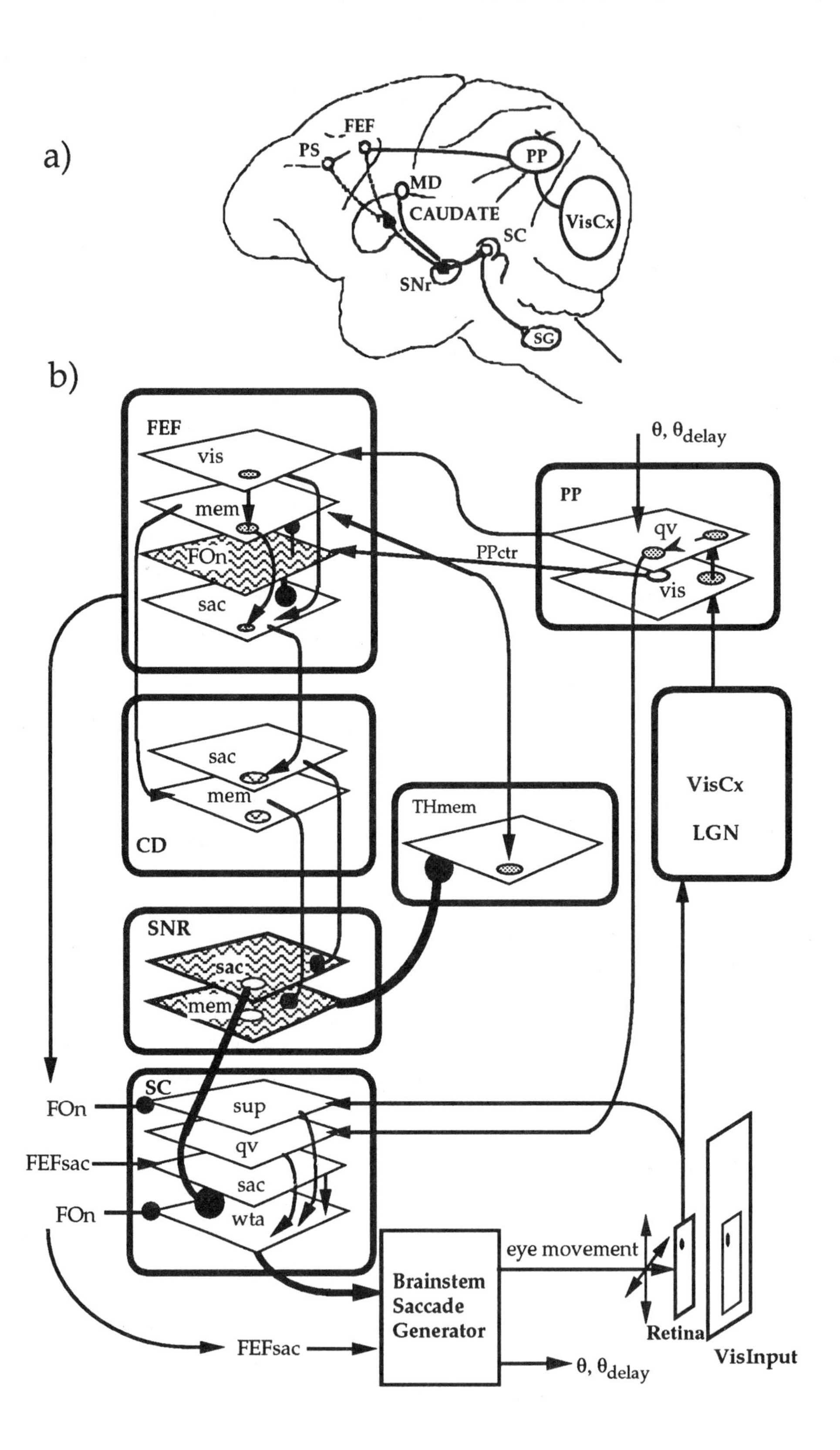
a)
FEF
PS
MD
CAUDATE
PP
VisCx
SC
SNr
SG
b)
FEF
vis
mem
FOn
sac
θ, θdelay
PP
qv
PPctr
vis
CD
sac
mem
THmem
VisCx
LGN
SNR
sac
mem
SC
FOn
sup
qv
FEFsac
sac
FOn
wta
eye movement
Brainstem
Saccade
Generator
Retina
VisInput
FEFsac
θ, θdelay

cells, which are active when a target occupies the fovea, and FEFsac (FEF cells coding saccades), which code the loci of possible targets.

Dominey and Arbib (1992) represent Fon cells as providing inhibition to SC to block saccades during active fixation, though a more realistic (but functionally equivalent) model would have these cells excite the cells in the rostral pole of SC which maintain fixation (Wurtz and Munoz, 1993). In either case, the crucial point for our further discussion is that FEF can encode the presence of a fixation target and accordingly stop SC from triggering a saccade.

In addition to the FEFsac cells, which show a transient response just prior to the onset of a saccade, FEF contains memory cells, FEFmem, which are active during the delay between the presentation of a stimulus and the initiation of a related response—with activity continuing even if the stimulus is no longer visible. In the model, it is a recurrent loop between the FEFmem array and mediodorsal thalamus (THAL) that maintains the activity that provides a working memory when a visual target is not visible (Alexander and Fuster, 1973). Through FEFmem, the FEFsac saccadic command can encode remembered targets as well as currently visible targets signaled via PP. When the fixation point goes off, the FEFsac activity to SC is no longer blocked by Fon activity, and movement can ensue. We postulate that if multiple targets are encoded by activity in the deep layers of SC, then a winner-take-all (WTA) mechanism will select the most salient peak of activity to provide the target for the ensuing saccade.

As this discussion shows, we may analyze two "channels," as exemplified by FEFsac and FEFmem in FEF, the phasic and the delay/maintained response channels. We may follow this through the basal ganglia: FEF excites the caudate (CD) which inhibits the substantia nigra pars reticulata (SNr), which in turn inhibits SC—and all these relations are topographical.[2] In both CD and SNr we have two channels, phasic and the delay/maintained, but the CD cells have a low background firing rate, while the SNr cells have a high rate of tonic activity. Thus the effect of FEF activation of CD is to lower the SNr inhibition of SC, thus *disinhibiting* SC at the corresponding locus. At first sight, this arrangement seems paradoxical: there is an apparent redundancy, since

FEF - Frontal Eye Fields
PP - Posterior Parietal cortex
CD - Caudate nucleus
SNR - Substantia Nigra pars Reticulatta
SC - Super Colliculus
TH - Thalamus (mediodorsal)
FOn - Fovea On (foveation)

vis - Visual response
mem - Memory (sustained)
sac - presaccadic (phasic)
qv - Quasi-Visual
wta - Winner Take All
PPctr - central element of PPqv

Figure 8.1 An overview of the Dominey and Arbib (1992) model of saccade generation.

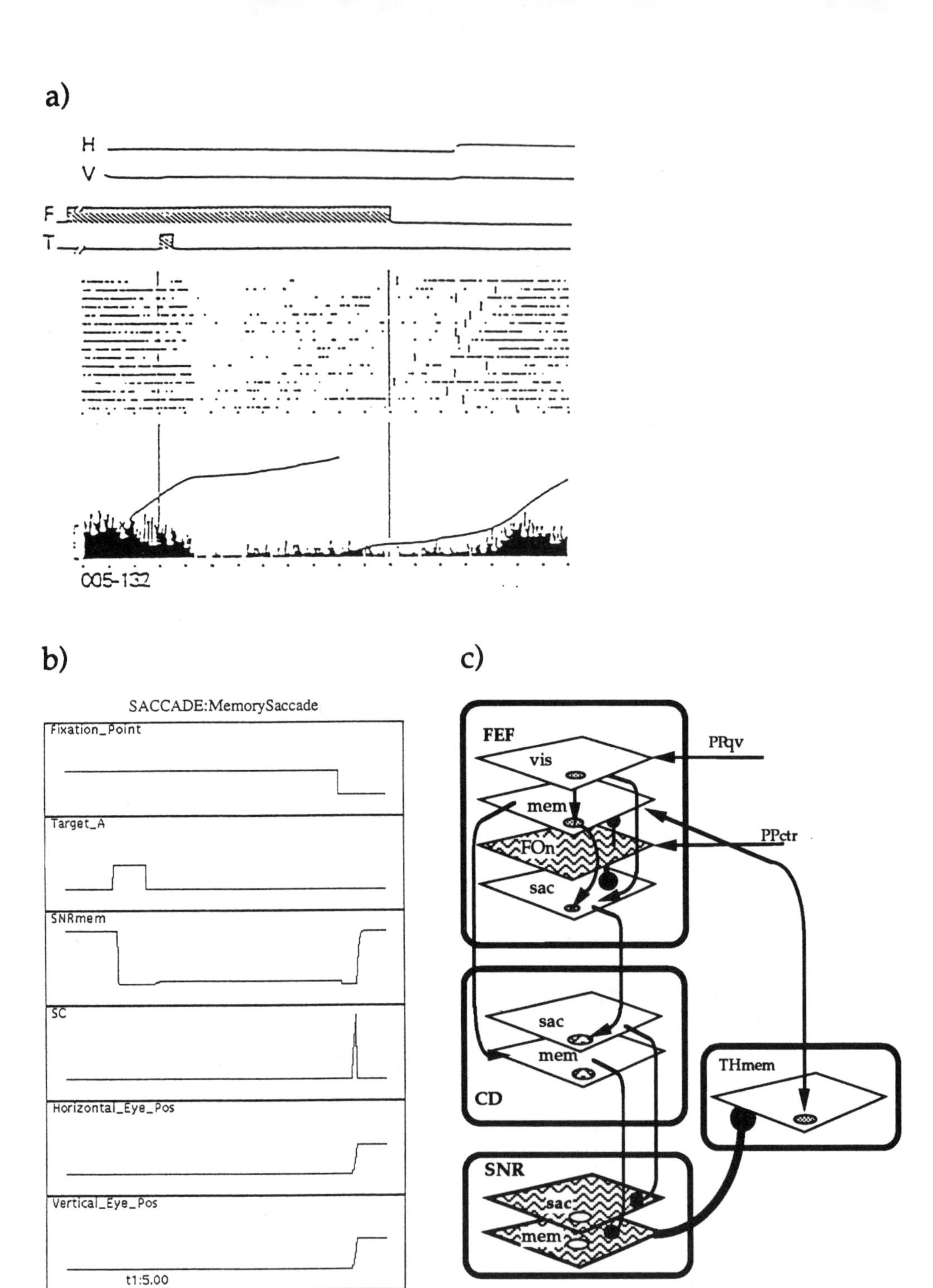
a)
H
V
F
T
005-132
b)
SACCADE:MemorySaccade
Fixation_Point
Target_A
SNRmem
SC
Horizontal_Eye_Pos
Vertical_Eye_Pos
t1:5.00
c)
FEF
vis
mem
FOn
sac
PPqv
PPctr
CD
sac
mem
THmem
SNR
sac
mem

FEF both signals the locus of a saccade by the excitation from FEFsac to SC, and also disinhibits the corresponding SC locus via the basal ganglia. The second section of this chapter provides one answer to this paradox, by showing how corticostriatal plasticity can allow the basal ganglia to not simply transmit the FEF pattern but to modulate it on the basis of experience—and this is of special significance when FEF codes the loci of multiple targets, rather than specifying a unique one. Here, we spell out our model of the cooperation of SC and SNr in managing the FEFmem ↔ THAL loop.

FEFmem cells excite CD memory (CDmem) cells, thus raising their activity during the delay period. CDmem cells inhibit SNr memory (SNrmem) cells, thus lowering their high tonic activity during the delay period (figure 8.2). This lowers the inhibitory drive from SNrmem to THAL, thus allowing the memory to be maintained. This virtuous circle is not broken until SC commands an eye movement. A nonspecific corollary discharge from SC then inhibits THAL, the loop is turned off, and activities in THAL, FEFmem, and CDmem return to their usual low levels, while SNRmem returns to its high tonic activity, inhibiting further loop activity until the next mnemonic episode is triggered. The memory period begins with the target onset during fixation, so in a saccade with overlap of target and fixation, we see some memory activity.

In the case of a double saccade, the activity from the first saccade erases the initial representation of the two targets and allows PP to install the new target map (the result of "dynamic remapping") in the FEFmem ↔ THAL loop.

Cerebellum and Metric Adaptation

We close by noting that specific choices have had to be made in the model which call for further experimental investigation. When cells with a particular property are found at several places in the brain, it may be (lacking adequate data to effect a discrimination) that (a) the property is independently computed in each place (possibly with cooperative

Figure 8.2 Substantia nigra pars reticulata (SNr) memory-contingent sustained response. (*a*) This cell shows sustained decrease in firing in response to a target in the contralateral visual field that is to be memorized for the delay saccade task. The decrease usually follows the target presentation and remains until the saccade onset. (From Hikosaka and Wurtz, 1983.) (*b*) Simulation of the memory-contingent sustained response (SNRmem). The sustained suppression of SNr allows the mediodorsal thalamus to participate in the reciprocal activity with frontal eye fields (FEF) that instantiates a spatial memory. This SNrmem is inhibited by caudate memory cells (CDmem), which in turn is tonically excited by FEFmem in the delay period of the memory saccade. (*c*) Corticostriatonigrothalamocortical path subserving spatial memory.

interactions to refine the computation), or (b) it may be computed in one place which transmits this information to other places, or (c) it may be a system property which depends on the interaction of the regions to be computed at all. In the present model, we chose option (b) for dynamic remapping (PP drives both FEF and QV cells of SC), while for working memory we chose (c), to explain the phenomenon via the FEFmem ↔ THAL loop. Clearly, these choices must be tested and the model refined. Conversely, we note that the role attributed to thalamus here is grossly simplified. Moreover, one can extend (and we extended) the model to take into account other brain regions which do not appear in figure 8.1. For example, we have also studied the role of the cerebellum in saccadic adaptation—where the concern is with the metric scaling of saccades, rather than their sequencing or working memory for target location. In our model (Schweighofer et al., submitted), when a saccade does not reach its target, there is an immediate corrective saccade, and it is the magnitude of this corrective saccade that supplies an "error signal" routed via inferior olive climbing fibers to that part of the cerebellum whose job is to modulate and coordinate the various saccade generators of the brainstem.

CORTICOSTRIATAL PLASTICITY

We now return to the paradox mentioned earlier, the apparent redundancy of the two pathways from FEF to SC: FEF both signals the locus of a saccade by direct excitation to SC and disinhibits the corresponding SC locus via the basal ganglia. We now suggest that an answer to this paradox lies in the use of corticostriatal plasticity to allow the basal ganglia to modulate the FEF pattern on the basis of experience. We thus turn from the question of working memory to that of long-term memory, and in particular to issues of visuomotor conditional learning, and sequential behavior.

We start by recalling our hypothesis that if the frontal eye field and direct visual input do not yield the encoding of a unique target in the deep layers of SC, then a WTA mechanism will "choose" one of the targets. To remove the paradox, we argue that experience based on inferior temporal or prefrontal information may provide contextual, learned information to bias activity in the basal ganglia to "tip the balance" to one "winner" or another. Our present model (Dominey et al., submitted) makes the very strong hypothesis that this learning is mediated by corticostriatal plasticity. Later versions can look at alternative sites (see discussion below), once we see how to martial the data to modify it. We first illustrate this for a model of visuomotor conditional learning, and then turn to sequence learning.

Visuomotor Conditional Learning

The paradigm for visuomotor conditional learning employed in our modeling, and in related monkey neurophysiological studies by Joseph and Dominey in Lyon, France, is to present two visual targets on either side of the fixation point, which is now a patterned stimulus such as a red circle or a blue triangle: (i) present the fixation point, which is also the cue for the appropriate response; (ii) present both targets with the cue; then (there is no delay in this paradigm) when the fixation point goes off, the monkey is to saccade to one of the two targets. Each cue codes for a particular direction; e.g., when the cue is the blue triangle, the monkey will always and only be rewarded for a saccade to the right target. In both experimental data and model simulation we see in the monkey's behavior a transition from a random response to a consistent response to the target associated with the current cue, with other cues reliably associated with the opposite response.

The basic idea of the model is illustrated in figure 8.3a. If we have two peaks in the FEF activity playing upon SC, and we feed them into a WTA with a little noise added, then it would be random as to which one of the peaks would be selected, and thus which target the system will saccade to. This is just what happens in the Dominey and Arbib (1992) model described above if we add a little bit of noise as the visual input passes through PP and FEF. But, if we add disinhibition to, say, the rightmost peak, then the WTA will select the rightmost target. What we would like to do is have a learning process which paired disinhibition with the appropriate cue. The way this is achieved is shown in figure 8.4.

Using the FEF as the center of the figure, the left-hand side of figure 8.4 can be recognized as a restructuring of figure 8.2, with PP driving FEF, which commands SC both directly and indirectly via CD and SNr.

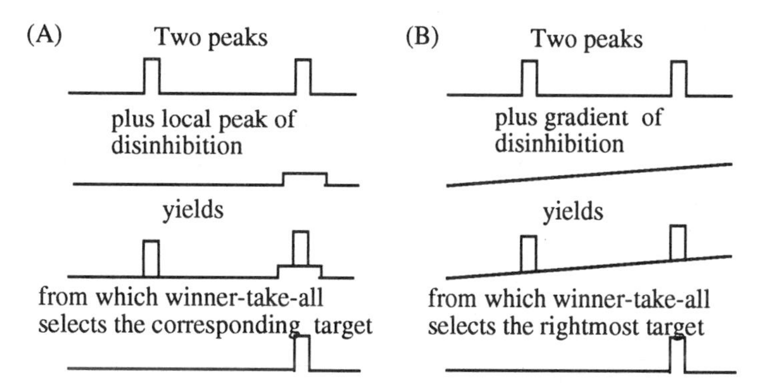

Figure 8.3 The second line of each diagram shows the peak of disinhibition that can tilt the balance in a winner-take-all mechanism to select (*A*) the peak that is in the corresponding position or (*B*) the rightmost of two peaks, whatever its position.

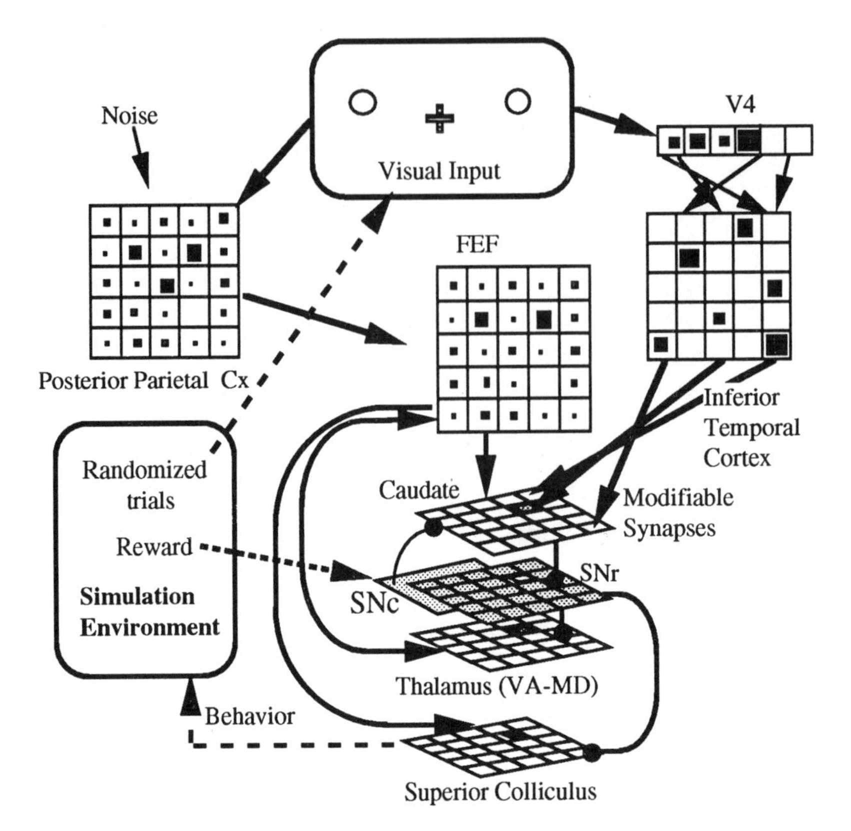

Figure 8.4 Schematic of visuomotor conditional learning model. Cue-related activity in inferior temporal cortex (IT) is produced from features extracted from visual cue input in V4. IT influences saccade production via its projections to caudate. Random noise added to the visual input in posterior parietal cortex (PP) breaks the symmetry between the two targets, providing a form of guessing. On correct cue-guided saccades, a reward contributes to strengthening synapses between IT cells activated by the cue and caudate cells participating in the saccade. After training, cue-driven activity in IT will preferentially drive the caudate cells involved in the correct saccade to overpower the noise in PP, yielding correct performance.

The new parts of the model are shown at right. While information about target location continues to be available via PP, information about the shape and color of the fixated cue travels via V4 to the inferior temporal cortex (IT) and thence projects onto the same caudate cells as those influenced by FEF so that visual information can pass through IT to project onto the same cells. At present, we assume no structure to these connections, only that each distinctive cue will yield a distinctive pattern of IT input to CD, and that the synapses carrying these patterns are uniformly distributed across CD. Our hypothesis is that if the monkey is correct it gets rewarded with a squirt of juice, and that the effect of reward is that dopamine (DA) is released which can modify the IT↔CD synapses according to a learning rule similar to that discussed by Wickens and Kötter in chapter 10. Basically, this is a reinforcement learning rule saying that at active CD cells, when DA is released, synapses from active IT cells will be strengthened (long-term potentiation, or LTP) while those for other IT cells will be weakened (long-term depression, or LTD) to yield a normalization effect with overall synaptic strength remaining the same. Formally, this is expressed by the equations:

$$w_{ij}(t+1) := w_{ij}(t) + \text{DA\_Modulation}*(\text{RewardContingency} - 1) *C1*F_i*F_j \quad (1)$$

$$w_{ij}(t+1) := w_{ij}(t+1)* \frac{\sum_j w_{ij}(t)}{\sum_j w_{ij}(t+1)} \quad (2)$$

where ":=" denotes assignment rather than equality. F_i and F_j are the firing rates of the IT and CD cells, respectively, while w_{ij} is the strength of the synapse connecting IT cell i to caudate cell j. The term DA_Modulation expresses a role of DA complementary to its role in learning, in a feedback loop that regulates the excitability of CD.[3] We simulate reward-related modulation (which we regard as mediated by DA release from substantia nigra pars compacta (SNc) by the term Reward-Contingency which is 1.5 for correct trials, and 0.5 for incorrect trials, and 1.0 when no reward or punishment is applied, corresponding to the increases and decreases in SNc activity for reward and error trials, respectively (Schultz, 1989). The term (DA_Modulation * (Reward-Contingency −1) will be positive on rewarded trials and negative on error trials. C1 is a constant that specifies the learning rate, and is set to $2.5e^{-5}$.

Weight normalization conserves the total synaptic weight that each IT cell can distribute to its striatal synapses [equation (2)]. If, after learning has occurred, one synapse from cell i to cell j was increased, then, since the total synaptic weight from cell i is conserved, the result of this increase is a small decrease in all other synapses from i. Similarly, when a weight is decreased due to an incorrect response, the other synapses from i are increased. Via this normalization, postsynaptic cells compete for influence from presynaptic cells, producing cue discrimination. Especially in the case of rewarded trials, equations (1) and (2) approximate how LTP may occur in the most active postsynaptic cells [via equation (1)] and LTD in the others [via normalization, equation (2)]. Figure 8.5 shows a simulated saccade preparation cell, trained by the above reinforcement learning procedure.

Eligibility The point about reinforcement learning is that the reward signal does not specify anything about the correct response. When a reward is given, there is hebbian learning—IT↔CD synapses that actively participate in the generation of the rewarded response are strengthened. However, when we change the sign of the reinforcement,

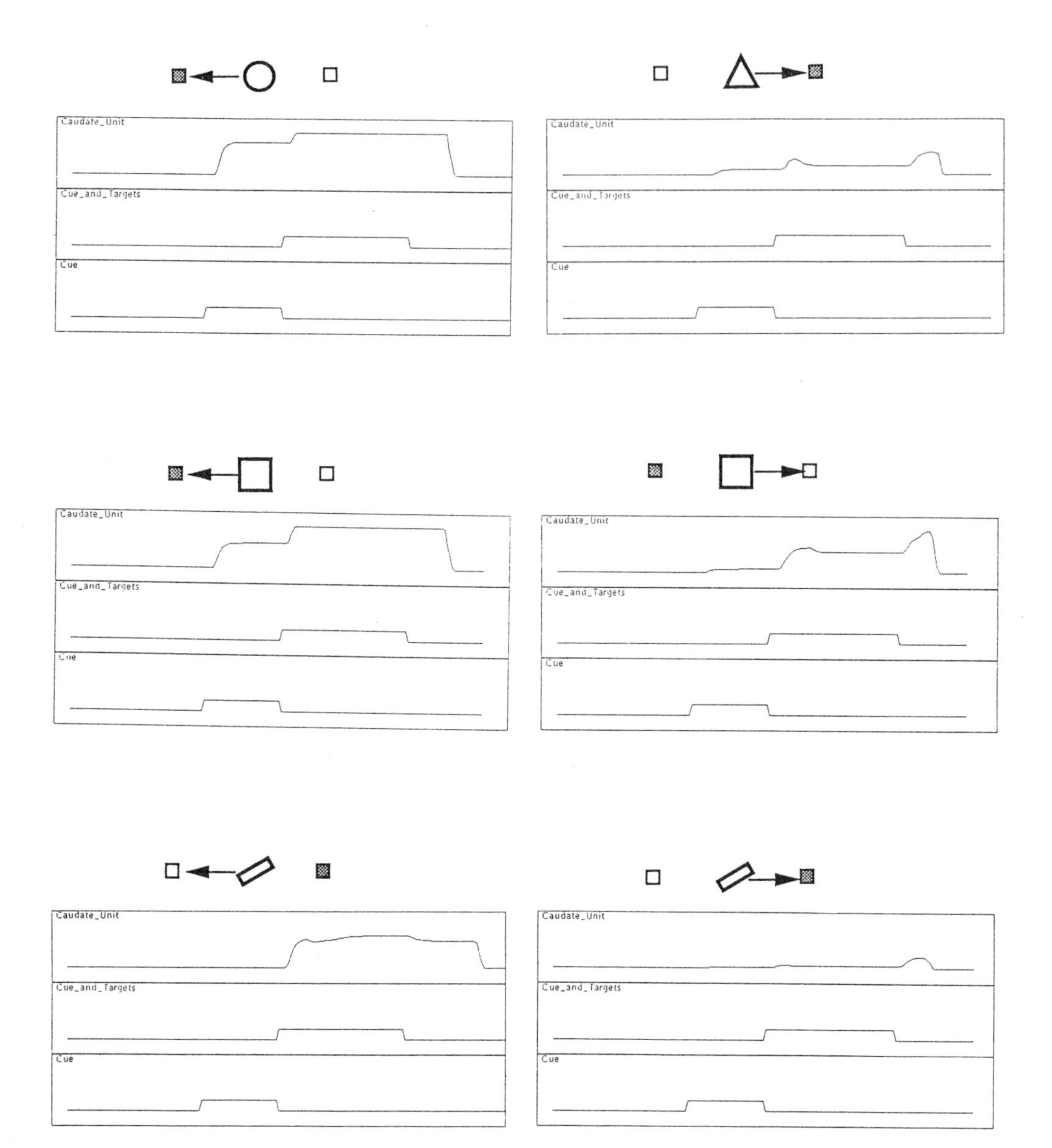

Figure 8.5 Simulated saccade preparation cell. Arrow indicates direction of saccade; dark square indicates correct direction. In each window the lower trace indicates when the cue alone appears, the middle trace shows when the cue and target overlap, and the upper trace shows the cell activity. All traces are from a cell that has preference for leftward saccades. (*Left*) Responses to cued leftward saccades. In the first and second rows, the cues are associated with a leftward saccade. The cell discharges with the cues, and increasingly with the targets through the saccade. This demonstrates the task-related activity and the equivalence of two cues associated with the same saccade. In the bottom row, an incorrect saccade is made to the left with a cue that is associated with a rightward saccade. The cell still discharges for its preferred direction. (*Right*) Rightward saccades are made both in correct and in incorrect situations, and in neither case is there significant activity.

we get an antihebbian rule, which weakens the synapses that were actively involved in generating an incorrect response.

Note, however, that equations (1) and (2) are set up on a trial-by-trial basis—they specify how synaptic weights are adjusted after each trial. However, we will in future need to address the real-time issue, namely that the release of DA by SNc occurs after the animal is rewarded, which occurs after the animal has acted, which occurs after the activity in the IT↔CD synapses which assisted the generation of the response. How then does such a synapse "know" that it should respond with LTD or LTP when the DA finally affects it? The answer offered (in the context of models of classical conditioning) by Klopf (1982) and Sutton and Barto (1981) was the notion of an *eligibility trace*—a synapse "remembers" that it was recently activated. We can now model this using second-messenger systems so that the reward will affect just those synapses whose eligibility is above zero. In equation (1), then, the time scale becomes much finer (say millisecond by millisecond, rather than trial by trial) and the term $F_i{}^*F_j$ is replaced by the eligibility of the synapse w_{ij}. We have an interesting wrinkle on eligibility in our companion study (Schweighofer et al., submitted) of the role of the cerebellum in metric adaptation of saccades. In such adaptation, as we have seen, the monkey's incorrect saccade is immediately followed by a corrective saccade. How, then, does the error signal affect those synapses involved in the incorrect saccade rather than those involved in the corrective saccade? The answer is to shape the eligibility trace so that it peaks only after a delay of about 200 ms. This ensures that when the error signal (climbing fiber input to a Purkinje cell) arrives, those plastic synapses (from parallel fibers to Purkinje cells) involved in the first saccade are at the peak of their eligibility, while those involved in the corrective saccade are not yet eligible.

This work suggests that the shaping of the eligibility signal may be task-dependent. It has thus become an important goal of work at the University of Southern California to try to bridge from this systems level of neural analysis to that of synaptic neurochemistry by trying to match (a) behavioral constraints on the dynamics of the eligibility trace with (b) the pharmacokinetics of the secondary messengers, etc., which may mediate it.

Spatial Generalization Returning to our study of the role of the basal ganglia in visuomotor conditional learning, we have been interested in spatial generalization. In figure 8.3A, we saw that a "bump" of cue-related activity placed by IT input in CD activity could cause the model to yield a saccade to a *specific* target associated with the fixation cue by reinforcement learning. But where do we put the "bump" if we want a given cue to signal that the animal should choose the rightmost, say, of two targets *no matter where the two targets are located*? The theoretical answer shown in figure 8.3B is to use not a bump but a gradient, sloping upward to the right, so that when it is added to twin peaks, the one on the right is going to become higher than the one on the left, and thus the WTA will always choose the rightmost one. To our delight, our reinforcement learning model "discovered" this solution, rather than having it inserted as a constraint of the model. If we present pairs of targets in many positions, but always reward, say, a saccade to the rightmost target when the cue is a blue triangle, then the incremental effect is that those IT↔CD synapses which are active for the blue triangle cue will be larger and larger for those CD cells associated with saccades further and further to the right.

Sequential Learning

Barone and Joseph (1989) studied a sequence learning paradigm. The monkey faced three keys displayed in a triangle around the fixation point. The monkey is repeatedly trained on sequences of one, then two, then three targets to respond to seeing the targets illuminated in a particular sequence while the fixation point is on (in this study, the fixation point itself has no cue properties) by pressing the targets in the same sequence in response to three successive "go" stimuli after the offset of the fixation point. This involves immense shaping. The monkey never "clicks" with the general idea of a sequence. Unlike a human, it cannot see a novel sequence and immediately repeat it; rather it must be trained again and again with a specific sequence before its visual presentation can act as the cue for its manual repetition. In fact, one monkey successfully learned to repeat five of the six possible three-element sequences, yet never achieved as low an error rate with the sixth!

The model for such sequence learning (Dominey et al., submitted) has a similar structure to that for visuomotor conditional learning in figure 8.3, but now the context dependence is provided by the input from prefrontal cortex (PFC) rather than IT. The self-loops in prefrontal cortex ensure that PFC activity changes as the visual inputs are received, and as subsequent saccades are triggered. Thus PFC input to CD is sufficiently diverse for the animal to eventually learn a set of sequences.

But Where Does Plasticity Really Reside?

In the two models presented here, we hypothesize that corticostriatal connections provide the crucial substrate for the two modes of learning under study.

Visuomotor Conditional Learning Our model posits the plasticity of a projection from IT to CD. The existence of such a projection is shown by Selemon and Goldman-Rakic (1985) but Goldman-Rakic (personal communication) has noted that this projection is sparse, and one should consider that the confluence might involve PP (rather than FEF) and IT. This alternative is functionally equivalent for the present model of activity in CD, SNr, and SC, but does require experimental resolution.

Sequential Learning Strick (personal communication) has questioned the extent of the convergence of PFC inputs (part of the prefrontal loop) with the region of caudate involved in the oculomotor loop as hypothesized in our model. With this in mind, the model could instead incorporate the known corticocortical connections from PFC to FEF, and plastic PFC↔FEF synapses (DA plays a role in cortical plasticity too!) could then bias the target representation in FEF—so that changes seen in CD in our preliminary experiments would then be the expression of FEF input, rather than the expression of corticostriatal plasticity. This alternative awaits simulation and the subsequent design of more refined experiments to test the new hypothesis.

NOTES

1. In this chapter, we discuss the structure of the model and highlight salient aspects of its behavior. The model has been fully formalized and simulated in the neural simulation language NSL (Weitzenfeld, 1991), with results in good general agreement with neurophysiological recordings (see Dominey and Arbib, 1992, for details).

2. In the present model, cells code for a given direction, and are quiet for other directions. In chapter 7 Goldman-Rakic shows that FEFmem topographical maps have an opponent organization, i.e., not only is a cell active for its preferred direction but is depressed below its resting rate when the opposite direction is encoded, and shows transient activity when the saccade occurs. We believe it would be easy to model this by simply adding inhibitory coupling between cells with opposed preferences—with the transient activity being the automatic consequence of postinhibitory rebound.

3. Release of DA in striatum decreases the excitability of striatal cells by corticostriatal afferents, and thus attenuates corticostriatal activity, allowing striatum once again to detect the effect of IT's bias. In our model, when strong cortical inputs overstimulate striatum SNr is strongly inhibited. This inhibition of SNr leads to the *dis*inhibition of SNc (Carlsson and Carlsson, 1990). The resulting increase in activity of SNc DA cells leads to the increased release of DA in striatum, which attenuates the corticostriatal signal (Calabresi et al., 1993), bringing the most active cells back to the maximum input capacity of striatum, and reducing the activity of the less active *surrounding* striatal cells, thus increasing the signal-to-noise ratio. See Dominey, et al. (submitted) for formalization and further details.

REFERENCES

Alexander, G. E., and Fuster, J. M. (1973) Effects of cooling prefrontal cortex on cell firing in the nucleus medialis dorsalis. *Brain Res.* 61:93–105.

Barone, P., and Joseph, J. P. (1989) Prefrontal cortex and spatial sequencing in macaque monkey. *Exp. Brain Res.* 78:447–464.

Calabresi, P., Mercuri, N. B., Sancesario, G., and Bernardi, G. (1993) Electrophysiology of dopamine-denervated striatal neurons: Implications for Parkinson's disease. *Brain* 116:433–452.

Carlsson, M., and Carlsson, A. (1990) Interactions between glutamatergic and monoaminergic systems within the basal ganglia—implications for schizophrenia and Parkinson's disease. *Trends Neurosci.* 13:272–276.

Dominey, P. F., and Arbib, M. A. (1992) A cortico-subcortical model for generation of spatially accurate sequential saccades. *Cereb. Cortex* 2:153–175.

Dominey, P. F., Arbib, M. A., and Joseph, J.-P. (submitted) A model of cortico-striatal plasticity for learning associations and sequences.

Funahashi, S., Bruce, C. J., and Goldman-Rakic, P. S. (1989) Mnemonic coding of visual space in the monkey's dorsolateral prefrontal cortex. *J. Neurophysiol.* 61:331–349.

Hikosaka, O., and Wurtz, R. (1983a) Visual and oculomotor functions of monkey substantia nigra pars reticulata. III. Memory-contingent visual and saccade responses. *J. Neurophysiol.* 49:1268–1284.

Hikosaka, O., and Wurtz, R. (1983b) Visual and oculomotor functions of monkey substantia nigra pars reticulata. IV. Relation of substantia nigra to superior colliculus. *J. Neurophysiol.* 49:1285–1301.

Klopf, A. H. (1982) *The Hedonistic Neuron: A Theory of Memory, Learning, and Intelligence.* Washington, D.C.: Hemisphere.

Schultz, W. (1989) Neurophysiology of basal ganglia. In D. B. Caine (ed.), *Handbook of Experimental Pharmacology,* Vol. 88. Heidelberg: Springer-Verlag, pp. 1–45.

Schweighofer, N., Arbib, M. A., and Dominey, P. F. (submitted) A model of the role of cerebellum in adaptive control of saccades.

Scudder, C. A. (1988) A new local feedback model of the saccadic burst generator. *J. Neurophysiol.* 59:1455–1475.

Selemon, L. D., and Goldman-Rakic, P. S. (1985) Longitudinal topography and interdigitation of corticostriatal projections in the rhesus monkey. *J. Neurosci.* 5:776–794.

Sutton, R. S., and Barto, A. G. (1981) Toward a modern theory of adaptive networks: Expectation and prediction. *Psychol. Rev.* 88:135–170.

van Gisbergen, J. A. M., Robinson, D. A., and Gielen, S. (1981) A quantitative analysis of generation of saccadic eye movements by burst neurons. *J. Neurophysiol.* 45:417–442.

Weitzenfeld, A. (1991) *NSL, Neural Simulation Language,* Version 2.1, Technical Report No. 91-02, Center for Neural Engineering, University of Southern California, Los Angeles.

Wurtz, R. H., and Munoz, D. P. (1993) Saccadic and fixation Systems of oculomotor control in monkey superior colliculus. In P. Rudomin, M. A. Arbib, F. Cervantes-Pérez, and R. Romo (eds.), *Neuroscience: From Neural Networks to Artificial Intelligence.* Heidelberg: Springer-Verlag, pp. 413–425.

9 A State-Space Striatal Model

Christopher I. Connolly and J. Brian Burns

This chapter considers the role of the striatum in computing purposeful state changes. Such state changes include, for example, movement through a maze, or an obstacle-avoiding arm reach. In each of these cases, the system (arm, body, etc.) must make its way from the current state to one or more goal states, avoiding undesirable states along the way. Certain properties of the striatum make it a strong candidate for a locus of computation for this problem.

Alexander et al. (1992) present a compelling view of the basal ganglia as being involved in parallel, functionally segregated circuits. These circuits appear to run from cortex, striatum, pallidum, thalamus, and back to cortex (Graybiel, 1991; see also Holsapple et al., 1991). Using this basic architecture, this chapter attempts to examine some striatal "cross sections" of these circuits. We postulate that these cross sections correspond to distinct, contiguous state spaces embedded in the striatal matrix, each of which is responsible for an aspect of organism function (e.g., arm reaching, egomotion, etc.). We also propose that the membrane potentials of cells in these regions comprise harmonic functions over each state space, and that the gradient descent of these functions produces trajectories—sequences of purposeful state changes.

Connolly and Burns (1993a) propose a striatal model which suggests that the striatum contains state-space representations. The model concentrates on the motor control aspects of basal ganglia function. This model is reviewed here, and some refinements of the model are considered. In particular:

• The model is further localized to the matrisomes of the striatum. These contiguous partitions of the matrix appear to be in the best position, both anatomically and pharmacologically, for performing the sort of motor control proposed here.

• The effect of continuous cortical firing (of the sort exhibited by pyramidal neurons of layer V) on medium spiny cells is considered.

• A more detailed (albeit speculative) picture is proposed for the communication of gradient (state change) information from the striatum to the pallidum.

The relevant mathematical description of a motor control scheme is reviewed first, followed by a description of the original striatal model. The chapter concludes by considering some extensions and suggested experiments for this model.

MOTOR CONTROL

An important problem in robotics concerns systems which are capable of deriving collision-free paths from one point to another (the path-planning problem). For most robot types, one can describe robot position and motion in terms of the robot's configuration space (Udupa, 1977; Lozano-Pérez, 1981). Configuration space is the space of parameters which can be used to describe the pose of the robot. For a small mobile robot, these parameters are simply the x and y positions of the robot on the floor. For a jointed robot arm, the parameters are the angles of each joint. Thus, such mechanical systems can be represented as single points in an appropriate configuration space. A change in configuration is represented by movement of the point. This scheme has been applied to a variety of mechanical systems.

Figure 9.1 provides an illustration of these concepts. On the left is a simulation of a two-link arm with revolute joints. The arm is executing an obstacle-avoiding reach to a goal region. Obstacles are shaded, and the goal region is black. On the right-hand side of this figure is the corresponding configuration space, which represents the two joint angles of the robot. Movement of the robot on the left corresponds to the

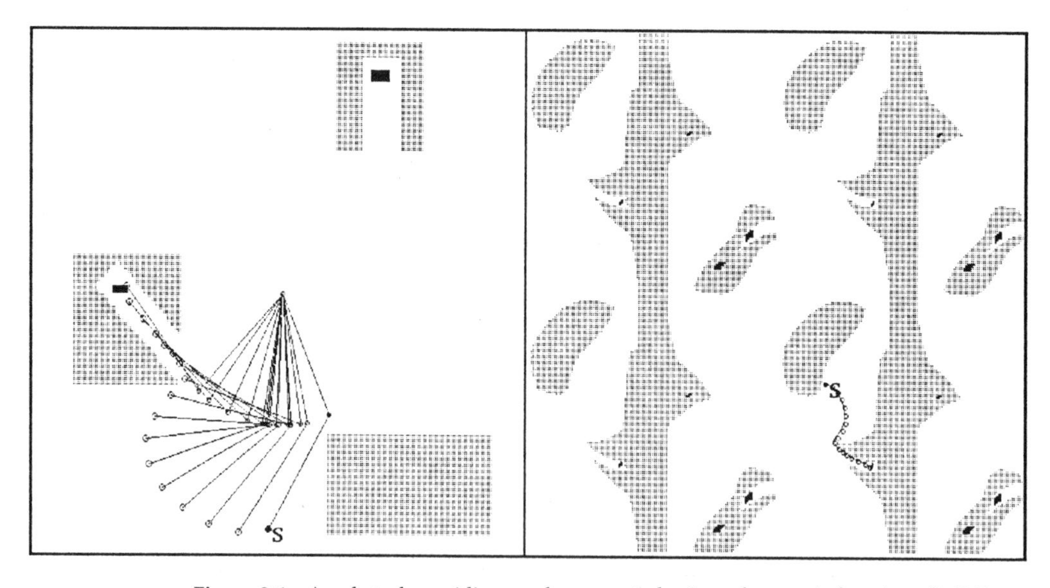

Figure 9.1 An obstacle-avoiding reach, generated using a harmonic function. (*Left*) Two-link arm avoiding obstacles, (*Right*) Configuration space trajectory.

movement of the point on the right. The shaded regions on the right correspond to obstacles.

Although several approaches have been taken for solving the path-planning problem in robotics (Schwartz and Sharir, 1983; Khatib, 1985; Arkin, 1989; Canny and Lin, 1990), we concentrate here on the use of harmonic functions (Connolly and Grupen, 1993). Harmonic functions are solutions u to Laplace's equation in n dimensions:

$$\sum_{i=0}^{n} \frac{\delta^2 u}{\delta x_i^2} = 0 \tag{1}$$

This equation describes several physical phenomena (e.g., ideal fluid flow, electrostatic potential) and says that all second derivatives of the harmonic function u sum to zero. The equation provides a model for many steady-state phenomena. The relationship of Laplace's equation to so many different physical phenomena, in concert with its utility for robot planning, makes this mathematical framework a possibility for planning in biological systems.

Analog computers (using resistive networks) have been constructed to compute solutions to this equation (McCann and Wilts, 1949). Since Kirchhoff's laws[1] represent a discrete form of Laplace's equation, such grids are able to compute discrete solutions to Laplace's equation very rapidly as voltages. Recently, Tarassenko and Blake (1991) employed a resistive grid to solve the path-planning problem for a mobile robot. Simply put, the scheme described here computes potentials governed by equation (1), then performs gradient descent on the potentials to generate trajectories in some state space.

State Spaces and Resistive Networks

For the purposes of this discussion, the following intuition may be useful: Consider a robot arm's configuration (joint angle) space. Divide this space into a discrete grid. Every grid point represents a particular arm configuration. Place resistors between pairs of adjacent grid nodes (as shown in figure 9.2). This grid can now be used to generate trajectories for the robot arm by manipulating voltages at certain nodes. Specifically, the potentials over this grid will be discrete solutions to equation (1).

In this framework, the setup for motor control is achieved by raising certain grid nodes to high potentials ("depolarization"), and lowering the potentials of other grid nodes ("hyperpolarization"). These potentials then become the boundary conditions for equation (1). The high potentials correspond to states which are to be avoided (e.g., states which would result in collision with obstacles). Low potentials correspond to goal states. Nodes which are not fixed are referred to as free nodes. To compute the potentials at free nodes, the mean-value prop-

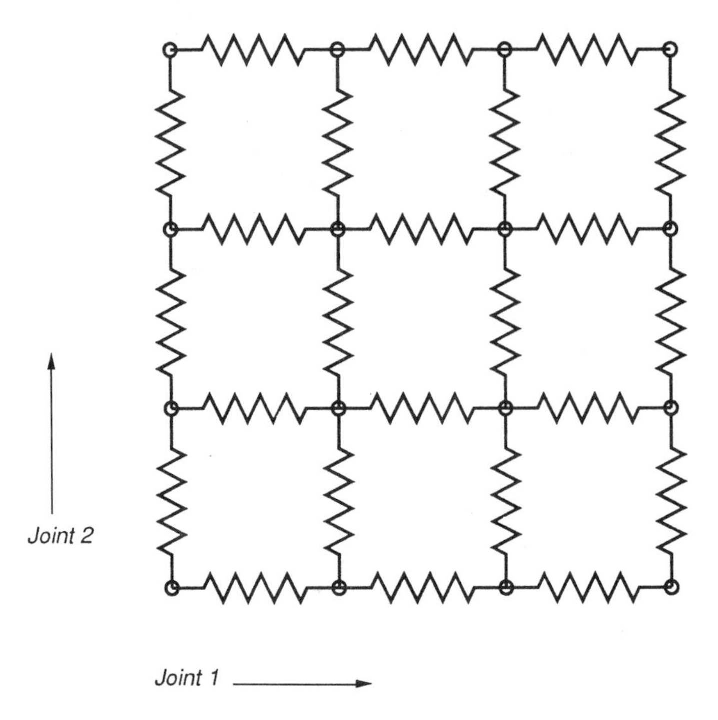

Figure 9.2 Resistive grid in configuration space.

erty of harmonic functions can be invoked (e.g., see Doyle and Snell, 1984, or Axler et al., 1991): The value of a harmonic function at a point in the grid is the average of the values found at grid neighbors[2] (figure 9.3). This also means that, unless the function is constant in the grid, every free node will be surrounded by nodes with both greater and lesser potentials. This shows directly that such functions are suitable for control based on gradient descent, since the only minima for the function will be at those states which have been held fixed at low potentials. Once such a resistive network has settled, obstacle-avoiding trajectories can be computed by using gradient descent. This basic scheme was used to generate the trajectory seen in figure 9.1.

Harmonic functions exhibit properties which are suitable for motor control. In particular, they are smooth, and decrease monotonically toward the designated state-space goal points. The smoothness property results in trajectories which are easily controlled. The monotonicity insures that gradient descent will always reach accessible goal points. A motor control scheme which relies on a resistive network with interpolation[3] (or any other mechanism for computing harmonic functions) is therefore sufficient for orchestrating smooth, goal-reaching state changes. Moreover, these state changes avoid designated undesirable states. The system to be controlled need only sample the harmonic function gradient at its current state-space point to generate the

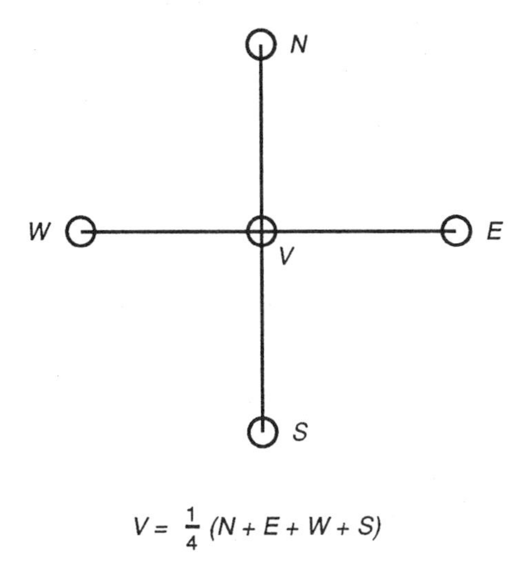

$$V = \frac{1}{4}(N + E + W + S)$$

Figure 9.3 Potential at the center is the average of neighbor potentials.

appropriate trajectory. If the computation method is sufficiently fast, then changes in goal and obstacle configuration will not affect the system's ability to reach its goal safely. Figure 9.4 illustrates these points. The shaded areas denote obstacles. This figure shows a series of gradient descent trajectories generated by the harmonic function corresponding to the obstacle-goal configuration. Each trajectory terminates at one of four goal points placed close to the obstacles. Note that these paths are all smooth and obstacle-avoiding.

PHYSIOLOGICAL CONSTRAINTS

The primary neurons of the striatum receive projections from the cortex, and these appear to emanate mainly (but not exclusively) from layer V (Kandel, et al., 1991). Three general cell types have been identified within the striatum (see Wilson, 1990): (1) giant aspiny, (2) medium aspiny, and (3) medium spiny. The giant aspiny cells are associated with Kimura's type I cells (Kimura et al., 1984), and are believed to be cholinergic. The medium aspiny cells are believed to be γ-aminobutyric acid (GABA)–ergic. The majority of striatal cells are medium spiny cells, which are also the output neurons of the striatum. The model presented here focuses on these cells.

Cortical axons synapse on the spines of medium spiny cells. The medium spiny cells, in turn, project to the globus pallidus and the GABAergic cells of the substantia nigra pars reticulata. The striatal architecture is monosynaptic from cortex to pallidum through the medium spiny cells (Goldman-Rakic and Selemon, 1990). The interneurons (aspiny cells) are much fewer in number and appear to serve a regu-

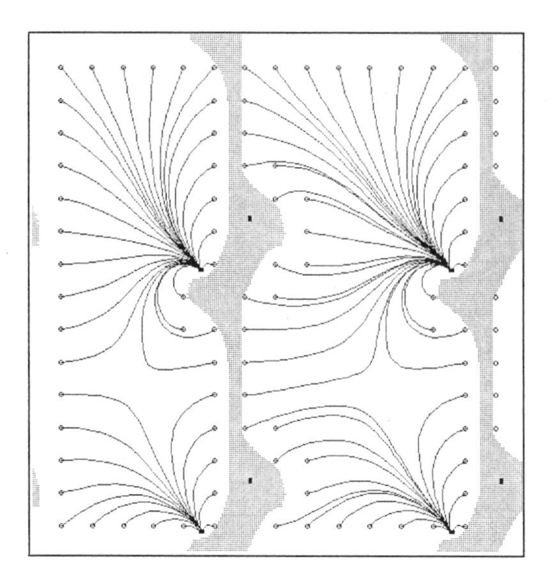

Figure 9.4 A sequence of trajectories generated by the gradient descent of a harmonic function.

latory function. The dendritic fields of the medium spiny cells are roughly spherical and overlapping. Recent research reports dopamine-dependent dye-coupling among medium spiny cells (Cepeda, et al. 1989; Walsh, et al., 1989; Cepeda et al., 1991; Walsh et al., 1991; O'Donnell and Grace, 1993), and some micrographic evidence which suggests the presence of gap junctions (Pasik et al., 1990). In addition, the striatum is divided into striosomes (regions that are poor in acetylcholinesterase [AChE]), and matrix. The matrix is further divided into contiguous matrisomes, which appear to have functional significance (Flaherty and Graybiel, 1991; Flaherty and Graybiel, 1993; see also chapter 5).

Williams and Millar (1990) report that the activity of medium spiny cells is dependent on dopamine (DA) concentrations. Specifically, DA at low concentrations (roughly 100nM) apparently induces cell firing, while higher DA concentrations (1–10μM) silences the cells. It is conceivable then that, among its other functions, DA can act as a trigger to induce striatal activity.

The type I cells described by Kimura et al. (1984) are tonically active, and become quiescent just before the onsent of voluntary movement. In addition, the striosomes are poor in AChE, the uptake enzyme for acetylcholine. It therefore seems reasonable to assume that changes in cholinergic cell activity will primarily affect the matrix (see also chapter 5). The matrix also receives a rich set of projections both from motor and sensory areas (Flaherty and Graybiel, 1993), and projects in turn to the globus pallidus (Graybiel, 1991).

THE PICTURE

In light of the algorithmic scheme described earlier and the physiological constraints mentioned in the previous section, an overall view of the striatopallidal system can be postulated. Figure 9.5 illustrates those components of the striatum and pallidum which are considered in this discussion.

Cortical neurons which project to the matrix are responsible for imposing constraints on this system. We propose that, at the very least, these neurons set the boundary conditions for striatal computation. The net effect of this activity is that certain striatal primary neurons will be depolarized, while others are hyperpolarized. The network of primary neurons represents a state space (e.g., arm joint angles) over which a potential function [determined by equation (1)] is passively computed, subject to the boundary conditions determined by corticostriatal control. The most direct and rapid possibility for passive computation of a harmonic function is through electrotonic coupling (figure 9.6). This would result in a resistive network within the striatum, probably regulated in part by the (dopaminergic) nigrostriatal system.

The potential function is therefore "sculpted" by the cortically imposed boundary conditions. One may envision this function as a thin rubber sheet, where some regions have been pulled up (obstacles, undesirable states), while others are pulled down (goal states). If the mechanism for computing this function is fast enough (e.g., a resistive network), then the function need not be completely static, but may change in response to external circumstances (e.g., movement of obstacles). The state change trajectory determined by this surface can be envisioned by imagining the motion of a ball rolling down the rubber sheet.

The gradient of this function must be extracted at the point corresponding to the current state. This provides the system with information about the next immediate state change. Connolly and Burns (1993a) postulate that a small region of cells (the current-state cluster) fires to deliver this information to the pallidum. Firing frequency presumably encodes the gradient information. Locally reduced DA concentrations may be responsible for this increased activity (Williams and Millar, 1990). This current-state cluster moves with the change in state, and essentially forms a feedback loop with the cortex providing current state information. In the case of arm movement, for example, the gradient induces a state change (through the pallidum, thalamus, and motor cortex) which changes the arm position. This change in state is manifested as a cortically induced change in the location of the current-state cluster within the striatum.

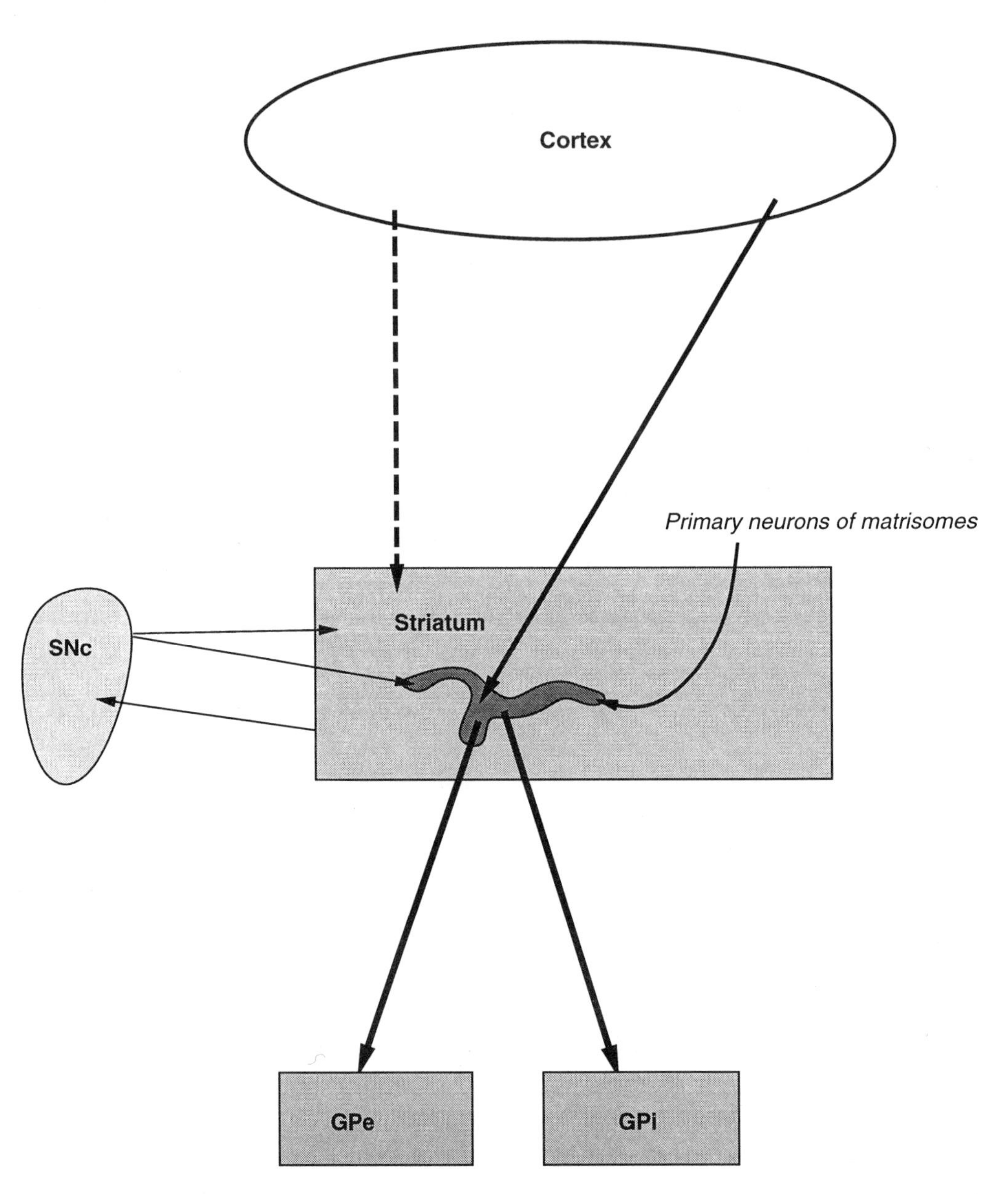

Figure 9.5 Schematic view of the system under consideration: the corticostriatopallidal pathway through the matrisomes. GPe, GPi, external and internal segments of globus pallidus.

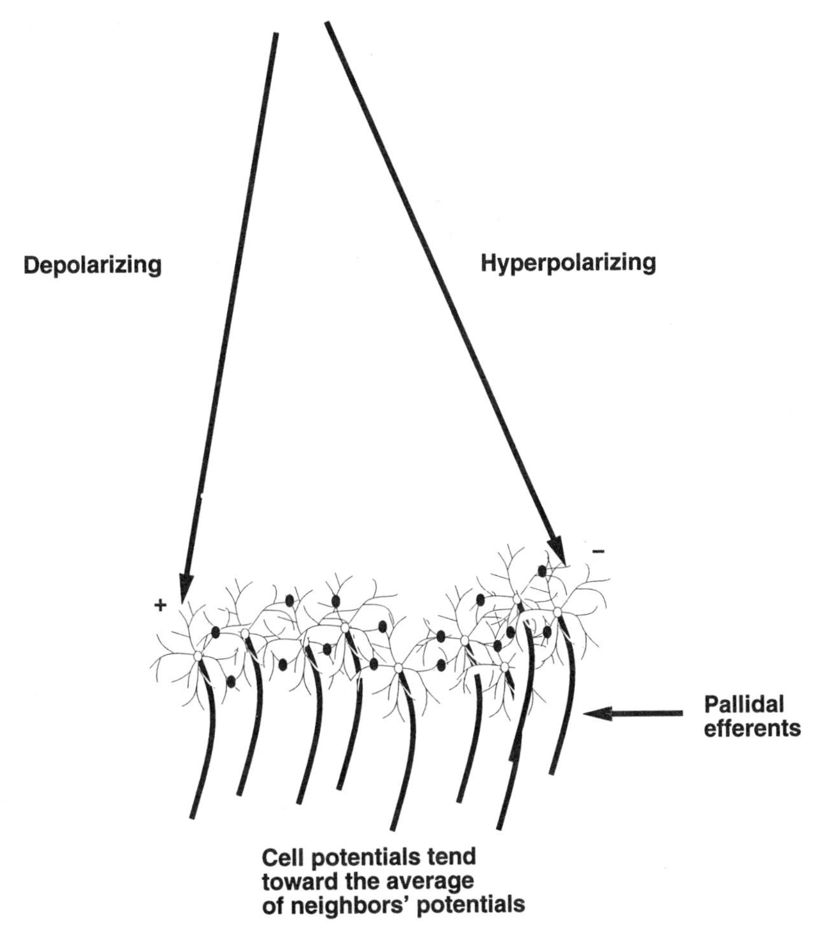

Figure 9.6 Coupled primary neurons: "free" cells' potentials will be the average of neighbor cells' potentials. Cells marked + and − are held at fixed potentials.

REFINEMENTS AND EXTENSIONS TO THE PICTURE

The medium spiny cells exhibit overlapping dendritic fields and local axonal arborization. Even so, there is currently no evidence for lateral inhibition among the network of medium spiny cells (see chapter 4). These cells are relatively quiet, and in concert with this their subthreshold activity suggests that the sort of passive computation described above could be taking place within the matrisomes. The proposed computation is a relaxation of potentials (Connolly and Burns, 1993a). We conjecture that the matrisomes correspond to the "state spaces" described above. These are, at least in a rough sense, striatal cross sections of the multiple parallel circuits described by Alexander et al. (1992). This localization is supported by the anatomical connections: afferents from the appropriate cortical sensory and motor areas, and efferents to the globus pallidus. In contrast, the striosomes project mainly to the substantia nigra pars compacta (Graybiel, 1991). The involvement of the globus pallidus in motor control (Kato and Kimura, 1992) and its apparent functional segregation (Hoover and Strick, 1993) lend some support to the view of matrisomes as functional units for direct motor control.

Pyramidal neurons of the cortex (some of which appear in layer V) have a regular firing pattern (Douglas and Martin, 1990) which is suitable for maintaining excitatory postsynaptic potentials (EPSPs) (and possibly inhibitory postsynaptic potentials [IPSPs]) in a finely graded manner over a period of time in striatal neurons (Wilson, 1990). In addition, striatal medium spiny cells exhibit a pronounced "up" state after coordinated corticostriatal excitation (see chapter 3), which could be a manifestation of cortically imposed boundary conditions.[4] In the context of the model described here, pyramidal neurons could represent a form of short-term state memory. Axonal arborization of corticostriatal fibers allows for a variety of possible mappings from cortical representations to striatal state-space representations. An example of such a mapping is the cartesian-to-joint-space transformation. In hardware, this requires nothing more than a massive boolean "OR" circuit which translates cartesian effector positions into corresponding joint angle "buckets" (Connolly and Burns, 1993a). A similar scheme has been used for performing obstacle mappings for robot systems (Connolly and Grupen, 1993). Figure 9.7 illustrates this mapping. All of the shaded cells in the left grid are mapped (through an "OR" gate) to one cell on the right. A two-link arm is superimposed on the left. The grid on the right represents joint angles (and thus all whole-arm configurations). In this way, obstacles which are sensed in cartesian coordinates can be mapped directly into a representation in joint space (figure 9.1 on the right shows the joint space mapping for a rectangular obstacle). It is

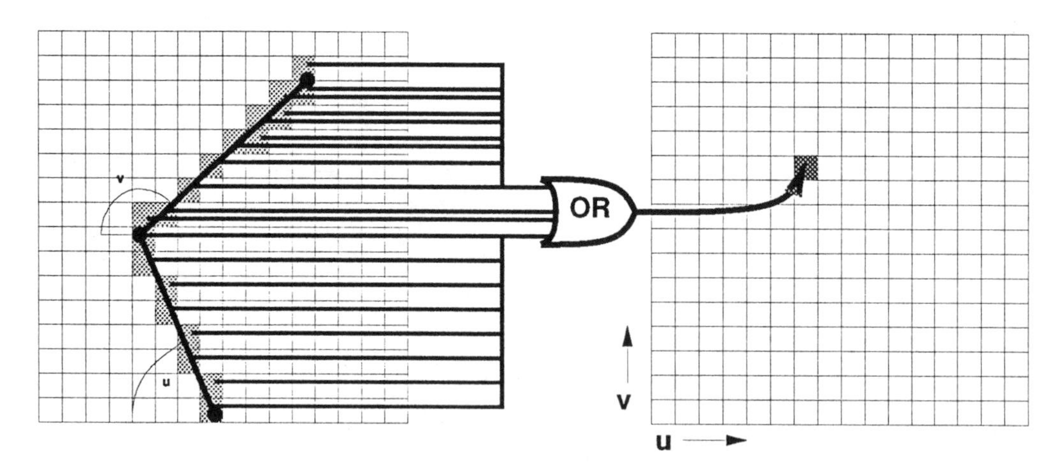

Figure 9.7 An example of a boolean "OR" circuit, which transforms cartesian obstacle information into joint angle space.

possible that this scheme has a counterpart in the corticomatrisomal projection.

Gradient Readout

The picture described above still begs the question: How is gradient information transferred from the striatum to the pallidum? Although the striatum is relatively quiet, certain cells do fire in bursts and in response to passive or active arm movement (Crutcher and DeLong, 1984; Kimura, 1990). This firing pattern could represent a "readout" of the information inherent in the distribution of membrane potentials in the matrisome network. The current-state cluster of firing cells should be able to transfer the potential gradient information from the current state cluster to the pallidum. To exemplify this gradient computation, the reduction of DA concentrations is approximated by a spatial gaussian distribution. Specifically, the increase in cell activity due to reduced DA concentration as a function of cell location is modeled here by:

$$G(x,y,z) = \alpha e^{-\beta(x^2+y^2+z^2)} \tag{2}$$

Here, x, y, and z represent distance from the centroid of the cluster. It is assumed here that the firing rate R of cells within the cluster will depend on the product of DA concentration and membrane potential:

$$R(x,y,z) = G(x,y,z)V(x,y,z) \tag{3}$$

This distribution of firing rates will result in a skewed gaussian. The position of the peak of this skewed gaussian is a function of the gradient of V, and thus represents an encoding of the gradient. Figure 9.8 shows how the slope of a function (top) results in such a skew in the distribution (bottom). The vertical line in the center of the figure shows the location of the peak of the original gaussian function. In general, only one extremum (a maximum) will occur in this function, since V is harmonic (thus has no extrema in the region of interest) and G is strictly positive with only one maximum. This relationship is determined by differentiating R and finding common zeros (i.e., the extrema of R):

$$R_x = G_xV + GV_x \tag{4}$$

$$= 2\alpha\beta x(x^2 + y^2 + z^2)GV + V_x[DA] \tag{5}$$

$$= G(V_x - 2\alpha\beta x(x^2 + y^2 + z^2)V) \tag{6}$$

$$= 0 \tag{7}$$

The function G is strictly positive. Thus, at the peak of the skewed gaussian, the following relationships will hold:

$$V_x = 2\alpha\beta x(x^2 + y^2 + z^2)V \tag{8}$$

$$V_y = 2\alpha\beta y(x^2 + y^2 + z^2)V \tag{9}$$

$$V_z = 2\alpha\beta z(x^2 + y^2 + z^2)V \tag{10}$$

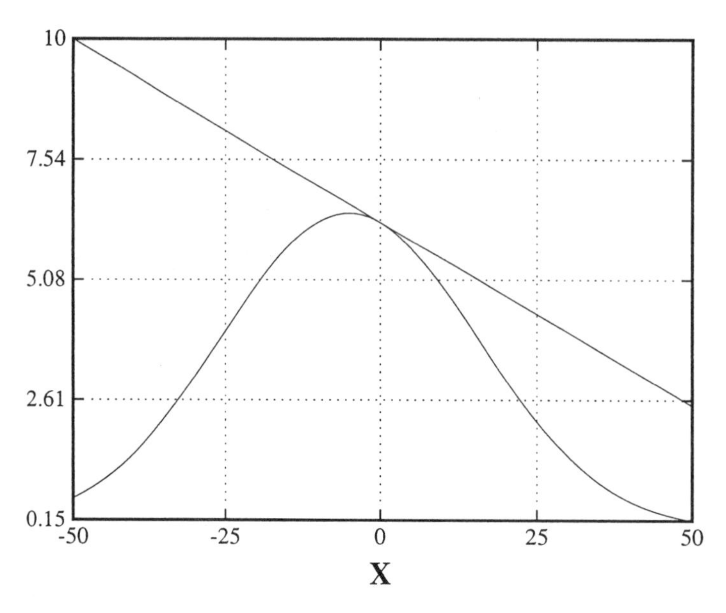

Figure 9.8 Function R (*lower curve*), which is the skewed gaussian resulting from the slope seen in the top curve. Note that the leftward shift in the peak is in the direction of the gradient.

Note also that at the origin, the function R_x is proportional through α to V_x (likewise for R_y and V_j, and R_z and V_z), thus implying that most of the mass of R is shifted in the direction of the gradient of V.

This model of striatopallidal firing frequency suggests that the potential gradient is directly related to the centroid of the resulting skewed gaussian. When the gradient of the potential function is zero, the gaussian is not skewed. This firing profile sends essentially the same information (a skewed gaussian through mass-frequency coding) from the striatum to both the internal and external segments of the globus pallidus. Both spines (Wilson, 1984) and intercellular resistances will result in a low-pass filtering for any changes in membrane potential. Because of this, it is unlikely that the action potentials (which are high-frequency signals) within this current-state cluster would have a significant effect on the potentials over the whole matrisomal network.

While this picture is extremely speculative, it is presented in order to give the reader at least one possible picture for the transferral of gradient information from striatal membrane potentials to the globus pallidus.

SUMMARY

A striatal model has been outlined here which proposes that the matrisomal networks of the striatum correspond to individual state spaces, each of which governs different aspects of organism behavior. The focus of this description has been on motor control, since this appears to be the most direct way of testing mathematical and algorithmic properties of the model. It is essential to be able to implement mathematical descriptions for striatal motor control in a straightforward manner on artificial mechanical systems. This ensures a certain *algorithmic* consistency to any proposed scheme.

The choice of matrisomes for the loci of computation is driven by their anatomical and pharmacological properties. These properties appear to be strongly related to a context-dependent motor control function. The similarities between putamen and caudate nucleus, and the existence of other functional units within the caudate nucleus (e.g., see Parthasarathy, et al., 1992) suggest that it should be possible to extend this model to nonmotor functions (see, e.g., Connolly and Burns, 1993b).

The model presented here is at best a schematic for one aspect of striatal motor control. In any event, it underscores the algorithmic and biological consistency required for successful models of basal ganglia function. As always, such models are most useful if they can predict new properties of the system being modeled. In the case of this state-space model, we can suggest the following as possibilities:

• Motor trajectories (e.g., purposful reaches) could be dependent on the presence and permeability of gap junctions.

• Membrane potentials of cells within the same matrisome should be related. Specifically, each cell's membrane potential should be a weighted average of neighboring cell potentials. This suggests that changes in membrane potential for one matrisomal spiny cell should affect the others in the network, to some degree.

• The observed spike duration of matrisomal spiny cells can be estimated and used to predict the intercellular resistances and capacitances which would be required for proper low-pass filtering of action potentials.

• The firing of a cluster of matrisomal cells should produce a stereotyped motor reaction. In the case of an arm, e.g., appropriate stimulation should result in a repeatable movement to a particular configuration.

• In patients with Huntington's disease, repeated and controlled arm movements (e.g., straight-line reaches to a goal point) should reveal "dead" zones, i.e., arm configurations which consistently result in deviations from the appropriate path.

NOTES

1. The governing laws for resistances in a circuit: Kirchhoff's current law states that current flowing into a junction must equal the current flowing out of a junction.

2. If the conductances ($1/R$) between nodes are allowed to vary, then the center value is the *weighted* average of the neighbors, where the weights are the internode conductances. Most of the principles described here still apply to this case.

3. Because of the mean-value property for harmonic functions, a simple gaussian weighted average will provide the required smooth interpolation.

4. It is interesting to note that the "up" state is just *below* the action potential threshold.

REFERENCES

Alexander, G. E., DeLong, M. R., and Crutcher, M. D. (1992) Do cortical and basal ganglia motor areas use "motor programs" to control movement? *Behav. Brain Sci.* 15:656–665.

Arkin, R. C. (1989) Motor schema–based mobile robot navigation. *Int. J. Robotics Res.* 8:92–112.

Axler, S., Bourdon, P., and Ramey, W. (1991) *Harmonic Function Theory,* Vol. 137: *Graduate Texts in Mathematics.* New York: Springer-Verlag,

Canny, J. F., and Lin, M. C. (1990) An opportunistic global path planner. In *Proceedings of the 1990 IEEE International Conference on Robotics and Automation,* Cincinatti, Ohio, May 1990, pp. 1554–1559.

Cepeda, C., Walsh, J. P., Hull, C. D., Howard, S. G., Buchwald, N. A., and Levine, M. S. (1989) Dye-coupling in the neostriatum of the rat: I. modulation by dopamine-depleting lesions. *Synapse* 4:229–237.

Cepeda, C., Walsh, J. P., Buchwald, N. A., and Levine, M. S. (1991) Neurophysiological maturation of cat caudate neurons: Evidence from in vitro studies. *Synapse* 7:278–290.

Connolly, C. I., and Burns, J. B. (1993a) A model for the functioning of the striatum. *Biol. Cybern.* 68:535–544.

Connolly, C. I., and Burns, J. B. (1993b) A new striatal model and its relationship to basal ganglia diseases. *Neurosci. Res.* 16:271–274.

Connolly, C. I., and Grupen, R. A. (1993) The application of harmonic functions to robotics. *J. Robotic Syst.* 10:931–946.

Crutcher, M. D., and DeLong, M. R. (1984) Single cell studies of the primate putamen II. Relations to direction of movement and pattern of muscular activity. *Exp. Brain Res.* 53:244–258.

Douglas, R. J., and Martin, K. A. C. (1990) Neocortex. In G. M. Shepherd (ed.), *The Synaptic Organization of the Brain.* New York: Oxford University Press, pp. 389–438.

Doyle, P., and Snell, J. L. (1984) *Random Walks and Electric Networks.* Carus Monographs in Mathematics. Washington, D.C.: American Mathematical Society, 1984.

Flaherty, A. W., and Graybiel, A. M. (1991) Corticostriatal transformations in the primate somatosensory system. Projections from physiologically mapped body-part representations. *J. Neurophysiol.* 66:1249–1263.

Flaherty, A. W., and Graybiel, A. M. (1993) Two input systems for body representations in the primate striatal matrix: Experimental evidence in the squirrel monkey. *J. Neurosci.* 13:1120–1137.

Goldman-Rakic, P. S., and Selemon, L. D. (1990) New frontiers in basal ganglia research. *Trends Neurosci.* 13:241–244.

Graybiel, A. M. (1991) Basal Ganglia—input, neural activity, and relation to the cortex. *Curr. Opin. Neurobiol.* 1:644–651.

Holsapple, J. W., Preston, J. B., and Strick, P. L. (1991) The origin of thalamic inputs to the "hand" representation in the primary motor cortex. *J. Neurosci.* 11:2644–2654.

Hoover, J. E., and Strick, P. L. (1993) Multiple output channels in the basal ganglia. *Science* 259:819–821.

Kandel, E. R., Schwartz, J. H., and Jessell, T. M. (eds.) (1991) *Principles of Neural Science.* New York: Elsevier, 1991.

Kato, M., and Kimura, M., (1992) Effects of reversible blockade of basal ganglia on a voluntary arm movement. *J. Neurophysiol.* 68:1516–1534.

Khatib, O. (1985) Real-time obstacle avoidance for manipulators and mobile robots. In *Proceedings of the 1985 IEEE International Conference on Robotics and Automation.* St. Louis, Missouri, March 1985, pp. 500–505.

Kimura, M. (1990) Behaviorally contingent property of movement-related activity of the primate putamen. *J. Neurophysiol.* 63:1277–1296.

Kimura, M., Rajkowski, J., and Evarts, E. (1984) Tonically discharging putamen neurons exhibit set-dependent responses. *Proc. Natl. Acad. Sci. U. S. A.* 81:4998–5001.

Lozano-Perez, T. (1981) Automatic planning of manipulator transfer movements. *IEEE Trans. Systems, Man, Dybernet.* 10:681–698.

McCann, G. D., and Wilts, C. H. (1949) Application of electric-analog computers to heat-transfer and fluid-flow problems. *J. Appl. Mechanics* 16:247–258.

O'Donnell, P., and Grace, A. A. (1993) Physiological and morphological properties of accumbens core and shell neurons recorded in vitro. *Synapse* 13:135–160.

Parthasarathy, H. B., Schall, J. D., and Graybiel, A. M. (1992) Distributed but convergent ordering of corticostriatal projections: Analysis of the frontal eye field and the supplementary eye field in the macaque monkey. *J. Neurosci.* 12:4468–4488.

Pasik, P., Pasik, T., and Holstein, G. R. (1990) The ultrastructural chemoanatomy of the basal ganglia: 1984–1989 I. The neostriatum. In G. Bernardi (ed.), *The Basal Ganglia III.* New York: Plenum Press, pp. 187–188.

Schwartz, J. T., and Sharir, M. (1983) On the piano movers' problem I: The case of a two-dimensional rigid polygonal body moving amidst polygonal barriers. *Commun. Pure Appl. Math.* 36:345–398.

Tarassenko, L., and Blake, A. (1991) Analogue computation of collision-free paths. In *Proceedings of the 1991 IEEE International Conference on Robotics and Automation,* Sacramento, California, April 1991, pp. 540–545.

Udupa, S. M. (1977) Collision detection and avoidance in computer controlled manipulators. *Proceedings of the Fifth International Joint Conference on Artificial Intelligence,* pp. 737–748.

Walsh, J. P., Cepeda, C., Hull, C. D., Fisher, R. S., Levine, M. S., and Buchwald, N. A. (1989) Dye-coupling in the neostriatum of the rat: II. Decreased coupling from between neurons during development. *Synapse* 4:238–247.

Walsh, J. P., Cepeda, C., Buchwald, N. A., and Levine, M. S. (1991) Neurophysiological maturation of cat substantia nigra neurons: Evidence from in vitro studies. *Synapse* 7:291–300.

Williams, G. V., and Millar, J. (1990) Concentration-dependent action of stimulated dopamine release on neuronal activity in rat striatum. *Neuroscience* 39:1–16.

Wilson, C. J. (1984) Passive cable properties of dendritic spines and spiny neurons. *J. Neurosci.* 4:281–297.

Wilson, C. J. (1990) Basal ganglia. In G. M. Shepherd (ed.), *The Synaptic Organization of the Brain.* New York: Oxford University Press, pp. 279–316.

Editors' Commentary on Part II

The emerging anatomical organization of the basal ganglia discussed in chapters 5 through 7 confirms the essential elements of the Alexander, DeLong, and Strick schema that was reviewed in chapter 1 (figure 1.1), while also demonstrating some interesting variations on this basic theme. When these findings are applied to the pattern recognition/ registration model discussed in chapter 1, there emerge implications for the kinds of contexts that might be recognized by the striatum and the role that pattern recognition might play in the planning and coordination of motor functions. We also consider relations to the two models of striatal function discussed in chapters 8 and 9.

CONVERGENCE PATTERNS

There is now agreement among anatomists that the input to a small patch within the striatum is derived from rather specific sites in the cerebral cortex. The patterns that Graybiel and Kimura describe in chapter 5 suggest that convergent inputs arrive preferentially from several homologous body part representations derived from different areas of the cerebral cortex. If, for example, the body part is the hand, this convergence might bring together phasic (area 1) and tonic (area 3b) cutaneous information, position (area 2) and velocity (area 3a) proprioceptive information, along with motor commands (area 4) that are being sent to hand muscles. Each spiny neuron that received this convergence would be exposed to a somatotopically specific "window" into a given task's sensorimotor context. Multiple expressions of these types of overlapping projections might provide different subgroupings of task-related inputs to spiny neurons across the striatum.

If, as in chapter 1, we assign the function of pattern recognition to individual spiny neurons, these cells would seem to be ideally situated for detecting patterns of input signifying key phases of sequential movements, which is consistent with the single unit data reviewed in chapter 6. As a hypothetical example, imagine that you are in a dark room and have decided to pick up a mug of coffee that you know is on a table directly in front of you. The first phase of this movement might involve ascertaining the current position of your arm. If your arm is initially resting on the table, this determination would require touch and joint position information from, perhaps, sensory cortex areas 3b and 2, respectively. From there, one strategy might be to search for the coffee mug by sliding your hand across the surface of the table. Recognition of this context might require combining slip detection information from cortical areas representing directionally selective cutaneous receptive fields (possibly area 1) with a representation of the motor command mediating this slip from area 4. Next, contact with the object could contribute to the context for signaling the discontinuation of the searching phase. Then, the identity of the object would need to be established through tactile apprehension of the object's size and shape (possibly area 2) and perhaps its texture (possibly area 3b). Similarly, various convergent combinations of cortical representations would be required for recognizing the remaining steps of this sequential task. While certainly an oversimplification, this example illustrates that striatal integration of the differential cortical representations of homologous body parts could provide the type of contextual recognition required for sequencing and/or controlling the subtasks of a particular movement.

TRANS-STRIATAL MOTOR LOOPS

Chapter 6 suggests that the patterns of convergence from homologous areas of motor cortex are not the same as those from homologous areas of somatosensory cortex. Instead, Strick and colleagues find a segregated pattern of closed trans-striatal loops interconnecting topographically specific regions of the basal ganglia with three motor areas of the cerebral cortex: primary motor cortex (M1), the supplementary motor area (SMA) and the ventral premotor region (PMv). Strick's loops, like the loops in the modules of the conceptual model described in chapter 1 (figure 1.2, loops through F neurons), consist of corticostriatal projections emanating from a given region of frontal cortex that return, via the pallidum and thalamus, back to their zone of origin in the frontal cortex. Each module (or, more appropriately, each array of modules) appears to be tethered by a primary loop through the frontal cortex, and its signaling would also be regulated by convergent projections from other cortical areas.

To summarize, it appears that the output of a module may provide one of the more prominent inputs to the same module. This suggests that a context detected by nonfrontal inputs to a module's spiny neurons could be reinforced by signal transmission through the module's trans-striatal loop. Returning to our coffee mug example, let us assume that the mug has been lightly grasped by your hand. Part of the context, which might consist of a combination of cutaneous and proprioceptive somatosensory information, could serve as a signal to increase gripping force in preparation for lifting the mug. Recognition of this context would release pallidal inhibition to a module's corticothalamic loop, promoting a weak reverberating signal. This signal would then be used to reinforce and prolong activity within the module's trans-striatal loop through the mechanism of disinhibition, a form of positive feedback. The result would be a full registration of the contextual pattern to which the module is tuned. Registration refers to the working memories that are discussed later.

The motor cortical areas as a group are special regions of frontal cortex in that they have direct projections to the spinal cord. In these instances, the resultant of basal ganglia processing could serve directly as a coordinating or commanding signal. Thus, as activity within an M1 loop increases, grip strength might be commanded to increase accordingly. In contrast, these same cortical-basal ganglionic mechanisms, when tethered to F neurons in prefrontal regions of the cerebral cortex, are more likely to function as planning signals for motor behavior, as discussed in chapter 7. These prefrontal outputs would have to project to other parts of the motor system where they would ultimately influence the generation of motor commands.

Application of these concepts onto some of the other anatomical patterns described in chapter 5 leads to interesting possibilities. Consider the convergence from three striatal foci onto a single pallidal patch illustrated in figure 5.4. According to the trans-striatal loop concept of chapter 6, the pallidal patch would project to a motor or prefrontal zone that would project back to each of the three striatal foci that provided the original pallidal input. This circuitry might provide a mechanism whereby a given motor signal, be it plan or command, might be instantiated by three quite different contexts. The three contexts would be detected by spiny neurons at the different striatal sites, which we assume would receive different constellations of convergent input, although some of these inputs could very well be similar (as was illustrated). Convergence from multiple striatal sites to a common pallidal focus might in some cases integrate information from opposite sides of the cerebral cortex, which could contribute to the planning of bimanual tasks. Divergence from a single cortical site to several striatal foci provides a useful mechanism for disseminating a given type of information for use in different motor plans. While the above ideas clearly involve speculation beyond our current anatomical knowledge, they give a taste of the considerable potential of this type of pattern recognition/registration architecture.

MEMORY-GUIDED MOTOR PERFORMANCE

The now considerable evidence that sustained activity in the frontal cortex represents working memories was reviewed in chapter 7. In exploring the behavioral significance of sustained activity, Goldman-Rakic found that individual neurons have receptive fields, which she termed memory fields, that are preferentially tuned for a particular target location or for a particular direction of a planned movement. Different neurons in the population are tuned to different directions. One can readily imagine arrays of pattern recognition/detection modules of the type discussed in chapter 1 controlling these prefrontal populations.

Clearly these memory signals should be useful in planning and controlling movements, and Arbib and Dominey developed this idea in a specific model of oculomotor saccade generation in chapter 8. This model uses working memories sustained by recurrent activity in corticothalamic loops through the frontal eye field (FEF) to provide potential saccade commands to the superior colliculus during tasks that require memories of potential targets. Unlike the context detection/registration model of chapter 1, this model uses long-term memory in the striatum primarily as a means for modulating the saccadic command as it traverses the striatum, as opposed to actually detecting the conditions for saccade selection and/or initiation. The modulated command is then compared in the superior colliculus to the unmodified command issued directly from the FEF. Within the colliculus, a competitive mechanism sees that one of several potential targets is selected for saccade implementation.

The anatomy of these brain regions identifies at least three loops that could participate in sustaining working memories: the corticothalamic loops mentioned in the previous paragraph, corticocortical loops that are a common feature of cortical architecture, and the trans-striatal loops mentioned earlier in this commentary. The relative importance of each of these loops remains unclear at present. Similarly, it is unclear whether sustained activity is initiated by excitatory corticocortical inputs to frontal neurons or by disinhibitory rebound in the pallidothalamic pathway. An attractive possibility is that these alternative anatomical routes represent operational alternatives that are available to the nervous system, each with particular specializations in function.

AN ALTERNATE STRIATAL MODEL

In chapter 9, Connolly's state-space model provided us with a completely different interpretation of striatal function. Rather than pattern classification, he suggested that the striatum might be involved in a computation that coordinates obstacle-avoiding limb movements. The striatum is thought of as an electrotonically conductive grid of spiny neurons, each representing a state in the limb's joint angle space. The model combines coordination and obstacle avoidance into the striatum by overlaying the position of salient environmental objects directly onto the limb's state representation through corticostriatal projections. Although the dual-state behavior of striatal neurons discussed in chapter 3 is not appropriate for the graded computations required in this model, the subthreshold fluctuations observed in spiny neurons might subserve this function. However, an alternative that is more consistent with the pattern detection/registration model would use spiny neurons for the detection of potential obstacles and prefrontal corticothalamic loops for sustaining working memories of the obstacles. The latter information could then be sent to the cerebellum for use in selecting movements that avoid the obstacles.

III Reward Mechanisms

10 Cellular Models of Reinforcement

Jeff Wickens and Rolf Kötter

The concept of reinforcement used in early formulations of learning theory was a psychological concept based on the experimental analysis of behavior. At the cellular level a similar concept—heterosynaptic plasticity—is also referred to as reinforcement. This chapter explores possible links between reinforcement on the level of behavior, and heterosynaptic plasticity mechanisms in the neostriatum.

The formal psychological concept of positive reinforcement probably originated with Thorndike (1911), who wrote: "Any act which in a given situation produces satisfaction becomes associated with that situation so that when the situation recurs the act is more likely than before to recur also." The effect of reinforcement (satisfaction) is to strengthen the association between the situation and the act. This formulation implicitly requires a mechanism capable of integrating three factors: situation, action, and reinforcement.

In heterosynaptic plasticity, the synaptic strength of one pathway is modified by activity in another pathway. For example, in *Aplysia*, synapses between sensory and motor neurons can be strengthened by being active in association with release of serotonin from a different set of neurons in response to noxious stimuli (Kandel and Tauc, 1965). In principle, such activity-dependent heterosynaptic plasticity has the potential to combine three factors (presynaptic and postsynaptic activity, and neurochemical reinforcement).

The facilitation of simple reflexes described in invertebrates is obviously different from the reinforcement learning which occurs in mammals. In the mammalian brain, the behavioral effects of reinforcement cannot be reduced to changes in the strength of connections between sensory and motor neurons. It is necessary to understand the effects of heterosynaptic plasticity in the context of the neural circuits within which the synapses are embedded. These circuits probably involve the connections between the cerebral cortex and the neostriatum, as originally proposed by Miller (1981).

The overall aim of this chapter is to consider synaptic modification mechanisms in the neostriatum, and their contribution to reinforcement learning. The specific proposals are:

1. Midbrain dopaminergic neurons projecting to the neostriatum are activated by behavioral reinforcement.

2. A major target of this dopamine reinforcement signal is the corticostriatal pathway.

3. The relevant effect of dopamine is to modify the synapses in the corticostriatal pathway in an activity-dependent way.

These proposals are considered in relation to the circulation of activity in circuits that connect the cerebral cortex and neostriatum in an array of parallel loops. It is argued that the effect of strengthened corticostriatal synapses is increased frequency of responses that previously led to reward being obtained. The increase in response frequency is brought about by a process of selective strengthening of corticostriatal connections. The strengthened connections produce positive feedback pathways which amplify the cortical antecedents of previously successful acts, making them more likely to occur in the future.

DOPAMINERGIC MECHANISMS IN REINFORCEMENT LEARNING

It has long been known that dopamine plays a key role in positive reinforcement, but the exact nature of this role has been—and continues to be—difficult to determine. At the behavioral level, dopamine has short-term motor-activating effects (which affect performance) as well as long-term reinforcing effects (which affect learning). Similarly, at the electrophysiological level, dopamine has immediate effects on neuronal activity which may be different from its long-lasting effects on synaptic efficacy. In reinforcement learning it is likely to be the long-term effects, which persist beyond the immediate period of dopaminergic activity, that are most relevant.

It is tempting to translate the psychological concept of positive reinforcement into physiological actions of dopamine. For example, the idea that reinforcement is a neurochemically specific process mediated by forebrain dopamine systems has served as a useful heuristic in psychopharmacology (Stein and Belluzzi, 1989). However, the psychological concept of reinforcement does not specify exactly how the neural substrate of reinforcement should behave. For example, it does not specify *when* the relevant neurons should fire in relation to primary reinforcement, or what activity they should have in relation to secondary reinforcement or to reinforcement that is expected but does not come. On the other hand, studies of the activity of dopamine neurons in behaving animals are very likely to contribute a great deal to the concept of reinforcement, but only if their design takes cognizance of the psychological issues at stake.

There is mounting evidence that dopaminergic neurons which project to the neostriatum may be a major branch of a "final common pathway" for positive reinforcement (Beninger, 1983; Wise and Bozarth, 1984). Other areas of the brain (such as the prefrontal cortex and ventral hippocampus) also receive a dopamine innervation from the midbrain (Lindvall et al., 1978; Verney et al., 1985). Restriction of the present discussion to dopaminergic mechanisms in the neostriatum is not intended to suggest a lesser role for these other areas in reinforcement learning. The main aim is to establish that striatal dopamine also makes a crucial contribution to reinforcement learning, in addition to its more generally accepted role in motor activation.

The literature implicating the mesostriatal dopaminergic pathway in reinforcement is vast, and not without controversy. There are several key pieces of evidence:

1. Direct electrical stimulation of certain sites in the brain (ESB) can produce conditioning effects similar to those produced by natural rewards in animals engaged in learning. The most effective ESB reward sites are the ones that directly or indirectly activate the dopaminergic neurons in the midbrain (Stellar and Stellar, 1985).

2. Injection of dopamine agonist drugs directly into the neostriatum can also produce positive reinforcement effects similar to those produced by natural rewards (Hoebel et al., 1983).

3. Dopamine antagonist drugs injected directly into the neostriatum can *attenuate* the positive reinforcing effects of ESB reward (Kurumiya and Nakajima, 1988).

4. The concentration of dopamine in the neostriatum is increased by both natural and ESB rewards (Hoebel et al., 1989) and decreased by aversive stimuli (Mark et al., 1991).

5. Dopamine neurons are activated by food and liquid rewards and by stimuli which predict such rewards (Ljungberg et al., 1992)

These pieces of evidence implicate mesostriatal dopamine pathways in positive reinforcement mechanisms. They are consistent with (but do not prove) the idea that the dopaminergic neurons projecting to the neostriatum carry a positive reinforcement command signal. However, while suggesting that the targets for the reinforcement signal may be found in the neostriatum, the evidence so far does not indicate what the effect of this signal is likely to be.

A crucial missing piece of the argument is what dopamine does. Although direct evidence on this point is still inconclusive, there are several important leads. These are considered in the following discussion. It is argued, by analogy with its reinforcing effects in other neural systems, that dopamine may produce heterosynaptic facilitation in the neostriatum.

Dopamine and Heterosynaptic Plasticity

Behavioral responses which have been strengthened by reinforcement persist for some time, even in the absence of continued reinforcement. If dopamine is a mediator of reinforcement, then it should be able to produce long-lasting changes in synaptic strength, outlasting the period of exposure to dopamine. Although there have been few studies of long-lasting effects of brief exposure to dopamine, there is evidence that dopamine has long-lasting effects on synaptic transmission at other synapses.

Monoamine-mediated heterosynaptic plasticity has been investigated in a variety of neural systems. These investigations have shown that release of monoamine neurotransmitters from one set of synapses can bring about long-lasting changes in the efficacy of transmission at another set of synapses, which use a different neurotransmitter. In many cases this heterosynaptic plasticity is brought about by a monoamine-mediated increase in the intracellular concentration of adenosine 3′:5′-cyclic phosphate (cAMP). In a variety of species, these actions of monoamines underlie the storage the memory in long-lasting changes in synaptic efficacy (Schwartz and Greenberg, 1987).

Mediation of heterosynaptic plasticity by cAMP occurs in many different species. For example, neurotransmitter-induced modulation of an electrotonic synapse has been described in the leech (*Hirudo medicinalis*) between an identified pair of electrically coupled neurons (Colombaioni and Brunelli, 1988). The transfer of electrical signals between these cells was reduced by perfusing the preparation with serotonin or dopamine. This action was reproduced by the iontophoretic injection of cAMP into one of the coupled neurons, suggesting again that these monoamines act by increasing the intracellular levels of cAMP, possibly by activating the adenylate cyclase system.

While cAMP-dependent mechanisms for synaptic modification have been most extensively studied in invertebrates, dopamine can act by similar mechanisms in the mammalian nervous system. For example, in rabbit sympathetic ganglia, a brief exposure to dopamine results in an enduring (hours long) enhancement of slow depolarizing responses to muscarinic agonists (Ashe and Libet, 1981). This long-term enhancement of the muscarinic response occurs via activation of a dopamine receptor that is coupled to adenylate cyclase and resembles the dopamine D1 receptors described in the brain (Ariano, 1987; Mochida et al., 1987). Activation of the dopamine D1 receptor increases cAMP. Thus dopamine in the mammalian nervous system, again by increasing cAMP, can bring about a heterosynaptic long-lasting increase in synaptic efficacy.

Similar effects of dopamine have been reported in the hippocampus. Gribkoff and Ashe (1984) found that temporary exposure to dopamine produced two effects: an initial suppression of the extracellularly recorded population responses, followed by "a profound potentiation of the population responses that can last for hours" (Gribkoff and Ashe, 1984). Furthermore, Frey et al. (1993) have shown that a brief application of a cAMP analog induces a long-lasting potentiation very similar to the late phases of hippocampal long-term potentiation (LTP).

In cellular operant-conditioning studies microinjections of reinforcing transmitters or drugs are applied directly to cells after bursts of neuronal firing. Stein and Belluzzi (1989) have demonstrated operant conditioning of single units in the hippocampus by this means. They used dopamine and various dopamine agonists as reinforcing neurotransmitters, and found that contingent application of these drugs increased firing, yet noncontingent application did not. Thus, at the cellular level, the frequency of certain neural events can be selectively increased by the application of reinforcing neurotransmitters, such as dopamine.

In summary, it has been shown in various neural systems that brief exposure to dopamine can produce long-lasting facilitation of synaptic transmission in nondopaminergic pathways. It may be argued, by analogy, that the reinforcing effects of dopamine in the neostriatum can be brought about in a similar way. In the following paragraphs, this argument is extended by reference to anatomical evidence showing that dopaminergic afferents converge with corticostriatal afferents in a spatial arrangement that is compatible with such a heterosynaptic influence.

Anatomical Substrates of Heterosynaptic Plasticity in the Striatum

A possible anatomical basis for heterosynaptic plasticity is found in the convergence of afferents from two sources upon the medium spiny neurons of the neostriatum: (1) excitatory, glutamatergic inputs from the cerebral cortex; and (2) dopaminergic inputs from the midbrain.

The inputs from the cerebral cortex to the neostriatum originate bilaterally from all major regions (McGeorge and Faull, 1989). The organization of the somatosensory projection has been studied in some detail. In cats and monkeys somatosensory information from different parts of the body is kept separate,while there is convergence of information concerning different modalities (Malach and Graybiel, 1988; Flaherty and Graybiel, 1991). There is an intricate pattern of convergence in which *different selections* of cortical inputs are *combined* in different ways at *different neostriatal sites.* This is an ideal arrangement for a heterosynaptic mechanism to select particular input-output combinations for reinforcement.

The specificity of connections suggested by the intricacy of the corticostriatal projection contrasts with the apparently more diffuse dopaminergic projection from the midbrain. The dopaminergic inputs to the neostriatum originate from the substantia nigra pars compacta (SNc) and the ventral tegmental area (VTA). The terminal fields of these inputs overlap to some degree, but the SNc input is directed mainly at the dorsal neostriatum, or caudate-putamen, while the VTA input is directed mainly at the ventral neostriatum, or nucleus accumbens (Parent, 1986).

Bouyer et al. (1984) showed that striatal afferents from the cerebral cortex and dopaminergic afferents from the substantia nigra frequently terminate in close proximity to each other, often synapsing on the same dendrite or dendritic spine. By a combination of Golgi impregnation, retrograde labeling from the substantia nigra, and tyrosine hydroxylase (TH) immunohistochemistry, Freund et al. (1984) showed that these dendrites and spines were located on striatal output neurons. They also showed that of 87 spines studied, 46 (52.9%) had only one asymmetrical synapse, 34 (39%) had *both* an asymmetrical (presumably cortical) *and* a TH-positive (dopaminergic) terminal, and (8.1%) received one asymmetrical and one unstained symmetrical terminal.

The convergence of dopaminergic and corticostriatal synapses upon individual neostriatal neurons provides a favorable substrate for a three-factor synaptic modification rule, because it brings together the processes of three groups of cells (cortical, striatal, and dopaminergic neurons).

The foregoing suggests that dopaminergic inputs to the neostriatum are in a position to control the efficacy of the corticostriatal synapses by a heterosynaptic mechanism. We next consider functional studies which suggest that dopamine increases striatal output by increasing the effectiveness of corticostriatal inputs.

Effects of Dopamine on Striatal Output

Functional studies of the effects of dopamine on the eventual output from the neostriatum have shown that dopamine agonists have non-uniform effects. The evidence lends itself to the interpretation that dopamine increases output from the most active cells (which are in the minority) and decreases output from the less active cells.

Studies of striatal output activity using regional cerebral glucose utilization suggest that overall output from the neostriatum is increased by dextroamphetamine (an indirect dopamine agonist), apomorphine (a direct-acting agonist), and levodopa (Wooten and Collins, 1983; Trugman and Wooten, 1986; Sirinathsinghji et al., 1988).

At the single-cell level, systematic apomorphine produces highly variable responses in the substantia nigra pars reticulata (SNr). These SNr cells receive inhibitory input from striatal output neurons. Many cells exhibit increased firing rates, some cells are markedly inhibited, and a large group show only modest or fluctuating changes in rate (Waszczak et al., 1984). The variability is reduced by lesions of the neostriatum, which suggests it is mediated by striatal projection neurons.

The cell-to-cell variability in the effects described can be explained in terms of an activity-dependent action of dopamine. West et al. (1986) and Haracz et al. (1993) found that amphetamine increased the activity of striatal units that were active in association with movements on the part of the animal, but inhibited units with little or no associated activity.

Since the action potential firing of striatal output neurons is largely due to excitatory input from the cerebral cortex, the changes in firing rates brought about by dopamine are probably due to changes in the effectiveness of the corticostriatal inputs. Consistent with this, Warenycia et al. (1987) found that systemic administration of amphetamine produced excitation in control animals, but produced inhibition in animals subjected to bilateral removal of the frontoparietal cortex. The authors concluded that the excitatory response depended on an intact cerebral cortex and required intact corticostriatal afferents.

In summary, the foregoing suggests that one effect of dopamine on striatal output is to increase the action potential firing rate of a small number of neurons and reduce the activity of a larger number. The basis for this nonuniform effect appears to be activity-related. The activity of active neurons is increased and the activity of less active neurons is decreased. The neurons that increase their output apparently do so because of strengthened excitatory inputs from the cortex. The results discussed are compatible with the hypothesis that dopamine brings about an activity-dependent enhancement of corticostriatal synapses. The association of this activity with movement on the part of the animal (Haracz et al., 1993) suggests that the increased firing of these cells may cause the increased behavioral activity brought about by dopamine. In the following discussion, the behavioral effects of dopamine are considered.

Effects of Striatal Dopamine on Behavior

Dopamine agonist drugs increase motor activity (Beninger, 1983). At low doses, exploratory behavior is increased. As doses increase, higher rates of activity occur in more and more limited categories of response (Lyon and Robbins, 1975). At very high doses behavior is reduced to repetitive performance of a limited range of behaviors such as gnawing, turning, and grooming movements.

McLennan et al. (1964) reported that electrical stimulation of the caudate nucleus of a cat caused it to raise and turn its head to look over its contralateral shoulder, as if it had just become aware of *something which it wished to see* but which was constantly just out of view behind it. These effects of direct electrical stimulation probably arise from activation of the dopaminergic terminals in the neostriatum. Similar effects can be produced by stimulation of the nigrostriatal system, and these can be blocked by dopamine antagonists (Arbuthnott and Ungerstedt, 1975; Phillips, 1979).

The curious behavior produced by the activation of dopaminergic terminals in the neostriatum is of particular interest because the increased activity is apparently due to an intensified response to sensory input (in this case, visual input). Similarly, Szechtman (1983) found that peripheral sensory input could direct apomorphine-induced circling in rats: bandaging of one side of the head yielded circling in the opposite direction in response to dopamine agonists.

In summary, dopamine agonists increase activity in limited categories of response. The increased activity appears to be directed by sensory input.

DOPAMINE-MEDIATED HETEROSYNAPTIC PLASTICITY IN THE STRIATUM

The evidence of the previous sections suggests that the activation of striatal neurons brought about by dopamine is due to increased efficacy of corticostriatal afferents (since it is blocked by cortical lesions). Similarly, the dopamine-induced increase in behavioral activity is due to increased responsiveness to peripheral stimuli. In both cases the increases are selective: a *minority* of striatal neurons are activated, and increased behavioral activity occurs in *limited* categories of response. At the neural level the basis for the selection appears to be activity-related: the activity of the most active striatal neurons is increased while that of the less active neurons is decreased.

These observations may be accounted for by a simplifying hypothesis which proposes that dopamine brings about activity-dependent heterosynaptic plasticity in the corticostriatal pathway. Such hypotheses have been proposed in the past (Miller, 1981; Beninger, 1983; Wickens, 1988; White, 1989). For example, Miller (1981) proposed that strengthening of corticostriatal synapses would occur when a conjunction of pre- and postsynaptic activity was followed by reward, mediated by dopamine. This three-factor rule was originally based on requirements for instrumental conditioning. Nonetheless, it also predicts that striatal neurons which were being caused to fire by activity in their corticostriatal afferents would increase their firing rate in response to dopamine. This would occur because according to the three-factor rule the combination of pre- and postsynaptic activity, plus dopamine, would strengthen the active corticostriatal synapses.

There are, of course, various possible combinations of the three factors: pre- and postsynaptic activity, and dopamine. At present, there is insufficient direct electrophysiological evidence to permit hard and fast rules to be defined concerning the effects of different combinations of the three factors.

In the following discussion, the limited electrophysiological evidence is considered in relation to the possible combinations of the three factors. Although the empirical evidence is incomplete it will be helpful to consider it in this framework.

Dopamine and Synaptic Plasticity in the Striatum: Electrophysiological Evidence

Table 10.1 presents an exhaustive list of the possible combinations of the three factors, followed by an indication of the results of electrophysiological experiments which have tested each combination. A number *1* in the top section of the table indicates either presynaptic activity (tetanization of the corticostriatal afferents); postsynaptic activity (action potential firing of the postsynaptic cell); or dopamine receptor activation. The arrows in the bottom section of the table indicate the direction of the change in response. A double arrow indicates both increases and decreases.

Condition A In the absence of any activity, the synaptic strength stays unmodified. Repeated measurements of the amplitude of striatal responses to test stimulation of the cortex are generally stable over time (Walsh, 1993).

Table 10.1 Effects of dopamine on responses of striatal cells to cortical stimulation

	Condition							
	A	B	C	D	E	F	G	H
Activity		One Factor			Two Factors			Three Factors
Presynaptic	0	1	0	0	1	1	0	1
Postsynaptic	0	0	1	0	1	0	1	1
Dopamine	0	0	0	1	0	1	1	1
Study								
Begg et al., 1993	⇔				⇓			⇑
Pennartz et al., 1993				⇕	⇕	⇕		
Haracz et al., 1993				⇓				⇑
Garcia-Munoz et al., 1992		⇕			⇕			
Calabresi et al., 1992		⇔			⇓			⇓
Schneider, 1991					⇓			
Chiodo and Berger, 1986								⇑
Hirata et al., 1984				⇕		⇕	⇕	⇕
Schneider et al., 1984						⇑	⇑	⇑

Condition B In order to produce this condition (presynaptic activity alone) it is necessary to repeatedly stimulate the cortex while preventing the postsynaptic cell from becoming depolarized. Calabresi et al. (1992) have shown that subthreshold tetanic stimulation of the cortex, or cortical tetanization in association with hyperpolarization of the postsynaptic neuron, produces neither LTP nor long-term depression (LTD). On the other hand, Garcia-Munoz et al. (1992) found long-term changes in terminal excitability following brief tetanization stimulation of the cortex. Since the changes in terminal excitability were also seen following destruction of the postsynaptic striatal neurons by kainic acid, these changes could presumably be induced by presynaptic factors alone.

Condition C Postsynaptic activity alone can be produced by injecting depolarizing current into the postsynaptic cell. This does not usually induce any changes in excitatory postsynaptic potential (EPSP) amplitude, but a systematic study of this condition has not been reported.

Condition D The effect of dopamine alone has been shown in numerous experiments to produce a decrease in EPSP amplitude as long as the exposure to dopamine continues (Mercuri et al., 1985; Calabresi et al., 1987, 1988). However, long-lasting effects, which persist beyond the period of exposure to dopamine, have seldom been studied. The few cases where the persisting effects of dopamine have been described will now be considered.

Hirata et al. (1984) studied the effects of repetitive stimulation of the substantia nigra, or iontophoretic application of dopamine agonists, on responses of striatal units to cortical stimulation. The responses of striatal cells to cortical simulation were attenuated in some cases (22 of 53 neurons) and enhanced in others (12 of 53 neurons). Although the data shown concerned only the effects on cortical stimuli delivered 100 ms after nigral stimulation, in the discussion section it was noted that "the attenuation and enhancement persisted for a period of time, up to several minutes, following nigral conditioning stimulation." It should be noted that the striatal cells recorded from were firing spontaneously. This spontaneous activity is probably due to activity in presynaptic afferents. Therefore pre- and postsynaptic activity were not controlled in these experiments. Thus, the effects may not all be due to dopamine alone, but may include cases in which a chance conjunction of pre- and postsynaptic activity occurred during the conditioning stimulation. However, the changes lasting several minutes are an important clue to the effects of dopamine on corticostriatal transmission.

Condition E Pre- plus postsynaptic activity has been shown to produce LTD in the corticostriatal pathway. Calabresi et al. (1992) have shown that *suprathreshold* tetanic stimulation of the cortex, or subthreshold stimulation in association with depolarization (sufficient to result in action potential firing of the postsynaptic cell) produces LTD.

Condition F The effect of presynaptic activity plus dopamine was studied by Pennartz et al. (1993) in the nucleus accumbens. They reported that dopamine did not modulate synaptic plasticity, because neither dopamine nor dopamine antagonists modified the amount of potentiation or depression seen. Their results are difficult to interpret in the present framework, however, since in intracellular experiments they found no overall evidence of synaptic plasticity in control solutions (average of 15 cells) or in dopamine-containing solutions (average of 14 cells). Since tetanization of afferents was not paired with postsynaptic current injection, these intracellular experiments show that presynaptic activity plus dopamine is not sufficient to produce LTP.

Condition G The effects of postsynaptic activity plus dopamine have not been tested.

Condition H The effect of pre- plus postsynaptic activity plus dopamine has been tested in a small number of striatal cells by Begg et al. (1993). In these experiments, pulsatile ejection of dopamine was timed to coincide with cortical tetanization and current injection into the postsynaptic cell. Conditioning with all three factors (presynaptic cortical stimulation plus postsynaptic action potential firing plus dopamine) brought about LTP (in seven out of seven cells), compared to controls (presynaptic plus postsynaptic, no dopamine) which showed LTD. These results are preliminary, however, and further experimental work is in progress to investigate the temporal requirements of the dopamine application.

The findings of Chiodo and Berger (1986) are also of interest. In their experiments, test pulses of L-glutamate were repeatedly applied by iontophoresis. In the presence of dopamine (when ejected at low iontophoretic currents), the responses to glutamate application *increased* with successive applications, and then decreased when dopamine was blocked by trifluoperizine (see their figure 1). The potentiation was reported to last for 3 to 10 minutes after dopamine ejection had ceased, and was probably longer because cells that did not exhibit recovery were not included in subsequent analyses.

Condition H assumes that the dopamine signal is a phasic increase in synaptic dopamine concentration. Bath-applied dopamine is not a sufficient test of this condition. Incubation of striatal slices with dopamine agonists has a desensitizing effect, expressed as a decrease in the amplitude of dopamine-stimulated adenylate cyclase activity (Memo et al., 1982). This effect is apparently mediated by persistent occupancy of the D1 receptor, and suggests that prolonged exposure to dopamine prior to a conjunction of presynaptic and postsynaptic activity would not produce the same effect as pulsatile exposure to dopamine at the time of the conjunction. Furthermore, there are some indications that the order in which different activators of adenylate cyclase act may have significant effects on peak activity levels. In *Aplysia* Yovell and Abrams (1992) showed that the amount of activation of adenylate cyclase produced by pairing of brief pulses of Ca^{2+} and serotonin depended on the sequence of the pulses: peak cyclase activation was greatest when the Ca^{2+} pulse immediately preceded the serotonin pulse.

Possible Mechanisms for Heterosynaptic Plasticity in the Neostriatum

Dopaminergic and corticostriatal synapses converging on medium spiny neurons interact via intracellular calcium- and cAMP-dependent mechanisms. The glutamate released from corticostriatal synapses can lead to an increase in intracellular calcium concentration, via *N*-methyl-D-aspartic acid (NMDA) receptors (Mayer and Westbrook, 1987), while stimulation of dopamine receptors leads to an increased concentration of cAMP (Azzaro et al., 1987). There are several ways in which interactions between calcium and cAMP can control the effectiveness of corticostriatal synapses:

First, adenylate cyclase in the striatum is regulated by calcium (Gnegy and Treisman, 1981). At low calcium concentrations (less than $1 \mu M$) the activity of adenylate cyclase increases up to twofold in a calcium-calmodulin–dependent manner (Kebabian et al., 1972). However, at higher concentrations (greater than $1 \mu M$) calcium inhibits adenylate cyclase (Piascik et al., 1981), as shown in figure 10.1A. Second, an increase in activation of adenylate cyclase can also be brought about by dopamine, acting via D1 receptors (Weiss et al., 1985). As illustrated in figure 10.1B, adenylate cyclase is stimulated up to twofold by dopamine (Gnegy and Treisman, 1981; Piascik et al., 1981). Dopamine D1 receptors are present at postsynaptic sites in dendrites and dendritic spines in the striatum (Huang et al., 1992).

Thus, in striatal neurons dopamine and glutamate can act as synergists to increase the concentration of cAMP, through the activities of dopamine-sensitive and calcium-sensitive adenylate cyclase systems (Piascik et al., 1981; Ahlijanian and Cooper, 1988; Natsukari et al., 1990). This synergy is a particular characteristic of the striatum, in contrast to the neocortex (Ahlijanian and Cooper, 1988; Natsukari et al., 1990).

The relative activity of adenylate cyclase calculated as a function of dopamine and calcium concentrations, is shown in figure 1C. The three-dimensional graph shows the complex interactions between dopamine and calcium in the activation of adenylate cyclase. The calculations were based on values obtained from many sources (Kebabian et al., 1972;

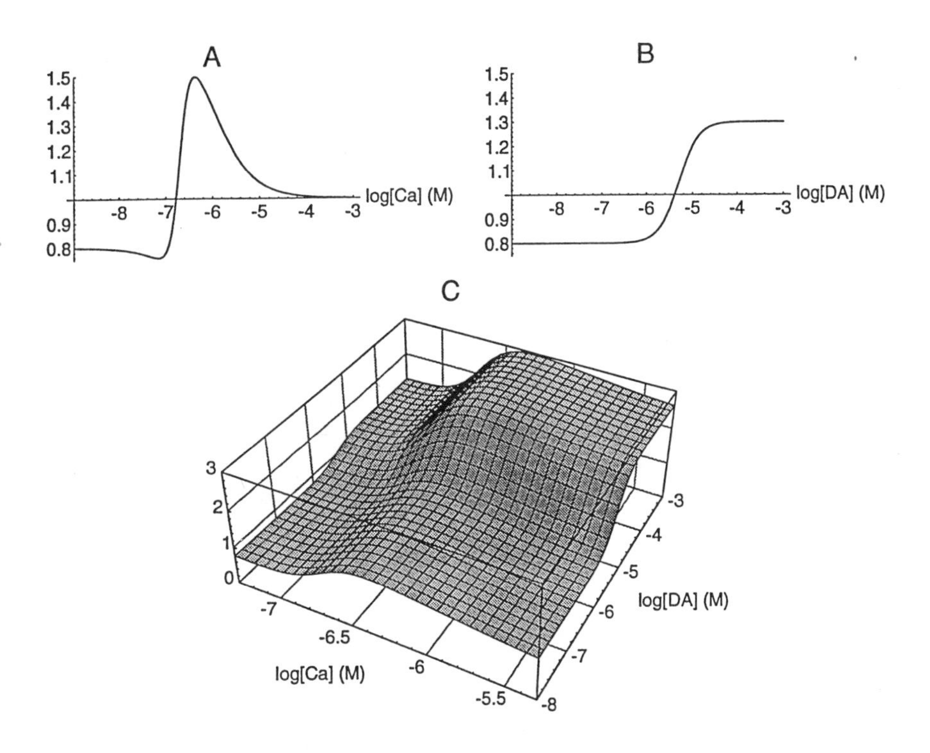

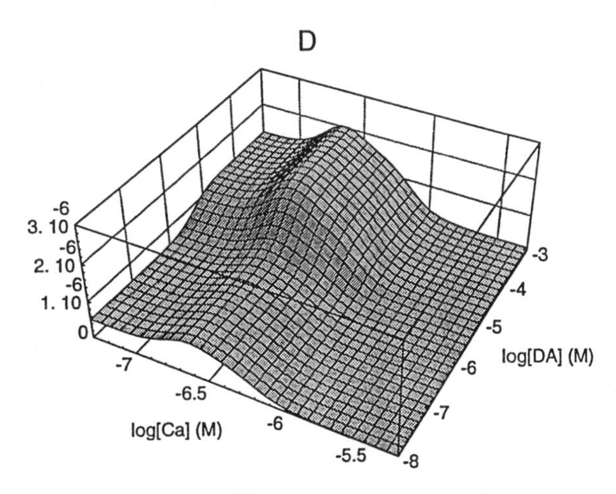

Figure 10.1 Interactions between Ca^{2+} and dopamine. (*A–C*) Model values for the relative activity of adenylate cyclase as judged from experimental data. For references, see text. (*D*) Estimated dependence of the concentration of threonine-phosphorylated DARPP (a dopamine- and cAMP-regulated phosphoprotein) on calcium and dopamine. The model includes effects of cAMP-dependent protein kinase and protein phosphatase 2B (Adapted from Kötter, submitted for publication.)

Piascik et al., 1981; Gnegy and Treisman, 1981; Weiss et al., 1985; Azzaro et al., 1987; Ahlijanian and Cooper, 1988; Natsukari et al., 1990) and are presented in full by Kötter (submitted for publication). For maximal effects simultaneous action of both dopamine and calcium is required.

The increased cAMP concentration brought about by stimulation of dopamine receptors increases the amount of phosphorylated DARPP, a dopamine- and cAMP-regulated phosphoprotein (Walaas et al., 1983). The highest concentrations of DARPP have been found in dopaminoceptive neurons in the striatum (Walaas and Greengard, 1984). DARPP is present in the dendrites and dendritic spines of striatal medium spiny neurons (Ouimet et al., 1984, 1992).

The phosphorylated form of DARPP is a potent inhibitor of protein phosphatase 1 (PP-1) (Hemmings et al., 1984). Since PP-1 reverses the effects of some of the protein kinases involved in LTP, dopamine might facilitate LTP in this indirect way.

The interaction between dopamine and glutamate is not straightforward, however. Maximal activation of NMDA receptors does not increase the proportion of phosphorylated DARPP but rather decreases it (Halpain et al., 1990). It is thought that this effect results from the activation of protein phosphatase 2B (PP-2B) which depends on calcium. In vitro, PP-2B dephosphorylates DARPP (Hemmings et al., 1984, 1990).

The foregoing shows that the phosphorylation of DARPP is influenced by both dopamine and calcium. The overall effect of interactions between dopamine and calcium in altering the phosphorylation state of phosphoproteins is very complicated. Thus, it is helpful to consider the behavior of a quantitative model of these interactions.

Kötter (submitted for publication) investigated a quantitative model in which the dependent variable was the concentration of phosphorylated DARPP. Dopamine and calcium concentrations were independent variables. The effects of dopamine and calcium on DARPP were estimated using simple equilibrium equations and data on intracellular concentrations and apparent *Km* values of intermediaries.

Figure 10.1D shows the concentration of phosphorylated DARPP predicted by the model, as a function of dopamine and calcium concentrations. Dopamine alone produces an increase in phosphorylated DARPP. A small increase in intracellular calcium concentration (to less than 1μM) has synergistic effects with dopamine, resulting in a greater increase in phosphorylated DARPP. Likewise, an increase in dopamine has synergistic effects with calcium. Thus, the model suggests that the simultaneous stimulation of glutamate and dopamine receptors should produce an increase in the amount of phosphorylated DARPP. However, it can also be seen from figure 10.1D that if the calcium concentration rises to much higher values (greater than 1μM) its effect is to oppose the phosphorylation of DARPP by dopamine.

Thus, the foregoing suggests a possible basis for interactions between dopamine and glutamate, expressed in terms of phosphorylated DARPP: a potential intermediary in synaptic plasticity. The activity of other phosphoproteins, including cytoskeletal proteins and membrane receptors, is also influenced by protein kinases and phosphatases in a similar way (Walaas and Greengard, 1991; Swope et al., 1992). Each of these phosphoproteins responds to specific concentrations of different enzymes. It is most likely that they have their own specific profiles of dopamine and calcium dependency. Estimates can be obtained for several protein kinases and phosphatases and for cAMP, DARPP, and microtubule-associated protein 2 (MAP2) (Kötter, submitted for publication).

At present the biochemical data available are too limited to work out precisely the total effects of dopamine and calcium on synaptic transmission. Nonetheless, it is already apparent that the direction of synaptic change is a sensitive function of both calcium and dopamine. It remains a major challenge to apply the insights gained from quantitative biochemical modeling to the interpretation of the electrophysiological studies presented in table 10.1. Furthermore, the biochemical calculations assumed equilibrium conditions and did not take account of temporal changes in intermediates. Temporal aspects of these mechanisms are considered below.

Temporal Requirements of Heterosynaptic Plasticity

In behavioral operant conditioning, the effectiveness of a reinforcing stimulus is reduced when its presentation is delayed for several seconds (Renner, 1964). When intracranial stimulation of the medial forebrain bundle is used in place of natural rewards, a delay of even 1 second markedly impedes the acquisition of self-stimulation behavior (Black et al., 1985). In the cellular operant-conditioning studies reported by Stein and Belluzzi (1989) reinforcement delays of the order of 100 ms considerably attenuated the reinforcing effect of the directly applied dopamine agonists. This suggests that the timing of the dopamine application in relation to the conjunction of pre- and postsynaptic activity could be critical.

If the heterosynaptic potentiating effect of dopamine can only take effect within a narrow time window of pre- and postsynaptic activity, then the dopamine signal must also be temporally precise. In vivo voltammetry can be used to measure the extrasynaptic concentration of dopamine. The maximum dopamine concentration increase in response to prolonged stimulation at ESB reward sites occurs 6 to 8 seconds after the onset of stimulation. These measurements reflect the overflow of dopamine from the synaptic cleft into the surrounding fluid. Synaptic dopamine concentration changes on a faster time scale. Calculations of synaptic concentration based on fast voltammetry measurements suggests that the concentration of dopamine within the synaptic cleft rises steeply with each presynaptic action potential, to a peak within *tens of milliseconds,* and then rapidly decays (Justice et al., 1988; Kawagoe et al., 1992).

In summary, the dopamine signal is pulsatile in nature, and the synaptic modification mechanism probably has strict temporal requirements.

Mechanisms for Dealing with Delayed Reinforcement

In real-world learning situations there is inevitably a delay between the neural events giving rise to a response, and reinforcement being obtained. The synaptic activity of relevance may have occurred several seconds prior to reinforcement and the subsequent release of dopamine. Some mechanism is required to bridge the gap between synaptic activation and dopamine release.

There are several potential mechanisms for dealing with delayed reinforcement. Two are considered here in relation to behavioral evidence and possible neural substrates. The first mechanism is based on the idea of an "eligibility trace" (Barto et al., 1981) or "state of readiness" (Miller, 1981). This suggests that the synapses involved in the behavior that led to reward being obtained maintain a trace of their activity. When reinforcement is given, it acts on the synapses that have been "marked" by the trace. The second mechanism is based on the idea that cues present in the learning situation can acquire secondary reinforcing power. In this scheme, the dopamine signal is controlled by an adaptive element which uses cues to predict reinforcement and activate the dopamine neurons in response to the earliest salient cues.

The two mechanisms proposed for dealing with delay in reinforcement operate on different time scales. The temporal properties of these mechanisms are the focus of the following discussion.

The Eligibility Trace or State of Readiness The dendritic spines on which the corticostriatal inputs terminate may function as chemically isolated compartments. Localized elevations of calcium concentration, restricted to individual dendritic spines, can be brought about by a conjunction of presynaptic and postsynaptic activity (Gamble and Koch, 1987; Wickens, 1988; Holmes and Levy, 1990; Regehr and Tank, 1990). Because of the chemical isolation brought about by the anatomical features of the spine neck (such as the spine apparatus) the elevated calcium concentration could persist for some time. In principle, the elevated spine calcium could serve as the physical basis of the eligibility trace (Wickens, 1990).

The calcium gradients brought about by a conjunction of cortical presynaptic and postsynaptic striatal activity should persist for at least tens or hundreds of milliseconds. By the mechanisms described in the previous section, if dopamine receptor activation occurred during this time, the synapse would be strengthened. However, if the dopamine signal failed to occur, the synapse would be weakened. This would result in extinction of the response. Figure 10.2 illustrates how calcium- and dopamine-activated mechanisms within the medium spiny neurons of the neostriatum would give rise to these effects.

It is unlikely that the proposed eligibility trace would persist for more than 1 second. However, in animal learning experiments it is possible to extend the temporal gradient of reinforcement over a much longer period, by holding certain cues constant during the learning situation. Some other mechanism is needed to bridge the gap between an eligibility trace operating on a 1-second time scale, and a much longer experimentally determined temporal gradient of reinforcement. This is now considered.

Prediction of Reinforcement Spence (1947) argued (on the basis of behavioral evidence) that all learning involving delay of primary reward results from the action of immediate secondary reinforcement which develops in the situation. According to Spence, "Such a hypothesis eliminates the necessity of explaining how reward seemingly acts backward over time to influence something which occurred earlier." The learning theory arguments behind this position are beyond the scope of this chapter, but there are some important clues from neurobiology which support this idea.

Ljungberg et al. (1992) studied the activity of dopamine neurons during several stages in the acquisition of a behavioral task. During acquisition, prominent activations of the dopamine neurons occurred in association with the delivery of primary liquid reward. However, after the behavior had been acquired the most prominent response was to a light signaling the start of the task. In related experiments, in trials in which reward was withheld because of errors, Schultz et al. (1993) described depression of activity in all dopamine neurons "at precisely the moment when reward would have been delivered." They suggested that the decrease in activity reflected a state of expectation of reward at the time of its usual delivery. It is particularly interesting that Schultz et al. (1992) described neural activity in the monkey ventral striatum that was time-locked to the delivery of reward. It is possible that the output from some such neurons might inhibit dopaminergic neurons from firing at the time of expected reward, while others (perhaps in the amygdala) might excite dopaminergic neurons to fire in an anticipatory fashion.

In artificial neural network models of learning, use of an internal reward signal controlled by an adaptive predictor of reinforcement has

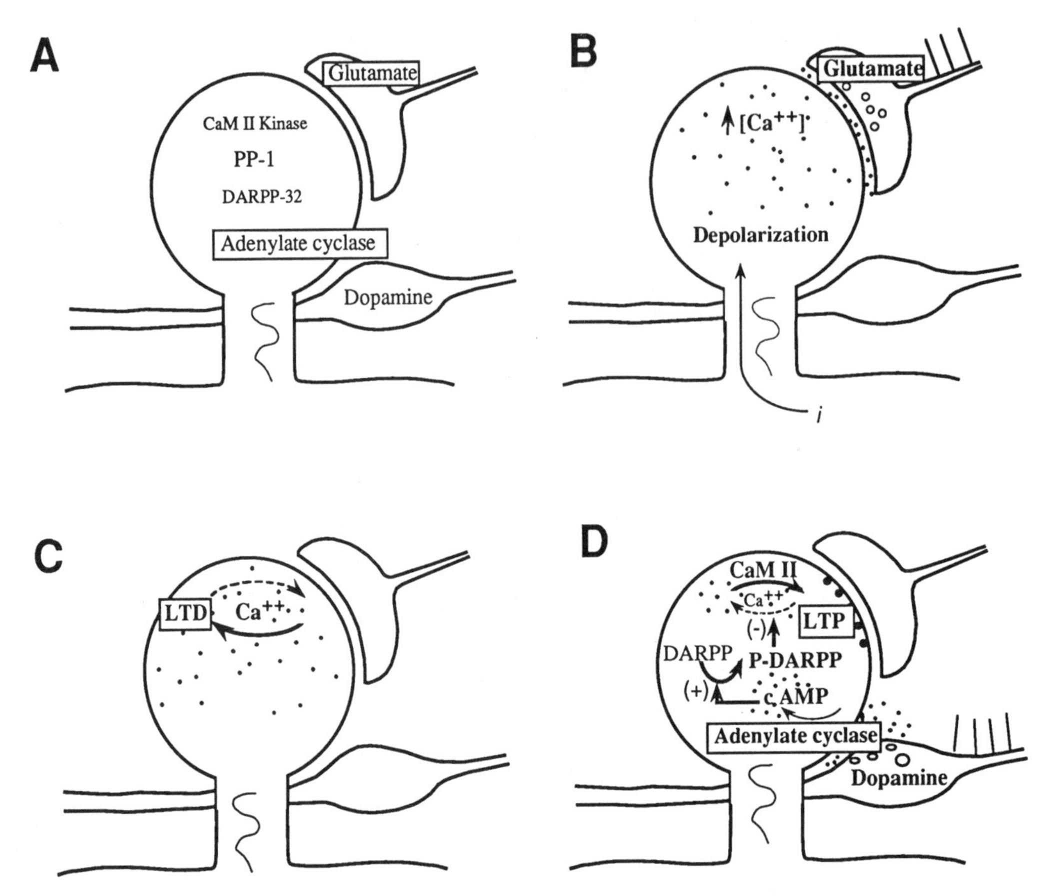

Figure 10.2 A synaptic mechanism for dealing with delayed reinforcement. (*A*) The processes of three different neurons meet at a dendritic spine which contains the biochemical mechanisms required for three-factor synaptic modification. (*B*) A conjunction of glutamate release from corticostriatal presynaptic terminals and postsynaptic depolarization of the striatal neuron leads to an increase in intracellular calcium concentration. This is localized to the dendritic spine concerned and persists as an "eligibility trace" lasting several hundred milliseconds. (C) If dopaminergic activity does not occur during the period of elevated calcium, long-term depression (LTD) is the result. (*D*) If dopaminergic activity occurs while the spine calcium concentration is still elevated, synergistic interactions between calcium and cAMP lead to long-term potentiation (LTP). CaMII Kinase, calcium-calmodulin–dependent protein kinase II; PP-1, protein phosphatase 1: DARRP, dopamine- and cAMP-regulated phosphoprotein. (Adapted from Wickens, 1990.)

been shown to be advantageous. When primary reinforcement is delayed, models using an internal reinforcement mechanism which functions predictively are able to learn more effectively than models without such a mechanism (Barto et al., 1983).

In summary, a heterosynaptic plasticity mechanism has been proposed which integrates three factors: pre- and postsynaptic activity, and a reinforcement signal mediated by dopamine. The conjunction of pre- and postsynaptic activity brings about a localized increase in spine calcium concentration. The effect of dopamine depends on timing. In order to produce potentiation, the dopamine reinforcement signal must act while the spine calcium concentration is elevated. Depression will result if the reinforcement is delayed. An internal control over the dopamine reinforcement signal provides a mechanism for learning on the basis of delayed but predictable reinforcement.

INTEGRATIVE ASPECTS OF REINFORCEMENT LEARNING

The capabilities of heterosynaptic plasticity mechanisms are determined by the circuits in which they are embedded. The mechanisms being proposed operate on the synapses connecting cortical afferents to medium spiny output neurons. The following discussion considers these mechanisms in relation to interactions among medium spiny neurons *within* the neostriatum, and then in relation to the interplay *between* the neostriatum and the cerebral cortex.

Local Circuit Interactions, Competition, and Reinforcement Learning

The efferent axons of the medium spiny neurons produce a number of collateral branches. The collaterals divide repeatedly within a volume of approximately 500μm diameter and synapse upon the proximal segments of the dendrites of neighboring medium spiny (and other) neurons. These synapses, and the neurons they terminate on, contain the inhibitory neurotransmitter γ-aminobutyric acid, or GABA (Kita and Kitai, 1988; Aronin et al., 1986; Pasik et al., 1988). This anatomical arrangement is consistent with an inhibitory relationship among striatal output neurons (Wilson and Groves, 1980; Groves, 1983; Rolls and Williams, 1986).

There is some electrophysiological evidence of inhibitory local interactions among striatal neurons. When caudate neurons are activated antidromically, other nearby spontaneously firing neurons show short latency suppression of activity (Katayama et al., 1981). This suppression is probably mediated by local axon collaterals of the medium spiny projection neurons, rather than an indirect effect transmitted via the cortex, because the suppression is not reduced when the cerebral cortex is removed.

On the basis of the limited available evidence, Wickens et al. (1991) proposed that a basic unit of striatal organization is an "inhibitory domain" comprising a set of several hundred mutually inhibitory neurons, occupying a spherical volume approximately 500μm in diameter. Computer simulations showed that in such a domain the prevailing dynamic is one of competition. *Competition* may be defined as a dynamic mode in which a small subset of neurons fire more than the rest, which are inhibited by the synapses of the local collateral branches of the axons of the active subset (Alexander and Wickens, 1993).

Competition among medium spiny neurons can improve the performance of the three-factor rule for synaptic modification. The improvement is due to the contrast-enhancing effect of competition on the spatial distribution of activity. The increased difference between neurons enhances learning by acting on the postsynaptic activity factor in the synaptic modification rule. Computer simulations of learning in a network model of the neostriatum show that competition facilitates learning of simple stimulus discriminations and response selections on the basis of differential reinforcements (Wickens, 1993). However, competition is not an absolute requirement for learning as it can still occur in models in which inhibitory interactions are reduced or absent (Borisyuk et al., 1994).

During *performance,* on the other hand, the significance of competition is that the neurons which receive the greatest weighted sum of excitation will respond most vigorously, and they will suppress the responses of their less strongly excited neighbors. Thus, the output from the neostriatum is activity in those striatal cells which most closely represent the current cortical activity pattern (or, in other words, whose synaptic strength vector most closely matches the cortical activity vector). These will be the ones that are associated with previously reinforced responses.

The foregoing suggests that the neostriatum contains a mechanism for making decisions on the basis of likelihood of reward.

Reinforcement Learning and Corticostriatal Interplay

The learning capabilities of the isolated neostriatum are limited in some important ways. In particular, the neostriatum is conceptually a single-layer network, because the cortical afferents terminate directly upon the striatal output neurons. This means that the discrimination abilities of the network *in isolation* will be limited to linearly separable sets of inputs, a limitation common to perceptron-like mechanisms (Minsky and Papert, 1969).

The neostriatum does not function in isolation, however, because it is connected with the cerebral cortex in an elaborate system of parallel loops (Alexander et al., 1986). The excitatory effect of the cerebral cortex on the striatum has already been discussed. Excitation of the neostriatum produces a perfectly time-locked increase of activity in a large number of thalamic cells projecting to the motor cortex (Deniau and Chevalier, 1985). Thus, the activity of neurons in the neostriatum may gain expression in the cortex by a process of disinhibition, as proposed by Kitai (1981). Overall, there is a positive-feedback pathway between the cerebral cortex and neostriatum.

Braitenberg (1978) suggested that a feedback loop between the cortex and neostriatum could be used to regulate the threshold of the cerebral cortex, and thereby regulate cortical cell-assembly formation. Progression from one cortical assembly to another could be controlled by the neostriatum. Increasing the threshold could "hold" a given cortical state. Decreasing the threshold could induce a change in cortical activation to a new pattern (Palm, 1987).

The primitive ability of the neostriatum to select output patterns that maximize the likelihood of reinforcement in a given cortical context can be used to considerable advantage in such a system. It provides a powerful way to search through all the cortical activity patterns that are possible (in a given situation) in order to select the one most likely to bring about further reinforcement.

In the proposed mechanism, the neostriatum is viewed as an amplifier of cortical activity patterns which works by positive feedback. In this view, the process of deciding to perform a particular action begins with activity in a loop involving the cortex and neostriatum. The activity is amplified and filtered as it cycles around this loop. A given pattern of striatal output (chosen on the basis of competitive interactions within the neostriatum) lowers the threshold for activation of a particular cortical cell assembly.

If, in the past, a dopamine reward signal had frequently followed the initial patterns of cortical activity, the associated corticostriatal synapses will be strong, and the activity will be amplified. Conversely, if the initial patterns of cortical activity had seldom been followed by reward, the associated corticostriatal synapses will be weak, and the activity will not be amplified. By this mechanism a cortical cell assembly that had led the organism toward a rewarding outcome in the past would become more attractive in the future. The neostriatum deepens the basins of attraction surrounding the states of the cortex that are most likely to lead to reinforcement.

The role of the neostriatum in this is to ensure that the assembly most likely to lead to a rewarding outcome will be amplified, while those less likely to lead to reward will be suppressed.

CONCLUSION

There is considerable evidence that the dopaminergic neurons which project from the midbrain to the neostriatum mediate an internal reinforcement signal. A major target of this signal is the synapses that connect the cerebral cortex to the neostriatum. However, the effect of dopamine on these synapses is not completely understood.

A useful heuristic has been that the corticostriatal synapses are modifiable according to a rule involving three factors: presynaptic activity in corticostriatal afferents; postsynaptic activity in striatal output neurons; and phasic activity in dopaminergic afferents. This rule provides a unifying hypothesis for the effects of dopamine on neostriatal output, motor activation, and learning.

Existing electrophysiological evidence for dopamine-mediated heterosynaptic plasticity neither confirms nor denies the postulated three-factor rule. In view of the neurochemical interactions involved, there are probably strict temporal requirements for the three interacting factors. Pre- and postsynaptic activity must coincide in order to bring about increases in intracellular calcium concentration, and synaptic dopamine release must occur *while* the calcium concentration is elevated. A quantitative approach indicates that different outcomes can occur depending on concentration and timing. Future experimental work should take these factors into account.

The corticostriatal pathway is a strategic location for modifiable synapses. The striatal neurons are favorably placed to evaluate cortical activity patterns. Since the strengths of the corticostriatal synapses may represent the accumulated sum of reinforcements, striatal neurons can compute a conditional expectation of reinforcement given the current cortical activity. The result of this computation can be used to select the cortical cell assemblies which maximize expected reinforcement. This is a simple mechanism for combining reinforcement learning with associative memory, to select and put to good use the knowledge stored in the form of cortical cell assemblies.

In general, neural networks using connections which are modified by reinforcement learning algorithms can operate very effectively to maximize reinforcement in complex environments. A model based on corticostriatal interactions, in which modifiable synapses are located in the corticostriatal pathway, promises to be a useful way of integrating cellular mechanisms of reinforcement into intelligent behavior.

ACKNOWLEDGMENTS

We thank Dr E. Gail Tripp for helpful comments. This work was supported by a grant from the Health Research Council of New Zealand.

REFERENCES

Ahlijanian, M. K., and Cooper, D. M. F. (1988) Distinct interactions between Ca^{2+} / calmodulin and neurotransmitter stimulation of adenylate cyclase in striatum and hippocampus. *Cell. Mol. Neurobiol.* 8:459–469.

Alexander, M. E., and Wickens, J. R. (1993) Analysis of striatal dynamics: The existence of two modes of behaviour. *J. Theor. Biol.* 163:413–438.

Alexander, G. E., DeLong, M. R., and Strick, P. L. (1986) Parallel organization of functionally segregated circuits linking basal ganglia and cortex. *Ann. Rev. Neurosci.* 9:357–381.

Arbuthnott, G. W., and Ungerstedt, U. (1975) Turning behavior induced by electrical stimulation of the nigro-neostriatal system of the rat. *Exp. Neurol.* 47:162–172.

Ariano, M. A. (1987) Comparison of dopamine binding sites in rat superior cervical ganglion and caudate nucleus. *Brain Res.* 421:245–254.

Aronin, N., Chase, K., and Difiglia, M. (1986) Glutamic acid decarboxylase and enkephalin immunoreactive axon terminals in the rat neostriatum synapse with striatonigral neurons. *Brain Res.* 365:151–158.

Ashe, J. H., and Libet, B. (1981) Modulation of slow postsynaptic potentials by dopamine in rabbit sympathetic ganglion. *Brain Res.* 217:93–106.

Azzaro, A. J., Liccione, J., and Lucci, J. (1987) Opposing actions of D-1 and D-2 dopamine receptor-mediated alterations of adenosine-3′,5′-cyclic monophosphate (cyclic AMP) formation during the amphetamine-induced release of endogenous dopamine *in vitro*. *Naunyn Schmiedebergs Arch. Pharmacol.* 336:133–138.

Barto, A. G., Sutton, R. S., and Brouwer, P. S. (1981) Associative search network: A reinforcement learning associative memory. *Biol. Cyber.* 40:201–211.

Barto, A. G., Sutton, R. S., and Anderson, C. W. (1983) Neuronlike elements that can solve difficult learning control problems. *IEEE Trans. Syst. Man Cyber.* 15:835–846.

Begg, A. J., Wickens, J. R., and Arbuthnott, G. W. (1993) A long-lasting effect of dopamine on synaptic transmission in the corticostriatal pathway, *in vitro*. Proceedings of the 11th International Australasian Winter Conference on Brain Research. *Int. J. Neurosc.*

Beninger, R. J. (1983) The role of dopamine in locomotor activity and learning. *Brain Res.* 287:173–196.

Black, J., Belluzzi, J. D., and Stein, L. (1985) Reinforcement delay of one second severely impairs acquisition of brain self-stimulation. *Brain Res.* 359:113–119.

Borisyuk, R. M., Wickens, J. R., and Kotter, R. (1994) Reinforcement learning in a network model of the basal ganglia. In R. Trappl (ed.), *Cybernetics and Systems '94, vol. II.* Proceedings of the Twelfth European Meeting on Cybernetics and Systems Research, Austrian Society for Cybernetic Studies, University of Vienna, Austria, 5–9 April 1994. Singapore, New Jersey, London, Hong Kong, World Scientific, pp. 1681–1686.

Bouyer, J. J., Park, D. H., Joh, T. H., and Pickel, V. M. (1984) Chemical and structural analysis of the relation between cortical inputs and tyrosine hydroxylase–containing terminals in rat neostriatum. *Brain Res.* 302:267–275.

Braitenberg, V. (1978) Cell assemblies in the cerebral cortex. In R. Heim and G. Palm (eds.), *Theoretical Approaches to Complex Systems.* Heidelberg: Springer-Verlag, pp. 171–188

Calabresi, P., Misgeld, U., and Dodt H. U. (1987) Intrinsic membrane properties of neostriatal neurons can account for their low level of spontaneous activity. *Neuroscience* 20:293–303.

Calabresi, P., Benedetti, M., Mercuri, N. B., and Bernardi, G. (1988) Endogenous dopamine and dopaminergic agonists modulate synaptic excitation in neostriatum: Intracellular studies from naive and catecholamine-depleted rats. *Neuroscience* 27:145–147.

Calabresi, P., Maj, R., Pisani, A., Mercuri, N. B., and Bernardi, G. (1992) Long-term synaptic depression in the striatum: Physiological and pharmacological characterization. *J. Neurosc.* 12:4224–4233.

Chiodo, L. A., and Berger, T. W. (1986) Interactions between dopamine and amino-acid induced excitation and inhibition in the striatum. *Brain Res.* 375:198–203.

Colombaioni, L., and Brunelli, M. (1988) Neurotransmitter-induced modulation of an electrotonic synapse in the CNS of *Hirudo medicinalis*. *Exp. Biol.* 47:139–144.

Deniau, J. M., and Chevalier, G. (1985) Disinhibition as a basic process in the expression of striatal functions. II. The striatonigral influence on thalamocortical cells of the ventromedial thalamic nucleus. *Brain Res.* 334:227–233.

Flaherty, A. W., and Graybiel, A. M. (1991) Corticostriatal transformations in the primate somatosensory system. Projections from physiologically mapped body-part representations. *J. Neurophysiol.* 66:1249–1263.

Freund, T. F., Powell, J. F., and Smith, A. D. (1984) Tyrosine hydroxylase–immunoreactive boutons in synaptic contact with identified striatonigral neurons, with particular reference to dendritic spines. *Neuroscience* 13:1189–1215.

Frey, U. U., Huang, Y.-Y., and Kandel, E. R. (1993) Effects of cAMP simulate a late stage of LTP in hippocampal CA1 neurons. *Science* 260:1661–1664.

Gamble, E., and Koch, C. (1987) The dynamics of free calcium in dendritic spines in response to repetitive synaptic input. *Science* 236:1311–1315.

Garcia-Munoz, M., Young, S. J., and Groves, P. M. (1992) Long-lasting changes in excitability of corticostriatal terminals following tetanic stimulation. In IBAGS, Proceedings of the Fourth International Basal Ganglia Society Meeting, Giens, France.

Gnegy, M., and Treisman, G. (1981) Effect of calmodulin on dopamine-sensitive adenylate cyclase activity in rat striatal membranes. *Mol. Pharmacol.* 19:256–263.

Gribkoff, V. K., and Ashe, J. H. (1984) Modulation by dopamine of population responses and cell membrane properties of hippocampal cal neurons in vitro. *Brain Res.* 292:327–338.

Groves, P. M. (1983) A theory of the functional organisation of the neostriatum and the neostriatal control of voluntary movement. *Brain Res. Rev.* 5:109–132.

Halpain, S., Girault, J. A., and Greengard, P. (1990) Activation of NMDA receptors induces dephosphorylation of DARPP-32 in rat striatal slices. *Nature* 343:369–372.

Haracz, J. L., Tschanz, J. T., Wang, Z., White, I. M., and Rebec, G. V. (1993) Striatal single-unit responses to amphetamine and neuroleptics in freely moving animals. *Neurosci. Biobehav. Rev.* 17:1–12.

Hemmings, H. C., Greengard, P., Tung, H. Y. L., and Cohen, P. (1984) DARPP-32, a dopamine-regulated neuronal phosphoprotein, is a potent inhibitor of protein phosphatase-1. *Nature* 310:503–505.

Hemmings, H. C., Nairn, A. C., Elliott, J. I., and Greengard, P. (1990) Synthetic peptide analogs of DARPP-32 (Mr 32,000 dopamine- and cAMP-regulated phosphoprotein), an inhibitor of protein phosphatase-1. Phosphorylation, dephosphorylation, and inhibitory activity. *J. Biol. Chem.* 265:20369–20376.

Hirata, K., Yim, C. Y., and Mogenson, G. J. (1984) Excitatory input from sensory motor cortex to neostriatum and its modification by conditioning stimulation of the substantia nigra. *Brain Res.* 321:1–8.

Hoebel, B. G., Monaco, A., Hernandes, L., Aulisi, E., Stanley, B. G., and Lenard, L. (1983) Self-injection of amphetamine directly into the brain. *Psychopharmacology* 81:158–163.

Hoebel, B. G., Hernandez, L., Schwartz, D. H., Mark, G. P., and Hunter, G. A. (1989) Microdialysis studies of brain norepinephrin, serotonin and dopamine release during ingestive behavior: Theoretical and clinical implications. *Ann. N. Y. Acad. Sci.* 575:171–191.

Holmes, W. R., and Levy, W. B. (1990) Insights into associative long-term potentiation from computational models of NMDA receptor-mediated calcium influx and intracellular calcium concentration. *J. Neurophysiol.* 63:1148–1168.

Huang, Q., Zhou, D., Chase, K., Gusella, J. F., Aronin, N., and DiFiglia, M. (1992) Immunohistochemical localization of the D1 dopamine receptor in rat brain reveals its axonal transport, pre- and postsynaptic localization, and prevalence in the basal ganglia, limbic system and thalamic reticular nucleus. *Proc. Nat. Acad. Sci. U. S. A.* 89:11988–11992.

Justice, J. B., Nicolaysen, L. C., and Michael, A. C. (1988) Modelling the dopaminergic nerve terminal. *J. Neurosci. Methods* 22:239–252.

Kandel, E. R., and Tauc, L. (1965) Mechanism of heterosynaptic facilitation in the giant cell of the abdominal ganglion of *Aplysia delipans. J. Physiol. (Lond.)* 181:28–47.

Katayama, Y., Miyazaki S., and Tsubokawa, T. (1981) Electrophysiological evidence favoring intracaudate axon collaterals of GABAergic caudate output neurons in the cat. *Brain Res.* 216:180–186.

Kawagoe, K. T., Garris, P. A., Wiedemann, D. J., and Wightman, R. M. (1992) Regulation of transient dopamine concentration gradients in the microenvironment surrounding nerve terminals in the rat striatum. *Neuroscience* 51:55–64.

Kebabian, J. W., Petzold, G. L., and Greengard, P. (1972) Dopamine-sensitive adenylate cyclase in caudate nucleus of rat brain, and its similarity to the "dopamine receptor." *Proc. Natl. Acad. Sci. U. S. A.*, 69:2145–2149.

Kita, H., and Kitai, S. T. (1988) Glutamate decarboxylase immunoreactive neurons in cat neostriatum: Their morphological types and populations. *Brain Res.* 447:346–352.

Kitai, S. T. (1981) Electrophysiology of the corpus striatum and brain stem integrating systems. In J. M. Brookhart, V. B. Mountcastle, V. B. Brooks, and S. R. Geiger (eds.), *Handbook of Physiology: The Nervous System,* Vol. 2. Bethesda, Md.: American Psychological Society, pp. 997–1013.

Kurumiya, S., and Nakajima, S. (1988) Dopamine D1 receptors in the nucleus accumbens: Involvement in the reinforcing effect of tegmental stimulation. *Brain Res.* 448:1–6.

Lindvall, O. A., Bjorklund, A., and Divac, I. (1978) Organization of catecholamine neurons projecting to frontal cortex in rat. *Brain Res.* 142:1–24.

Ljungberg, T., Apicella, P., and Schultz, W. (1992) Responses of monkey dopamine neurons during learning of behavioral reactions. *J. Neurophysiol.* 67:145–163.

Lyon, M., and Robbins, T. (1975) The action of central nervous system stimulant drugs: A general theory concerning amphetamine effects. *Curr. Dev. Psychopharmacol.* 2:79–163.

McGeorge, A. J., and Faull, R. L. M. (1989) The organization of the projection from the cerebral cortex to the striatum in the rat. *Neuroscience* 29:503–537.

McLennan, H., Emmons, P. R., and Plummer, P. M. (1964) Some behavioral effects of stimulation of the caudate nucleus in unrestrained cats. *Can. J. Physiol. Pharmacol.* 42:329–339.

Malach, R., and Graybiel, A. M. (1988) Mosaic architecture of the somatic sensory–recipient sector of the cat's striatum. *J. Neurosc.* 6:3436–3458.

Mark, G. P., Blander, D. S., and Hoebel, B. G. (1991) A conditioned stimulus decreases extracellular dopamine in the nucleus accumbens after the development of a learned taste aversion. *Brain Res.* 551:308–310.

Mayer, M. L., and Westbrook, G. L. (1987) Permeation and block of *N*-methyl-D-aspartic acid receptor channels by divalent cations in mouse cultured central neurons. *J. Physiol.* 394:501–527.

Memo, M., Lovenberg, W., and Hanbauer, I. (1982) Agonist-induced subsensitivity of adenylate cyclase coupled with a dopamine receptor in slices from rat corpus striatum. *Proc. Nat. Acad. Sci.* U. S. A. 79:4456–4460.

Mercuri, N., Bernardi, G., Calabresi, P., Cotugno, A., Levi, G., and Stanzione, P. (1985) Dopamine decreases cell excitability in rat striatal neurons by pre- and postsynaptic mechanisms. *Brain Res.* 358:110–121.

Miller, R. (1981) *Meaning and Purpose in the Intact Brain.* Oxford: Oxford University Press.

Minsky, M. L., and Papert, S. A. (1969) *Perceptrons: An Introduction to Computation Geometry. Expanded Edition.* Cambridge, Mass.: MIT Press.

Mochida, S., Kobayashi, H., and Libet, B. (1987) Stimulation of adenylate cyclase in relation to dopamine-induced long-term enhancement (LTE) of muscarinic depolarization, in the rabbit superior cervical ganglion. *J. Neurosc.* 7:311–318.

Natsukari, N., Hanai, H., Matsunaga, T., and Fujita, M. (1990) Synergistic activation of brain adenylate cyclase by calmodulin, and either GTP or catecholamines including dopamine. *Brain Res.* 534:170–176.

Ouimet, C. C., Miller, P. E., Hemmings, H. C., Walaas, S. I., and Greengard, P. (1984) DARPP-32, a dopamine and adenosine 3′:5′-monophosphate–regulated phosphoprotein enriched in dopamine-innervated brain regions. *J. Neurosc.* 4:111–124.

Ouimet, C. C., Lamantia, A. S., Goldman-Rakic, P., Rakic, P., and Greengard, P. (1992) Immunocytochemical localization of DARPP-32, a dopamine and cyclic-AMP-regulated phosphoprotein, in the primate brain. *J. Comp. Neurol.* 323:209–218.

Palm, G. (1987) Associative memory and threshold control in neural networks. In J. L. Casti and A. Kalqvist (eds.), *Real Brains, Artificial Minds.* New York: North-Holland, pp. 165–179.

Parent, A. (1986) *Comparative Neurobiology of the Basal Ganglia.* New York: John Wiley & Sons.

Pasik, P., Pasik, T., Holstein, G., and Hamori, J. (1988) GABAergic elements in the neuronal circuits of the monkey neostriatum: A light and electron microscopic immunocytochemical study. *J. Comp. Neurol.* 270:157–170.

Pennartz, C. M. A., Ameerun, R. F., Groenewegen, H. J., and Lopes da Silva, F. H. (1993) Synaptic plasticity in an *in vitro* slice preparation of the rat nucleus accumbens. *Eur. J. Neurosci.* 5:107–117.

Phillips, A. G. (1979) Electrical stimulation of the neostriatum in behaving animals. In I. Divac and R. G. E. Öberg (eds.), *The Neostriatum.* New York: Pergamon Press, pp. 183–194.

Piascik, M. T., Piascik, M. F., Hitzemann, R. J., and Potter, J. D. (1981) Ca^{2+}-dependent regulation of rat caudate nucleus adenylate cyclase and effects on the response to dopamine. *Mol. Pharmacol.* 20:319–325.

Regehr, W. G., and Tank, D. W. (1990) Postsynaptic NMDA receptor-mediated calcium accumulation in hippocampal slices. *Nature* 345:807–810.

Renner, E. K. (1964) Delay of reinforcement: A historical review. *Psychol. Bull.* 61:341–361.

Rolls, E. T., and Williams, G. V. (1986) Sensory and movement-related neuronal activity in different regions of the primate striatum. In J. S. Schneider and T. I. Lidsky (eds.), *Basal Ganglia and Behavior: Sensory Aspects of Motor Functioning*. Bern: Hans Huber, pp. 37–60.

Schneider, J. S. (1991) Responses of striatal neurons to peripheral sensory stimulation in symptomatic MPTP-exposed cats. *Brain Res.* 544:297–302.

Schneider, J. S., Levine, M. S., Hull, C. D., and Buchwald, N. A. (1984) Effects of amphetamine on intracellular responses of caudate neurons in the cat. *J. Neurosci.* 4:930–937.

Schultz, W., Apicella, P., Scarnati, E., and Ljungberg, T. (1992) Neuronal activity in monkey ventral striatum related to the expectation of reward. *J. Neurosci.* 12:4595–4610.

Schultz, W., Apicella, P., and Ljungberg, T. (1993) Responses of monkey dopamine neurons to reward and conditioned stimuli during successive steps of learning a delayed response task. *J. Neurosci.* 13:900–913.

Schwartz, J. H. and Greenberg, S. M. (1987) Molecular mechanisms for memory: Second-messenger induced modification of protein kinases in nerve cells. *Annu. Rev. Neurosci.* 10:459–467.

Sirinathsinghji, D. J. S., Dunnett, S. B., Isacson, O., Clarke, D. J., Kendrick, K., and Bjorklund, A. (1988) Striatal grafts in rats with unilateral neostriatal lesions. II *In vivo* monitoring of GABA release in globus pallidus and substantia nigra. *Neuroscience* 24:803–811.

Spence, K. W. (1947) The role of secondary reinforcement in delayed reward learning. *Psychol. Rev.* 54:1–8.

Stein, L., and Belluzzi, J. D. (1989) Cellular investigations of behavioral reinforcement. *Neurosci. Biobehav. Rev.* 13:69–80.

Stellar, J. R., and Stellar, E. (1985) *The Neurobiology of Motivation and Reward*. Heidelberg: Springer-Verlag.

Swope, S. L., Moss, S. J., Blackstone, C. D., and Huganir, R. L. (1992) Phosphorylation of ligand-gated ion channels: A possible mode of synaptic plasticity. *FASEB J.* 6:2514–2523.

Szechtman, H. (1983) Peripheral sensory input directs apomorphine-induced circling in rats. *Brain Res.* 264:332–335.

Thorndike, E. L. (1911) *Animal Intelligence*. New York: Macmillan.

Trugman, J. M., and Wooten, G. F. (1986) The effects of L-DOPA on regional cerebral glucose utilization in rats with unilateral lesions of the substantia nigra. *Brain Res.* 379:264–274.

Verney, C., Baulac, M., Berger, B., Alvarez, C., Vignay, A., and Helles, K. B. (1985) Morphological evidence for a dopaminergic terminal field in the hippocampal formation of young and adult rat. *Neuroscience* 14:1039–1052.

Walaas, S. I., and Greengard, P. (1984) DARPP-32, a dopamine- and adenosine 3′:5′-monophosphate–regulated phosphoprotein enriched in dopamine-innervated brain regions. *J. Neurosci.* 4:84–98.

Walaas, S. I., and Greengard, P. (1991) Protein phosphorylation and neuronal function. *Pharmacol. Rev.* 43:299–349.

Walaas, S. I., Aswad, D. W., and Greengard, P. (1983) A dopamine- and cyclic AMP–regulated phosphoprotein enriched in dopamine-innervated brain regions. *Nature* 301:69–71.

Walsh, J. P. (1993) Depression of excitatory synaptic input in rat striatal neurons. *Brain Res.* 608:123–128.

Warenycia, M. W., McKenzie, G. M., Murphy, M., and Szerb, J. C. (1987) The effects of cortical ablation on multiple unit activity in the striatum following dexamphetamine. *Neuropharmacology* 26:1107–1114.

Waszczak, B. L., Lee, E. K., Ferraro, T., Hare, T. A., and Walters, J. R. (1984) Single unit responses of substantia nigra *pars reticulata* neurons to apomorphine: Effects of striatal lesions and anesthesia. *Brain Res.* 306:307–318.

Weiss, S., Sebben, M., Garcia-Sainz, J. A., and Bockaert, J. (1985) D2-dopamine receptor-mediated inhibition of cyclic AMP formation in striatal neurons in primary culture. *Mol. Pharmacol.* 27:595–599.

West, M. O., Michael, A. J., Knowles, S. F., Chapin, J. K., and Woodward, D. J. (1986) Striatal unit activity and the linkage between sensory and motor events. In J. S. Schneider and T. I. Lidsky (eds.), *Basal Ganglia and Behaviour: Sensory Aspects of Motor Functioning.* Bern: Hans Huber, pp. 27–35.

White, N. M. (1989) A functional hypothesis concerning the striatal matrix and patches: Mediation of S-R memory and reward. *Life Sci.* 45:1943–1957.

Wickens, J. R. (1988) Electrically coupled but chemically isolated synapses: Dendritic spines and calcium in a rule for synaptic modification. *Prog. Neurobiol.* 31:507–528.

Wickens, J. R. (1990) Striatal dopamine in motor activation and reward-mediated learning: Steps towards a unifying model. *J. Neural Transm.* 80:9–31.

Wickens, J. R. (1993) *A Theory of the Striatum.* New York: Pergamon Press.

Wickens, J. R., Alexander, M. E., and Miller, R. (1991) Two dynamic modes of striatal function under dopaminergic-cholinergic control: Simulation and analysis of a model. *Synapse* 8:1–12.

Wilson, C. J., and Groves, P. M. (1980) Fine structure and synaptic connections of the common spiny neuron of the rat neostriatum. A study employing intracellular injection of horseradish peroxidase. *J. Comp. Neurol.* 194:599–615.

Wise, R. A., and Bozarth, M. A. (1984) Brain reward circuitry: Four circuit elements "wired" in apparent series. *Brain Res. Bull.* 297:265–273.

Wooten, G. F., and Collins, R. C. (1983) Effects of dopaminergic stimulation on functional brain metabolism in rats with substantia nigra lesions. *Brain Res.* 263:267–275.

Yovell, Y., and Abrams, T. W. (1992) Temporal assymetry in activation of *Aplysia* adenylyl cyclase by calcium and transmitter may explain temporal requirements of conditioning. *Proc. Nat. Acad. Sci. U. S. A.* 89:6526–6530.

11 Adaptive Critics and the Basal Ganglia

Andrew G. Barto

One of the most active areas of research in artificial intelilgence is the study of learning methods by which "embedded agents" can improve performance while acting in complex dynamic environments. An agent, or decision maker, is embedded in an environment when it receives information from, and acts on, that environment in an ongoing closed-loop interaction. An embedded agent has to make decisions under time pressure and uncertainty and has to learn without the help of an ever-present knowledgeable teacher. Although the novelty of this emphasis may be inconspicuous to a biologist, animals being the prototypical embedded agents, this emphasis is a significant departure from the more traditional focus in artificial intelligence on reasoning within circumscribed domains removed from the flow of real-world events. One consequence of the embedded-agent view is the increasing interest in the learning paradigm called *reinforcement learning* (*RL*). Unlike the more widely studied *supervised learning* systems, which learn from a set of examples of correct input-output behavior, RL systems adjust their behavior with the goal of maximizing the frequency or magnitude, or both, of the reinforcing events they encounter over time.

While the core ideas of modern RL come from theories of animal classical and instrumental conditioning (although the specific term *reinforcement learning* is not used by psychologists), the influence of concepts from artificial intelligence and control theory has produced a collection of computationally powerful learning architectures. Despite similarities between some of these architectures and the structure and function of certain brain regions, relatively little effort has been made to relate these architectures to the nervous system (but see Klopf, 1982; Werbos, 1987; Wickens, 1990; Houk 1992). In this chapter I describe the RL system called the *actor-critic* architecture, giving enough detail so that it can be related to basal ganglionic circuits and dopamine neurons. Specifically, I focus on a component of this architecutre called the *adaptive critic,* whose behavior seems remarkably similar to that of the dopamine neurons projecting to the stiatum and frontal cortex (see chapter 12). In chapter 13 of this book, Houk et al. present a hypothesis about how the actor-critic architecture might be implemented by the circuits of the basal ganglia and associated brain structures. My explanation of the adaptive critic largely follows that of Sutton (1984, 1988).

The adaptive critic is a device that learns to *anticipate reinforcing events* in a way that makes it a useful conjunct to another component, the *actor,* that adjusts behavior to maximize the frequency or magnitude, or both, of reinforcing events. The adaptive critic also forms the basis of the *temporal difference model* of classical, or Pavlovian, conditioning (Sutton and Barto, 1987, 1990) which extends the Rescoral-Wagner model (Rescoral and Wagner, 1975) to take into account some of the fine temporal structure of conditioning. The learning rule used by the adaptive critic is due to Sutton, who was developing it as part of his Ph.D. dissertation (Sutton 1984) when it was used in the pole-balancing system of Barto et al. (1983). Sutton (1988) developed this class of learning algorithms further, calling them temporal difference (TD) methods.

This line of research work began with the exploration of Klopf's (1972, 1982) idea of *generalized reinforcement,* which emphasized the importance of sequentiality in a neuronal model of learning. An earlier precursor, however, is the technique used by Samuel (1963) in his learning program for the game of checkers. Current research on the adaptive critic focuses on its relationship to an optimization technique known as dynamic programming used for solving control problems. This connection follows the research of Werbos (1977, 1987) and Watkins (1989). Barto et al. (1990, 1994) provide detailed accounts of this perspective. A remarkable demonstration of the power of the actor-critic architecture is provided by Tesauro's (1992) backgammon playing program, which used an actor-critic architecture to learn how to play world-class backgammon.

REINFORCEMENT LEARNING

Following the basic idea of Thorndike's "law of effect" (Thorndike 1911), the simplest RL algorithms are based on the commonsense idea that if an action is followed by a satisfactory state of affairs, or an improvement in the state of affairs, then the tendency to produce that action is strengthened, i.e., reinforced. Although this is often called "trial-and-error" learning, I prefer to call it learning based on the "generate-and-test" procedure: alternatives are generated, evaluated by testing them, and behavior is directed toward the better alternatives. The reason for my preference is that it is too easy to confuse trial-and-error learning with supervised learning.

For example, an artificial neural network trained using the well-known supervised learning method of error backpropagation (e.g., see Rumelhart et al., 1986) produces an output, receives an error vector, and adjusts the network's weights to reduce the magnitude of the error.

This is a kind of trial-and-error learning, but it differs from the kind of learning Thorndike had in mind. Error vectors in supervised learning are derived from standards of correctness: the "target" responses of supervised learning. In contrast, RL emphasizes response-dependent feedback that *evaluates* the learner's performance by processes that do not necessarily have access to standards of correctness. Evaluative feedback tells the learner whether or not, and possibly by how much, its behavior has improved; or it provides a measure of the "goodness" of the behavior; or it just provides an indication of success or failure. Evaluative feedback does not directly tell the learner what it *should* have done, and although it is sometimes the *magnitude* of an error vector, it does not include *directional* information telling the learner how to change its behavior, as does the error vector of supervised learning. Although evaluative feedback is often called *reinforcement* feedback, it need not involve pleasure or pain.

Instead of trying to *match* a standard of correctness, an RL system tries to *maximize* the goodness of behavior as indicated by evaluative feedback.[1] To do this, the system has to probe the environment—perform some form of *exploration*—to obtain information about how to change its behavior. It has to actively try alternatives, compare the resulting evaluations, and use some kind of selection mechanism to guide behavior toward the better alternatives. I discuss the distinction between reinforcement and supervised learning in more detail elsewhere (Barto, 1991, 1992).

To actually build an RL system, one has to be more precise about the objective of learning. What does evaluative feedback evaluate? If the learning system's life consisted of nothing but a series of discrete trials, each consisting of a discrete action followed by an evaluation of that, and only that, action, the situation would be simple. But actions can have delayed as well as immediate consequences, and evaluative feedback generally evaluates the consequences of all of the system's past behavior. How can an RL system deal with complex tangles of actions and their consequences occurring throughout time? This has been called the *temporal credit assignment problem.* The concept of an adaptive critic is one way to approach this problem: the critic learns to provide useful *immediate* evaluative feedback based on predictions of future reinforcement. According to this approach, RL is not only the process of improving behavior according to given evaluative feedback; it also includes learning to improve evaluative feedback.

THE ACTOR-CRITIC ARCHITECTURE

The actor-critic architecture is usually viewed within the framework of control theory. Figure 11.1A is a variation of the classical control system

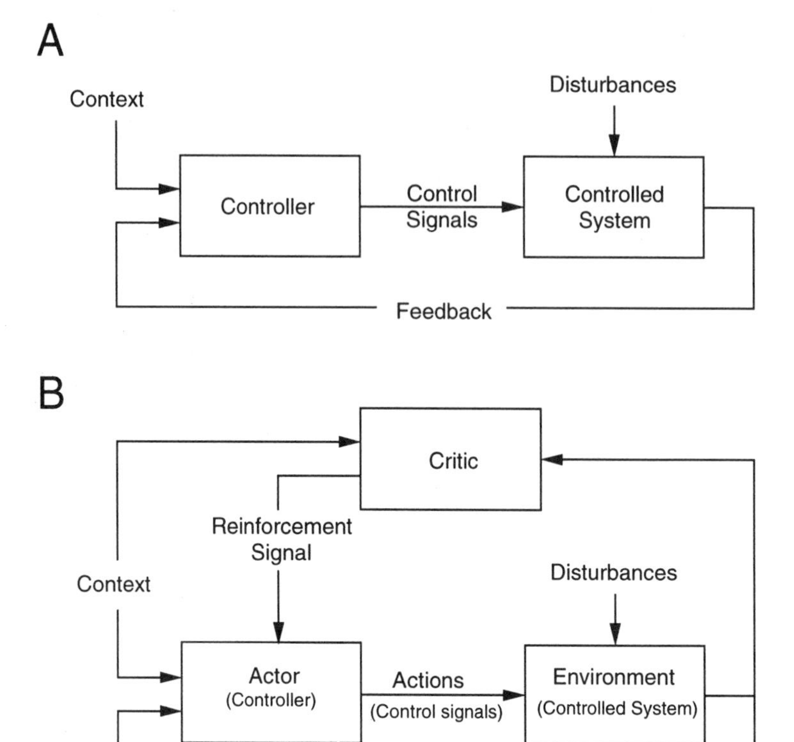

Figure 11.1 The actor-critic architecture as a controller. (*A*) A basic control loop. A controller provides control signals to a controlled system, whose behavior is influenced by disturbances. Feedback from the controlled system to the controller provides information on which the control signals can depend. The context inputs provide information pertinent to the control task's objective. (*B*) The actor-critic architecture. A critic provides the controller with a reinforcement signal evaluating its success in achieving the control objectives.

block diagram. A controller provides control signals to a controlled system. The behavior of the controlled system is influenced by disturbances, and feedback from the controlled system to the controller provides information on which the control signals can depend. The controller inputs labeled "context" provide information pertinent to the control task's objective. You might think of the context signals as specifying a "motivational state" that implies certain control goals.

Figure 11.1B extends the block diagram of figure 11.1A to the actor-critic architecture. Another feedback loop is added for providing evaluative feedback to the controller, now called the actor. The critic produces evaluative feedback, or reinforcement feedback, by observing the consequences of the actor's behavior on the controlled system, now called the environment. The critic also needs to know the motivational context of the task because its evaluations will be different depending on what the actor should be trying to do. The critic is an abstraction of the process that supplies evaluative feedback to the learning mechanism responsible for adjusting the actor's behavior. In most artificial RL systems, the critic's output at any time is a number that scores the actor's immediately preceding action: the higher the number, the better the action.

The actor-critic architecture is an abstract learning system, and care must be taken in relating it to animals and their nervous systems. It can help us in thinking about animal reinforcement learning, but it also can be misleading if it is taken too literally. Specifically, it is deceptive to identify the actor with an animal and the environment with the animal's environment. It is better to think of the actor-critic architecture as a model of any reinforcement learning component, or subsystem, of an animal. There are probably many such subsystems, only some of them directly controlling the animal's overt motor behavior.

Figure 11.2 elaborates the actor-critic architecture to emphasize this point. Think of the shaded box in this figure as an animal. The actor is not the same as the entire animal, and its actions are not necessarily motor commands. Furthermore, the critic (perhaps one of many) is in the animal. I have split the environment box of figure 11.1 into an internal and external environment to emphasize that the critic evaluates the actor's behavior on the basis of both its internal and external consequences. The internal environment can contain other actor-critic architectures, some of which do generate overt external behavior. In suggesting how the actor-critic architecture might be related to the basal ganglia, Houk et al. (see chapter 13) suggest that the actions of the

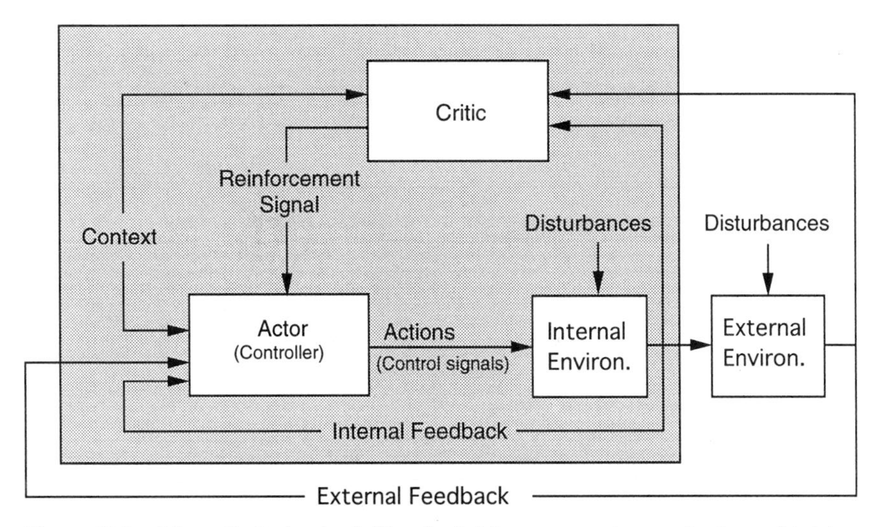

Figure 11.2 A hypothetical animal. The shaded box represents an animal, emphasizing that it is misleading to identify the actor with an entire animal and the critic with an external agent.

relevant actor are the signals influencing frontal cortex. Both the frontal cortex and cerebellum are components of this actor's internal environment.

Figure 11.3 shows the critic in more detail. It consists of a fixed and an adaptive critic. We think of the fixed critic as assigning a numerical *primary reinforcement value*, r_t, to the sensory input (both internal and external) received by the critic at each time instant t; r_t summarizes the strength of that input's primary reinforcing effect, i.e., the reinforcing effect that is wired in by the evolutionary process, not learned through experience. Although in animals this reinforcing effect depends on motivational state, we simplify things by assuming a fixed motivational state (so figure 11.3 does not show the context input to the critic). The adaptive critic assigns a different reinforcement value to the sensory input via an adaptive process. The output of the critic at t is the *effective reinforcement signal* sent to the actor. We label it $\hat{r}_t$.

IMMINENCE WEIGHTING

The basic objective of the actor-critic architecture is to learn to act so as to produce sensory input for which the primary reinforcement value is maximized. But because behavior continues over time, producing sensory input over time, the learning system has to maximize some measure of the entire time course of its input. This measure has to take into account the fact that actions can have long-term as well as short-term consequences on reinforcement, and that sometimes it is better to forgo short-term reward in order to achieve more reward later. Most RL researchers have adopted a measure based on the theory of optimal control. Although mathematical simplicity is its main advantage, this measure has some plausibility for animal learning as demonstrated by the TD model of classical conditioning (Sutton and Barto 1987, 1990). According to this measure, the objective of learning is to act at each

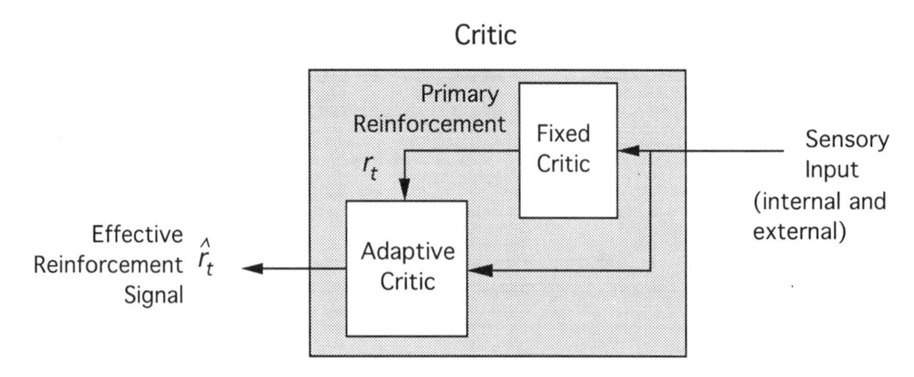

Figure 11.3 Elaboration of the critic. It contains fixed and adaptive components.

time instant so as to maximize *a weighted sum of all future primary reinforcement values.* It is plausible to weight immediate primary reinforcement more strongly than slightly delayed primary reinforcement, which should be mroe strongly weighted than long-delayed reinforcement. Sutton and Barto (1990) call this *imminence weighting* and suggest that the adaptive critic attempts to predict theimminence-weighted sum of future primary reinforcement.

Figure 11.4 illustrates the idea of imminence weighting for the particlar time course of primary reinforcement, shown in panel (A). One can think of this as a sequence of unconditioned stimuli (hence its labeling as *US*/λ, where λ is a normalization factor that we do not discuss here). (b) shows a particular imminence weighting function, which specifies how the weight given to primary reinforcement falls off with delay with respect to a particular time t. (c) shows how the primary reinforcement signal is transformed by the imminence weighting function applied at time t to give reduced weight to delayed primary reinforcement. The quantity the adaptive critic is trying to predict at time t is the area under this curve. To obtain the correct predictions for other times, this weighting function is slid along the time axis so that its base starts at the time in question, the primary reinforcement signal is reweighted according to the new position, and the new area is calculated. An example for another time, t', is shown in (D) and (E). By repeating this process for every time, one obtains the sequence of correct predic-

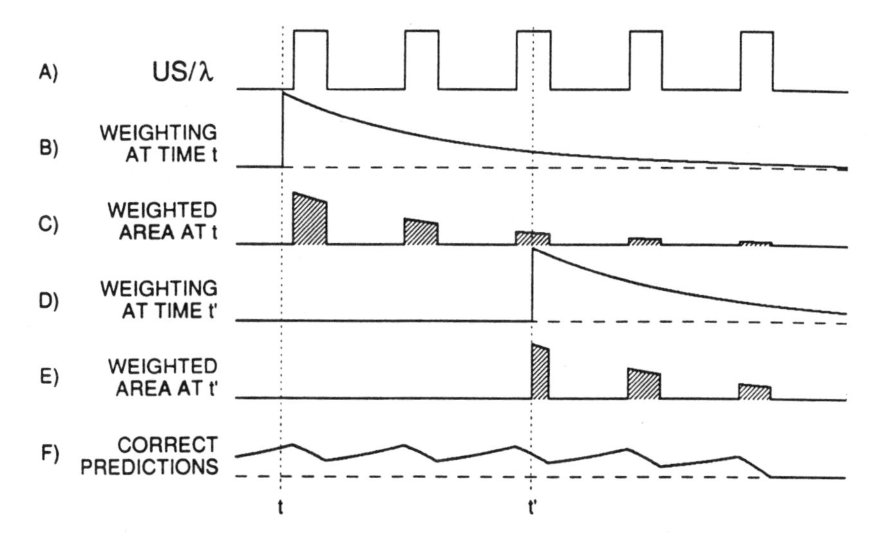

Figure 11.4 Imminence weighting. (*A*) A primary reinforcement signal representing a sequence of unconditioned stimuli (US/λ). (*B*) An imminence weighting function. (*C*) Primary reinforcement weighted by the imminence weighting function. The correct prediction at time t is the area under this curve. (*D* and *E*) Imminence weighting for time t'. (*F*) The correct predictions at each time. The heights at times t and t' equal the total areas in (*C*) and (*E*). (From Sutton and Barto, 1990.)

tions shown in (F). If the adaptive critic is correctly predicting the imminence-weighted sum of future primary reinforcement for the primary reinforcement signal of (A), its predictions should look like (F).

The simplest way to explain how it is possible to predict an imminence-weighted sum of future primary reinforcement is to adopt a *discrete-time* model of the learning process. Consequently, suppose t takes on only the integer values 0, 1, 2, . . . , and think of the time interval from any time step t to $t + 1$ as a small interval of real time. I make the additional assumption, again for simplicity, that *at minimum* it takes one time step for an action to influence primary reinforcement. This is the basic delay through the environment and the critic. Hence, by the *immediate primary reinforcement* for an action taken at time t, I mean r_{t+1}, and by the *immediate effective reinforcement*, I mean $\hat{r}_{t+1}$. Of course, this minumum delay can be different for different actor-critic systems in the nervous system.

Using a discrete-time version of imminence weighting, the objective of the actor is to learn to perform the action at each time step t that maximizes a weighted sum of the primary reinforcement values for time step $t + 1$ and all future times, where the weights decrease with decreasing imminence of the primary reinforcement value:

$$\alpha_1 r_{t+1} + \alpha_2 r_{t+2} + \alpha_3 r_{t+3} + \cdots , \tag{1}$$

with $\alpha_1 > \alpha_2 > \alpha_3 > \ldots$. Typically, these weights are defined in terms of a *discount factor*, γ, with $0 \leq \gamma < 1$, as follows:

$$\alpha_i = \gamma^{i-1},$$

for $i = 1, 2, \ldots$. Then the imminence-weighted sum of equation (1) is

$$r_{t+1} + \gamma r_{t+2} + \gamma^2 r_{t+3} + \cdots = \sum_{i=1}^{\infty} \gamma^{i-1} r_{t+i}.$$

The discount factor determines how strongly future primary reinforcement should influence current actions. When $\gamma = 0$, the imminence-weighted sum is just the immediate primary reinforcement r_{t+1} (because $0^0 = 1$). In this case, the desired actions maximize only the immediate primary reinforcement. As γ increases toward 1, future primary reinforcement becomes more significant. Here we think of γ as being fixed close to 1, so that the long-term consequences of actions are important and the adaptive critic plays an essential role in learning.

If actions were reinforced by immediate primary reinforcement only, learning would depend only on the short-term consequences of actions. This learning objective, which has been called a *tactical* objective (Werbos 1987), ignores the long-term consequences of actions. Since immediate primary reinforcement is usually lacking entirely (formalized by letting $r_t = 0$ for those times when there is no primary reinforcement), a purely tactical learning system cannot learn how to manipulate its environment in order to bring about future primary reinforcement. Even worse, acting only to attain immediate primary reinforcement can disrupt, or even preclude, attaining better primary reinforcement in the future. A *strategic* objective (Werbos, 1987), on the other hand, takes into account long-term as well as short-term consequences.

How tradeoffs between consequences at different times are handled is determined by exactly how one defines the strategic objective, imminence weighting by means of a discount factor being one definition. With discounting, any amount of primary reinforcement that is delayed by one time step is worth a fraction (γ) of that same amount of undelayed primary reinforcement. As γ increases toward 1, the delay makes less and less difference, and the objective of learning becomes more strategic.

The idea of the adaptive critic is that it should learn how to provide an effective reinforcement signal so that when the actor learns according to the tactical objective of maximizing immediate effective reinforcement, it is actually learning according to the strategic objective of maximizing a long-term measure of behavior. Here, the long-term measure is the imminence-weighted sum of future primary reinforcement. In order to do this, the adaptive critic has to predict the imminence-weighted sum of future primary reinforcement, and these predictions are essential in forming the effective reinforcement, as discussed below. Because effective reinforcement incorporates these predictions, the actor only needs to perform tactical leearning with respect to the effective reinforcement signal: it is geared so that the action at time t is always reinforced by the immediate effective reinforcement $\hat{r}_{t+1}$.

AN INPUT'S VALUE

We call the imminence-weighted sum of the primary reinforcement values from $t + 1$ into the future the *value* of the sensory input (internal and external) at time t. Let V_t denote this value; that is,

$$V_t = \sum_{i=1}^{\infty} \gamma^{i-1} r_{t+i}. \tag{2}$$

The objective of learning, then, is to learn to influence the environment so that the sensory inputs received by the learning system have the highest possible values. The job of the adaptive critic is to estimate these values and use them to determine the immediate effective reinforcement.

For the sake of brevity, I have to disregard a lot of important technical details about how this is even possible, but let me give some hints about these details. Because an estimate of an input's value is a prediction of the future, how can these estimates be made? Does not an input's value depend on the course of action the learning system will take in the future? Indeed, does it not depend on all kinds of unpredictable aspects of the environment? First, prediction is possible if one assumes that environmental situations tend to recur, so that a prediction is really a kind of recollection of what happened in the same situation, or in similar situations, in the past. The critic's sensory input must be rich enough to allow the detection of situations having the same or similar futures (formalized as *states* of a dynamic system). Second, for much of our discussion, we assume that the learning system's policy of acting stays fixed throughout the prediction process. This does not mean that the actor always produces the same action, but that it always responds the same way whenever the same situation recurs: its response rule, or policy, is fixed. Of course, because the whole point of RL is to change this response rule, this is only a subproblem of the entire RL problem. Finally, when the sensory input cannot resolve all the unpredictable aspects of the environment (i.e., when what *appears* to be a previously sensed situation is followed by a different course of events), probability theory is invoked. By the value of an input we really mean the *expected* imminence-weighted sum of future primary reinforcement values: the average over all the possible future scenarios.

So the adaptive critic is supposed to estimate the values of sensory inputs so it can compute a suitable effective reinforcement, $\hat{r}_t$. Let the critic's estimate of V_t be denoted P_t; it is a *prediction* of the imminence-weighted sum of future primary reinforcement. Then from equation (2) we would like the following to be true:

$$P_t \approx V_t = r_{t+1} + \gamma r_{t+2} + \gamma^2 r_{t+3} + \cdots, \tag{3}$$

where $\approx$ means "approximately equal."

LEARNING TO PREDICT

It is relatively easy to devise a supervised learning system for learning to predict the future values of specific signals. For example, suppose we wanted to have a prediction at any time t of the primary reinforcement signal at $t + 1$; that is, suppose we want $P_t = r_{t+1}$ for all t. This is a one-step-ahead prediction problem, and the usual kind of error-driven supervised learning system (e.g., Rumelhart et al. 1986) can be used to solve it. This system would need, at each time t during learning, an error between its actual prediction, P_t, and the prediction target (the quantity being predicted), r_{t+1}. It can obtain this error simply by computing P_t, *waiting one time step* while remembering P_t, then observing the actual r_{t+1}. It also has to remember for one time step the sensory input on which the prediction was based in order to update its prediction function.

For example, suppose P_t is the output of a simple linear connectionist unit:

$$P_t = \sum_{i=1}^{m} v_t^i x_t^i,$$

where v_t^i and x_t^i, for $i = 1, \ldots, m$, are respectively the connection weights and input activations at time t. Then the standard delta learning rule for one-step-ahead prediction is

$$v_{t+1}^i = v_t^i + \eta[r_{t+1} - P_t]x_t^i, \tag{4}$$

where $\eta > 0$ is the learning rate. If we think of this update equation being applied at time $t + 1$, then P_t is the *remembered* prediction, x_t^i is the *remembered* input activation, and r_{t+1} is the *currently observed* primary reinforcement value.

This is perhaps clearer if we rewrite the learning rule as it would appear if it were applied at time t instead of $t + 1$:

$$v_t^i = v_{t-1}^i + \eta[r_t - P_{t-1}]x_{t-1}^i. \tag{5}$$

This form, equivalent to equation (4), makes it clear that the previous prediction and the previous input activations have to be remembered as well as the previous connection weights. Following Klopf (1972, 1982), we say that input activity at $t - 1$ (i.e., $x_{t-1}^i \neq 0$) makes the connection weight v^i *eligible* for modification at t. In neural terms, eligibility would be a synaptically local memory for storing information about the past activity of the presynaptic fiber. Houk et al. (see chapter 13) postulate that this notion of eligibility is implemented by a period of high receptivity of spiny neuron synapses to the reinforcing effects of dopamine.

Of course, for this learning rule to work there must be information in the input stream that is predictively useful. If one wanted to predict more than one time step into the future, the procedure would be essentially the same except that it would need to remember the predictions and the input activations for the entire time interval until the actual predicted value becomes available. Consequently, for a prdiction interval of k time steps, the procedure would need to keep in memory k past predictions and k past input activations. Any kind of eligibility mechanism for this situation would have to be much more complicated than the simple period of receptivity mentioned above. One of the advantages of the adaptive critic is that it can learn to predict many time steps into the future without the need for a more complicated eligibility mechanism.

THE ADAPTIVE CRITIC LEARNING RULE

The adaptive critic learning rule begins with the one-step-ahead prediction method described in the previous section. However, for the critic the prediction targets are the *values* of inputs, which involve all future primary reinforcement, not just the primary reinforcement at the next time step. Extending the one-step-ahead method to this situation in the most obvious way would require an infinite amount of storage and the weights could not be updated until an infinite amount of time had passed.

The adaptive critic learning rule rests on noting that correct predictions must satisfy a certain consistency condition which relates the predictions at adjacent time steps. Moreover, it is true that any predictions that satisfy this consistency condition for all time steps must be correct. (This is a result from the theory of optimal control that is not particularly obvious.) Supose that the predictions at any two adjacent time steps, say steps $t - 1$ and t, are correct. This means that

$$P_{t-1} = r_t + \gamma r_{t+1} + \gamma^2 r_{t+2} + \cdots \tag{6}$$

$$P_t = r_{t+1} + \gamma r_{t+2} + \gamma^2 r_{t+3} + \cdots \tag{7}$$

Now note that we can write P_{t-1} as follows:

$$P_{t-1} = r_t + \gamma(r_{t+1} + \gamma r_{t+2} + \cdots).$$

But this is exactly the same as

$$P_{t-1} = r_t + \gamma P_t.$$

This is the consistency condition that is satisfied by the correct predictions. The error by which any two adjacent predictions fail to satisfy this condition is called the *temporal difference error* (TD error) by Sutton (1988):

$$r_t + \gamma P_t - P_{t-1}. \tag{8}$$

The adaptive critic uses the TD error to update its weights. The term *temporal difference* comes from the fact that this error essentially depends on the difference between the critic's predictions at adjacent time steps.

The adaptive critic therefore adjusts its weights according to the following modification of the the one-step-ahead learning rule of equation (5):

$$v_t^i = v_{t-1}^i + \eta[r_t + \gamma P_t - P_{t-1}]x_{t-1}^i. \tag{9}$$

This rule adjusts the weights to decrease the magnitude of the TD error. Note that if $\gamma = 0$, this is equal to the one-step-ahead learning rule [equation (5)].

In analogy with the one-step-ahead learning rule [equation (5)], we can think of $r_t + \gamma P_t$ as the prediction target: it is the quantity that each P_{t-1} should match. The adaptive critic is therefore trying to predict the next primary reinforcement, r_t, *plus its own next prediction* (discounted), γP_t. On the surface it is not clear that this would work: it is like the blind leading the blind. How can an incorrect prediction be improved by moving it toward another incorrect prediction? The key observation, however, is that the target $r_t + \gamma P_t$ tends to be more accurate than the prediction P_{t-1} because it includes the additional data provided by r_t. It is more like the blind being led by the slightly less blind. Although this method is very simple computationally, it actually converges to the correct predictions under fairly general conditions.

EFFECTIVE REINFORCEMENT

The output of the adaptive critic at time t is the effective reinforcement $\hat{r}_t$, which reinforces the action made at $t - 1$. For the actor-critic architectures with which we have the most experience, the effective reinforcement is same as the TD error:

$$\hat{r}_t = r_t + \gamma P_t - P_{t-1}. \tag{10}$$

Effective reinforcement is therefore the sum of the primary reinforcement, r_t, and the term $\gamma P_t - P_{t-1}$, which corresponds to *secondary reinforcement*. To understand why this makes sense, one has to consider how the learning rule of the actor works.

THE ACTOR LEARNING RULE

The basic idea of the actor learning rule is that if an action produced in response to a sensory input has the *expected* consequences, then that response tendency remains unchanged. On the other hand, if its consequences are better (worse) than expected, the response tendency is strengthened (weakened) (cf. the Rescorla-Wagner model; Rescorla and Wagner, 1972). When the TD error equals zero, the consequences are as they were predicted by the critic, and no learning should occur. When the TD error is positive (negative), consequences are better (worse) than predicted so that the response tendency should be strengthened (weakened).

Suppose the actor makes decisions by comparing the activities of a collection of linear connectionist units, where there is one unit for each possible action. The action selected is the one whose unit has the most vigorous activity. Let a_t denote the activity at time t of the unit corresponding to action a and suppose that

$$a_t = \sum_{i=1}^{m} w_t^i x_t^i,$$

where w_t^i and x_t^i, for $i = 1, \ldots, m$, are respectively the weights and input activations at time t. The following learning rule for this unit is applied *only if action* a *was selected for execution at time* $t - 1$:

$$w_t^i = w_{t-1}^i + \zeta \hat{r}_t x_{t-1}^i. \tag{11}$$

where $\zeta > 0$ is the learning rate. The weights of the units for the unselected actions remain unchanged.

Owing to the definition of the effective reinforcement $\hat{r}_t$ [equation (10)], this rule is almost identical to the adaptive critic learning rule [equation (9)]. It possibly has a different learning rate and, more important, it applies only to the weights for the action a that was selected at $t - 1$.

NEURAL IMPLEMENTATION

Figure 11.5 illustrates how the actor-critic architecture could be implemented by a neural network. Both the actor units and the predictor unit use the same learning rule and the same modulatory signal, $\hat{r}_t$, as a factor in updating their synaptic weights. All the modifiable synapses require local memory to implement the necessary eligibility mechanism. The only difference between these units is that the actor units compete with one another so that only one unit wins the competition, and the learning rule applies only to the winning unit. This could be implemented by suitable lateral inhibition and a slightly different eligibility mechanism for the actor units. Whereas the eligibility mechanism of the predictor unit remembers only past presynaptic activity, the eligibility mechanism of an actor unit would have to remember past *conjunctions of pre- and postsynaptic activity* in such a way that if it were not selected, none of its synapses could become eligible for modification. Consequently, the modifiable synapses of the prediction unit must use a two-factor learning rule, whereas those of the actor units must use a

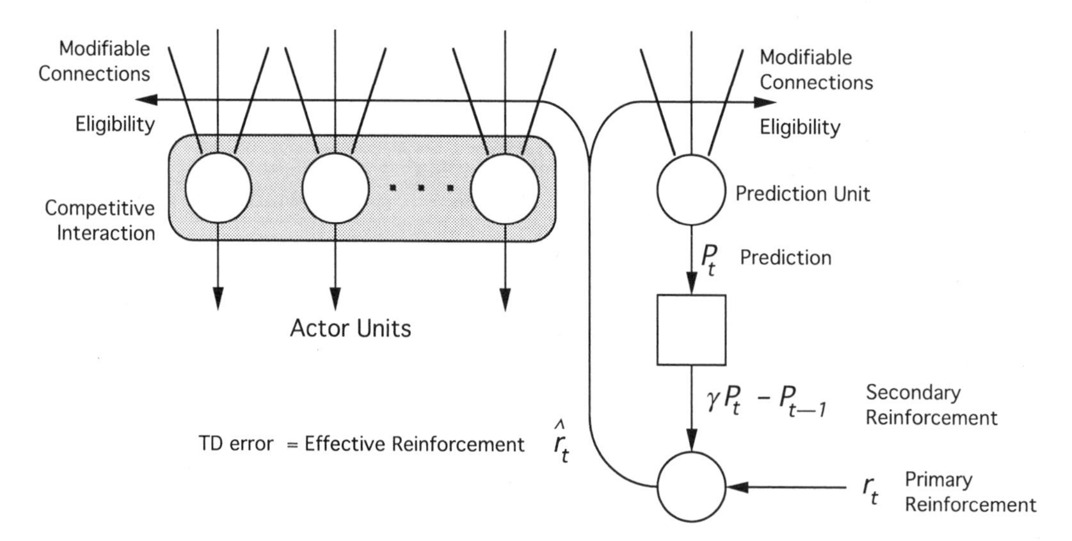

Figure 11.5 Network implementation of the actor-critic architecture. Both the actor units and the predictor unit use the same learning rule and the same modulatory signal, $\hat{r}_t$, as a factor in updating their synaptic weights. All the modifiable synapses require local memory to implement the necessary eligibility mechanism. The actor units compete with one another to determine the action, and the learning rule is applied only to the winning actor unit.

three-factor learning rule. Finally, some mechanism is required to compute the secondary reinforcement signal $\gamma P_t - P_{t-1}$. This could be accomplished by a kind of neural differentiator that is shown by the box in figure 11.5. Houk et al. elaborate this basic network in relation to the circuitry and intracellular chemistry of the basal ganglia and dopamine neurons (see chapter 13).

THE CASE OF TERMINAL PRIMARY REINFORCEMENT

Most relevant to animal learning experiments are cases in which a sequence of actions has to be accomplished before a primary reinforcing event occurs (e.g., a monkey reaching, picking up a food morsel, and transferring it to its mouth; see chapters 2 and 12. In this case, during each trial $r_t = 0$ for all t except when the food is actually tasted, at which time, say T, it is some positive number, say 1; so that $r_T = 1$. Suppose that the discount factor γ is very nearly 1 so that we can effectively ignore it, and further suppose that the adaptive critic starts out by producing $P_t = 0$ throughout the first trial.

Then until the first occurrence of the terminal primary reinforcing event at time T of some trial (due to the accidental execution of the right action sequence), all the TD errors, and hence all the effective reinforcements, are zero. At time T of this first successful trial, the TD error and the effective reinforcement are 1. This positive effective reinforcement causes the actor to increase its tendency to produce the immediately preceding response, and the positive TD error causes the adaptive critic to adjust its weights so that when the stimulus at time $T - 1$ of this successful trial recurs in a later trial, the critic will predict that the immediately following stimulus will have positive value. That is, P_{T-1} will be greater than 0 at time $T - 1$ of a trial in which the stimulus at $T - 1$ is the same as (or similar to) the stimulus at $T - 1$ of the first successful trial.

Now things become more complicated but also more closely related to the observed responses of dopamine neurons. Consider the next successful trial. In addition to the events of the first successful trial happening again, so that the actor's response tendency and the critic's prediction become stronger, the fact that P_{T-1} is positive has two additional consequences:

1. The TD error and the effective reinforcement at time $T - 1$ will now be positive. This quantity is

$$r_{T-1} + \gamma P_{T-1} - P_{T-2},$$

and since both r_{T-1} and P_{T-2} are zero,[2] this equals $\gamma P_{T-1} > 0$. Just as the positive TD error at time T of the first successful trial caused the critic to make a positive prediction at time $T - 1$ of later trials, this positive TD error at $T - 2$ will cause the critic to make a positive prediction at time $T - 2$ of later trials. Also, as effective reinforcement, this positive quantity causes the actor to increase its tendency to produce the response it made at $T - 2$ of this successful trial.

2. The TD error and the effective reinforcement at time T will *decrease*. This quantity is

$$r_T + \gamma P_T - P_{T-1}.$$

r_T is still 1 since this is a successful trial; P_T is still zero because it is predicting that zero primary reinforcement occurs *after* the trial[3]; and P_{T-1}, which is positive, is being subtracted. Thus, the TD error and effective reinforcement at time T will be smaller than in earlier trials.

With continued successful trials, which become increasingly likely owing to the actor's changing response rule, the TD errors and effective reinforcements propagate backward in time: the activity transfers from later to earlier times within trials. Learning stops when these quantities all become zero, which happens only when the adaptive critic correctly predicts the values of all stimuli, i.e., when all expectations are met by actual events (which requires certain assumptions about the regularity of the environment and the richness of the sensory input), and the actor always produces the correct actions.

CONCLUSION

The actor-critic architecture implements one approach to learning when actions have delayed consequences. It has a well-developed theoretical basis, works well in practice, and makes strong contact with animal learning through the TD model of classical conditioning. The adaptive critic computes an effective reinforcement signal such that the action-selection subsystem achieves long-term goals while learning only on the basis of immediate effective reinforcement. The TD error used by the adaptive critic's learning mechanism is the same as the effective reinforcement supplied to the action-selection subsystem. When primary reinforcement occurs only after a sequence of correct actions, the adaptive critic's activity parallels that observed in dopamine neurons during similar animal learning experiments. This suggests the hypothesis that the activity of dopamine neurons plays the dual roles of TD error and effective reinforcment in a neural implementation of the actor-critic architecture. Houk et al. explore this hypothesis in more detail in chapter 13.

NOTES

1. Klopf's (1972, 1982) theory of *heterostasis,* in contrast to *homeostasis,* emphasizes the significance of the difference between seeking to match and seeking to maximize.

2. Actually, P_{T-2} might not be zero because it might have increased in intervening unsuccessful trials in which the animal made a mistake only on the last move. But it will be small enough so that the TD error at $T - 1$ will still be positive.

3. Actually, P_T might be non-0 because it is really predicting the imminence-weighted sum of future primary reinforcement, which includes primary reinforcement obtained in later successful trials. However, it is only with considerable experience that P_T takes on significant positive value due to the presumably long duration of the intertrial interval.

REFERENCES

Barto, A. G. (1991) Some learning tasks from a control perspective. In L. Nadel and D. L. Stein (eds.), *1990 Lectures in Complex Systems*. Redwood City, Calif.: Addison-Wesley, pp. 195–223.

Barto, A. G. (1992) Reinforcement learning and adaptive critic methods. In D. A. White and D. A. Sofge (eds.), *Handbook of Intelligent Control: Neural, Fuzzy, and Adaptive Approaches*. New York: Van Nostrand Reinhold, pp. 469–491.

Barto, A. G., Sutton, R. S., and Anderson, C. W. (1983) Neuronlike elements that can solve difficult learning control problems. *IEEE Trans. Syst. Man Cyber.* 13:835–846; reprinted in J. A. Anderson and E. Rosenfeld, (eds.) (1988) *Neurocomputing: Foundations of Research*. Cambridge, Mass.: MIT Press.

Barto, A. G., Sutton, R. S., and Watkins, C. J. C. H. (1990) Learning and sequential decision making. In M. Gabriel and J. Moore (eds.), *Learning and Computational Neuroscience: Foundations of Adaptive Networks*, Cambridge, Mass.: MIT Press, pp. 539–602.

Barto, A. G., Bradtke, S. J., and Singh, S. P. (1994) Learning to act using real-time dynamic programming, *Artifi. Intell. J.* in press. [Also available as *Computer Science Technical Report 93-02*, University of Massachusetts, Amherst.]

Houk, J. C. (1992) *Learning in Modular Networks. NPB Technical Report 7*, Northwestern University Medical School, Department of Physiology, Ward Building 5-342, 303 E. Chicago Ave., Chicago, IL 60611-3008.

Klopf, A. H. (1972) *Brain Function and Adaptive Systems—A Heterostatic Theory*, Air Force Cambridge Research Laboratories Technical Report AFCRL-72-0164, Bedford, Mass.

Klopf, A. H. (1982) *The Hedonistic Neuron: A Theory of Memory, Learning, and Intelligence*. Washington, D.C.: Hemisphere.

Rescorla, R. A., and Wagner, A. R. (1972) A theory of Pavlovian conditioning: Variations in the effectiveness of reinforcement and nonreinforcement. In A. H. Black and W. F. Prokasy, (eds.), *Classical Conditioning II*. New York: Appleton-Century-Crofts, pp. 64–99.

Rumelhart, D. E., Hinton, G. E., and Williams, R. J. (1986) Learning internal representations by error propagation. In D. E. Rumelhart and J. L. McClelland (eds.), *Parallel Distributed Processing: Explorations in the Microstructure of Cognition*, Vol. 1: *Foundations*. Cambridge, Mass.: Bradford Books/MIT Press.

Samuel, A. L. (1963) Some studies in machine learning using the game of checkers, in E. A. Feigenbaum and J. Feldman (eds.) (1963) *Computers and Thought*. New York: McGraw-Hill.

Sutton, R. S. (1984) *Temporal Credit Assignment in Reinforcement Learning*, Ph.D. Dissertation, University of Massachusetts, Amherst.

Sutton, R. S. (1988) Learning to predict by the method of temporal differences. *Machine Learning*, 3:9–44.

Sutton, R. S., and Barto, A. G. (1981) Toward a modern theory of adaptive networks: Expectation and prediction. *Psychol. Rev.* 88:35–170.

Sutton, R. S., and Barto, A. G. (1987) A temporal-difference model of classical conditioning. In *Proceedings of the Ninth Annual Cnoference of the Cognitive Science Society.* Hillsdale, N.J.: Lawrence Erlbaum Associates.

Sutton, R. S., and Barto, A. G. (1990) Time-derivative models of Pavlovian reinforcement. In M. Gabriel and J. Moore (eds.), *Learning and Computational Neuroscience: Foundations of Adaptive Networks.* Cambridge, Mass.: MIT Press, pp. 497–537.

Tesauro, G. J. (1992) Practical issues in temporal difference learning. *Machine Learning* 8:257–277.

Thorndike, E. L. (1911) *Animal Intelligence.* Darien, Conn.: Hafner.

Watkins, C. J. C. H. (1989) *Learning from Delayed Rewards,* Ph.D. Dissertation, Cambridge University, Cambridge, England.

Werbos, P. J. (1977) Advanced forecasting methods for global crisis wraning and models of intelligence. *Gen. Syst. Yearbook* 22:25–38.

Werbos, P. J. (1987) Building and understanding adaptive systems: A statistical/numerical approach to factory automation and brain research. *IEEE Trans. Syst. Man Cyber.* 17:7–20.

Wickens, J. (1990) Striatal dopamine in motor activation and reward-mediated learning: Steps toward a unifying model. *J. Neural Transm.* 80:9–31.

12 Reward-related Signals Carried by Dopamine Neurons

Wolfram Schultz, Ranulfo Romo,
Tomas Ljungberg, Jacques Mirenowicz,
Jeffrey R. Hollerman, and Anthony Dickinson

Impaired neurotransmission of dopamine neurons projecting from the ventroanterior midbrain to the striatum and frontal cortex results in a multitude of severe behavioral impairments comprising motor, cognitive, and motivational processes. However, several lines of anatomical, physiological, and behavioral evidence suggest that dopamine neurons are not able to encode all details of the behaviors in which they are apparently involved. In order to assess more closely their function, we investigated the activity of dopamine neurons in monkeys during the performance of various behaviors.

SELECTIVE RESPONSES TO A LIMITED RANGE OF STIMULI

Primary Rewards

The optimal stimulus for activating dopamine neurons consists of a phasically occurring unpredicted food and liquid reward. Responses occur when animals touch a small morsel of hidden food during exploratory movements or when receiving a drop of liquid at the mouth outside of any behavioral task or while learning a task (figure 12.1 top) (Romo and Schultz, 1990; Ljungberg et al., 1992; Mirenowicz et al. unpublished observation). The responses do not occur when similarly shaped nonfood objects are touched or when a fluid valve is operated without actually delivering liquid.

Reward responses are abolished when the time of reward delivery is predicted by conditioned phasic stimuli, whereas the general reward prediction provided by known behavioral contexts alone does not abolish them. Thus, responses to reward encountered during spontaneous movements disappear when the same movement is performed in response to a conditioned stimulus. In a learning situation, reward responses disappear when a stimulus has become a valid reward predictor. Learning of a more complex delayed response task via several subtasks is accordingly accompanied by reward responses during the learning phase of each subtask which disappear when each learning curve reaches an asymptote.

The neurophysiological reasons for the disappearance of the response to predicted reward are unclear but might consist in an active inhibition. Omission of expected reward following an error of the animal or experimenter interaction phasically depresses the activity of dopamine neurons exactly at the time where reward would have occurred (Romo and Schultz, 1990; Ljungberg et al., 1991), even with a reward normally delivered 0.5 second after the last external stimulus (Schultz et al., 1993).

Conditioned Stimuli

Dopamine neurons are activated by conditioned visual or auditory stimuli that have become valid reward predictors (figure 12.1, middle) (Schultz, 1986; Schultz and Romo, 1990; Ljungberg et al., 1992). These stimuli are generally less effective than primary rewards, both in terms of response magnitude in individual neurons and in terms of fractions of neurons activated. Instruction and discriminative cues, which are primarily occasion-setting stimuli rather than predictors of reward, are less effective for activating dopamine neurons. Responses of dopamine neurons to a conditioned light have also been observed in haloperidol-treated rats (Miller et al., 1981). Dopamine neurons show minor or no activations prior to and during the execution of movements and are not activated during the delays of typical frontal cognitive tasks (De Long et al., 1983; Schultz et al. 1983, 1993; Romo and Schultz, 1990).

The large majority of responses specifically occur to appetitive stimuli, and only a few neurons are also activated by a conditioned aversive light or sound stimulus in an air puff or a saline avoidance task, and in response to other arousing stimuli (Mirenowicz and Schultz, unpublished observation).

As dopamine neurons respond to a conditioned, reward-predicting stimulus, they stop responding to the reward which is now predicted by the stimulus (see figure 12.1, middle). This response transfer occurs in single dopamine neurons tested both with unpredicted rewards and with reward-predicting conditioned visual or auditory stimuli in stable behavioral situations (Romo and Schultz, 1990). Likewise, the learning of an instrumental lever-pressing task is accompanied by a transfer of the response from the unpredicted reward to the reward-predicting conditioned stimulus (Ljungberg et al., 1992). The transfer is not always complete; the conditioned stimulus is usually somewhat less effective in activating dopamine neurons than the unpredicted primary reward, whereas the primary reward response is virtually always abolished in these situations.

Responses to conditioned stimuli also depend on whether that stimulus itself is unpredicted and are reduced when it is signaled by a preceding cue, to which dopamine neurons then may respond (figure 12.1, bottom) (Schultz et al., 1993). Extensive overtraining also atten-

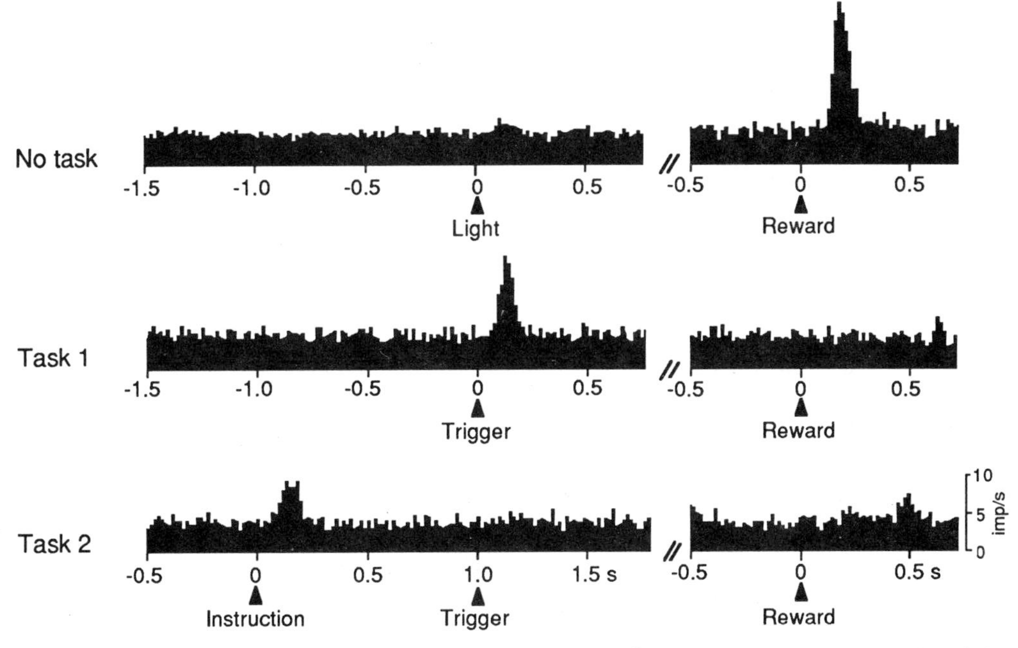

Figure 12.1 Responses of dopamine neurons to unpredicted primary reward and the transfer of this response to progressively earlier reward-predicting stimuli. All displays show population histograms obtained by averaging the normalized perievent time histograms of all dopamine neurons recorded in the indicated behavioral situations, independent of the presence of a response. (*Top*) In the absence of any behavioral task, there was no population response in 44 neurons tested with a small light (data from Ljungberg et al., 1992), but an average response in 35 neurons to a drop of liquid delivered at a spout in front of the animal's mouth (unpublished data). (*Middle*) Response to a reward-predicting trigger stimulus in a spatial choice reaching task, but absence of response to reward delivered during established task performance (23 neurons; data from Schultz et al,. 1993).(*Bottom*) Response to an instruction cue preceding by a fixed interval of 1 second the reward-predicting trigger stimulus in an instructed spatial reaching task (19 neurons; data from Schultz et al., 1993). The time base is interrupted because of varying intervals between the conditioned stimuli and the reward.

uates responses to conditioned stimuli (Ljungberg et al., 1992), probably because the animal comes to anticipate the stimulus on the basis of the events in the preceding trial in the highly stereotyped and automated task. Moreover, the conditioned stimulus may lose some of its reward-predicting incentive properties when the performance is established as an overtrained habit, where each response is no longer explicitly executed in order to receive an individual reward (Dickinson, 1980).

Dopamine neurons discriminate between reward-predicting and nonpredictive stimuli as long as the stimuli are sufficiently dissimilar (Ljungberg et al., 1992), in particular when different modalities are used (Mirenowicz and Schultz, unpublished observation). By contrast, dopamine neurons respond to both rewarded and unrewarded stimuli when they are physically very similar. A small box that opens rapidly in front of the animal does not by itself activate dopamine neurons. However, responses occur to every opening if the box on some trials contains a visible morsel of food (Ljungberg et al., 1992). Also, responses occur indiscriminately to two identical boxes placed side by side despite the fact that only one of them contains food (Schultz and Romo, 1990). Animals perform an indiscriminate ocular orienting response to each opening, but approach only the baited box with their hand. Similarly, dopamine neurons continue to respond to conditioned stimuli even when these stimuli are used in a different task and thus have lost their original reward prediction property (Schultz et al., 1993).

Novel Stimuli

Unexpected novel stimuli are effective in activating dopamine neurons as long as they elicit behavioral orienting reactions (e.g., ocular saccades). Neuronal responses subside together with orienting reactions after several stimulus repetitions (Ljungberg et al., 1992). Similar responses of dopamine neurons to unconditioned high-intensity or novel stimuli in parallel with whole-body orienting reactions have been described in cats (Steinfels et al., 1983).

Common Response Characteristics

The responses of dopamine neurons to these different stimuli are remarkably similar. They are phasic and occur with latencies of 50 to 120 ms, last less than 200 ms, and are composed of occasionally a single impulse or, more often, a short burst of a few impulses. They are polysensory and independent of the side of visual, auditory, or somatosensory stimulus presentation relative to the body axis. Effective stimuli in most situations activate the majority of the population of dopamine neurons, and different dopamine neurons in groups A8, A9, and A10 respond to the same stimuli in a similar manner, such that clearly separate response types cannot be associated with different populations of neurons. This homogeneity of neuronal responses suggests that dopamine neurons respond in parallel as a population rather than displaying differential response profiles.

NATURE OF STIMULI EFFECTIVE FOR ACTIVATING DOPAMINE NEURONS

Predominance of Reward

The major stimuli for dopamine neurons are reward-related events, both primary rewards and conditioned, reward-predicting stimuli. Exceptions to this are novel or arousing stimuli and conditioned aversive stimuli. Novel stimuli are potential rewards or reward predictors and might be included in the class of reward-related events. Aversive responses are found in relatively few dopamine neurons (<20%) which apparently belong to a less specific subpopulation responding to a larger spectrum of behaviorally important stimuli.

Salient Stimuli

The events to which dopamine neurons respond belong to the most important, salient external stimuli to which a subject needs to react in order to not miss an important object. Salient stimuli are unconditioned rewards and aversive stimuli, conditioned stimuli predicting rewards or punishment, and high-intensity, surprising, novel stimuli. These stimuli alert the subject, which interrupts its ongoing behavior, orients to the stimulus, and processes it with high priority (Schultz, 1992). However, most dopamine neurons respond best to only a subset of salient stimuli, namely primary rewards and conditioned reward-predicting stimuli. This stimulus repertoire suggests a more circumscript coding of reward-related stimulus properties.

Characteristics of Reward-related Stimuli

In summary, the most potent stimuli for activating dopamine neurons are primary rewards and conditioned stimuli that signal such rewards. The maintenance of the dopamine response depends, however, upon ensuring that the occurrence of the reward or conditioned stimulus is not itself predicted by another phasic stimulus. With increasing experience of a signaling relation between a predictive stimulus and a primary reward, the dopamine response to reward is attenuated as that to the conditioned stimulus develops. In order for this conditioned attenuation to occur, it appears to be necessary that the stimulus itself provide information about the time of occurrence of reward, for in its absence the dopamine response to the reward is maintained across repetitive presentations in a constant, tonic environmental context. Furthermore, establishing a predictive relation between a stimulus and reward does not produce a general depression of the dopamine response to rewards for, if after such training, the reward is presented unexpectedly in the absence of the stimulus, a full dopamine response is elicited. Finally, it should be noted that the dopamine responses do not encode the particular physical properties of the eliciting event; both primary rewards and conditioned stimuli with differing sensory characteristics elicit a common dopamine response.

This characterization of the responses of dopamine neurons to primary rewards suggests that their activity may be related to the reinforcing function of rewards. In general terms, rewards appear to play two roles in the process of conditioning. The first is their function in bringing about the learning of a new behavioral response, whereas the second reflects the capacity of a reward to maintain an established behavioral response. In the absence of an effective reward, an established conditioned response will undergo extinction. Given this distinction, the response characteristics of dopamine neurons to rewards suggest a relationship to *learning* of new behavior rather than to *maintaining* an established behavior. Contemporary learning theories in general assume that only unexpected or surprising rewards bring about the acquisition of a new conditioned response (Dickinson, 1980). Thus, the critical feature of dopamine neurons that suggests a role in learning is their sensitivity to unexpected rewards. Dopamine neurons do not respond to fully predicted rewards, just as such rewards, although capable of maintaining already established conditioned behavior, are not effective in bringing about the acquisition of a new behavioral response. The finding that stimuli predicting primary rewards also activate dopamine neurons is compatible with this parallel for it is well established that such stimuli can act themselves as conditioned reinforcers in the acquisition of a new behavioral response.

In contrast to the responses of dopamine neurons to unexpected rewards, neurons in the striatum and in structures projecting to the striatum, such as the amygdala and orbitofrontal and anterior cingulate cortex, respond to primary rewards in well-established behavioral tasks and are not affected by the uncertainty of reward (Niki and Watanabe, 1979; Thorpe et al, 1983; Nishijo et al., 1988; Apicella et al., 1991). This suggests that different aspects of the reward signal are distributed over different neuronal systems in the brain and that the function of reward signals accordingly may vary among these systems. The neuronal response to reward in striatal, amygdalar, and cortical neurons during well-established behaviors suggests a function in maintaining rather than bringing about learning. Other neurons in the striatum are activated for several seconds during the expectation of reward following a well-established conditioned stimulus (Apicella et al., 1992), a further argument for a role of the striatum in the maintenance of learned behaviors.

Error Signals

The hypothesis that dopamine neurons may mediate the role of unexpected rewards to bring about learning can be refined by reference to the concept of an *error* signal. Contemporary learning theories characterize learning as acquisition of associative strength by a conditioned stimulus (Dickinson, 1980). The increment in associative strength in each learning episode in which the conditioned stimulus is paired with a reward is determined by the discrepancy Δ which equals $(\lambda - V)$ where V is the current associative strength of the conditioned stimulus on that episode and λ is the maximum associative strength that could be sustained by the reward (Rescorla and Wagner, 1972; Mackintosh, 1975; Pearce and Hall, 1980). Thus, this discrepancy reflects the extent to which the conditioned stimulus has already been established as a predictor of reward. When Δ is zero, $V=\lambda$ which reflects the fact that the associative strength of the conditioned stimulus is sufficient to predict the occurrence of the reward, and little or no further learning (i.e., no increments in V) will occur. By contrast, when $\lambda > V$, the associative strength of the stimulus does not fully predict the reward and the Δ term is positive and leads to further increments in associative strength. In this sense, the Δ term represents the extent to which the animal has failed as yet to learn the full predictive relation between the conditioned stimulus and the reward, and for this reason it is referred to as an error signal. Learning algorithms that serve to minimize the Δ term can be considered as error correcting. Recent work has suggested that this learning rule is formally equivalent to the delta rule (Widrow and Hoff, 1960) of artificial neuronal networks (Sutton and Barto, 1981), thus allowing single neuronal elements, at least theoretically, to implement this learning rule. That temporal aspects in the unpredictedness of events are important for learning is suggested by the fact that the temporal variation in reinforcer occurrence constitutes an element of unpredictedness that allows learning (Dickinson et al., 1976).

Within this analysis of learning, the activity of dopamine neurons would appear to be a good candidate for representing the error signal or Δ term and also possibly applying it as a teaching signal controlling the changes in neuronal structures representing the associative strength of a conditioned stimulus or response. The error signal should be large on a learning episode when the reward is unexpected and a corresponding dopamine response is observed. By contrast, the error signal should be small when a predicted reward is presented and, in accord with the hypothesis, little or no activity is recorded in dopamine neurons in response to this event. Moreover, the depressant response of dopamine neurons to the unexpected omission of a reward following a well-established reward-predicting stimulus suggests a mechanism by which the Δ term may be computed at a neuronal level. Thus, it may well be that a predictive stimulus for a reward serves to inhibit the dopamine response that would otherwise be elicited by the reward.

In stressing another parallel between dopamine activity and the error signals governing learning, it should be noted that any error signal represented by the dopamine response can only relate to the salient, affective, or hedonic properties of the reward and not to any particulars of the reward objects, for, as we have noted, the same neuronal response is elicited by a variety of different rewards, as well as by conditioned, reward-predicting stimuli of various sensory modalities. This assumption is, however, in accord with accounts of behavioral deficits observed following interference with dopamine neurotransmission (Wise et al., 1978; Robbins and Everitt, 1992; Robinson and Berridge, 1993).

IMPACT OF THE DOPAMINE MESSAGE ON POSTSYNAPTIC PROCESSING

Further clues to the function of phasic dopamine responses to reward-related stimuli can be obtained by comparing them with task-related activities in the target structures, and by taking into account the circuitry and neuronal mechanisms influenced by dopamine neurons. There could hardly be a greater difference of behavior-related activity than that seen between dopamine and striatal or cortical neurons. In contrast to the rather homogeneous responses of dopamine neurons to a narrow range of stimuli, striatal neurons show highly differentiated activity related to various stimuli; to the preparation, initiation, and execution of limb and eye movements; to the expectation of known behavioral events; and to the expectation and reception of reward.

Large Circuits

Dopamine neurons innervate basically the dorsal striatum, the ventral striatum (the nucleus accumbens of rodents), and the frontal cortex. These structures globally subserve the organization of behavioral output. The striatum is a part of loops involving the frontal cortex and in addition receives input from postcentral sensory and association cortex and from limbic cortical and subcortical structures (figure 12.2). Besides the loops with the cortex, basal ganglia output is directed to limbic structures and to the superior colliculus, which are involved in affective behavioral components, memory, and simple motor output. Thus, dopamine neurons are in a position to influence the control of behavioral output by their widespread projections to structures involved in highly differentiated information processing. The limited spectrum of their largely homogeneous activity suggests that dopamine neurons contribute a particular component of neuronal processing that is commonly important for their target structures.

Local Circuitry in Target Structures

The basic arrangement of synaptic influences of dopamine neurons on striatal and frontal cortex neurons of rats and monkeys consists of three elements: the dendritic spine of the postsynaptic neuron, the dopamine presynaptic varicosity contacting the stem of the dendritic spine, and

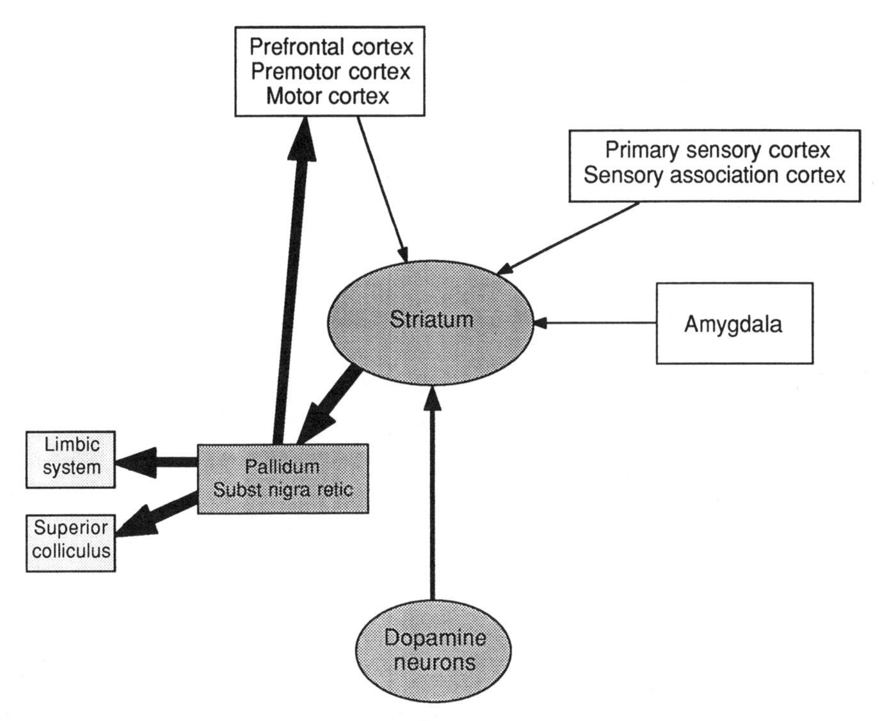

Figure 12.2 Highly simplified schema showing how dopamine neurons act on the striatum, which is linked through major circuits to other brain centers controlling behavioral output. The striatum denotes both the dorsal caudate and putamen and the ventral striatum including the nucleus accumbens. For reasons of simplicity, the dopamine innervation to the frontal lobe (*upper left: prefrontal cortex, premotor cortex, motor cortex)* has been omitted.

an excitatory cortical terminal at the tip of the dendritic spine (figure 12.3) (Freund et al., 1984; Goldman-Rakic et al., 1989; Smith et al., 1993). Every medium-sized striatal neuron receives about 500 to 5000 dopaminergic synapses at its dendritic spines and about 5000 to 10,000 cortical synapses (Doucet et al. 1986; for more details, see chapters 3 and 4). It is unclear whether all cortical inputs to a single striatal neuron could come from the same cortical column and rather unlikely that they originate from the same cortical neuron. It would be interesting to know which cortical areas may possibly converge onto striatal neurons. Although current schemes of basal ganglia connections suggest a maintained segregation of cortical inputs into the striatum (Alexander et al., 1986), it is conceivable that some functionally related, homotopical somatosensory and motor cortical areas project into common regions of the striatum (Flaherty and Graybiel, 1993). Such specific patterns of convergence would allow dopamine neurons to select the synaptic effects of cortical inputs at individual striatal neurons.

In a reduced model of such an anatomical arrangement, let A and B be two inputs converging on a single striatal neuron I, each of which contacts a dendritic spine of that neuron (see figure 12.3). The stems of the same spines are indiscriminately contacted by dopaminergic input *X*. When a reward-related signal is encountered, both inputs *X* and *A* are activated. Dopamine neuron *X* transmits the message that a reward-related event has occurred, without giving further details. At the same time, cortical input *A* carries detailed information about the same reward-related event, including its modality, body side, color, texture, position, surrounding, and whether it is food, fluid, or a conditioned sound or light, and may also code the details of an approach movement (different aspects of the same event being encoded by specific activity in different inputs *A*). Input *B*, encoding a different event, remains inactive. Through the simultaneity or near simultaneity of activity in *A*

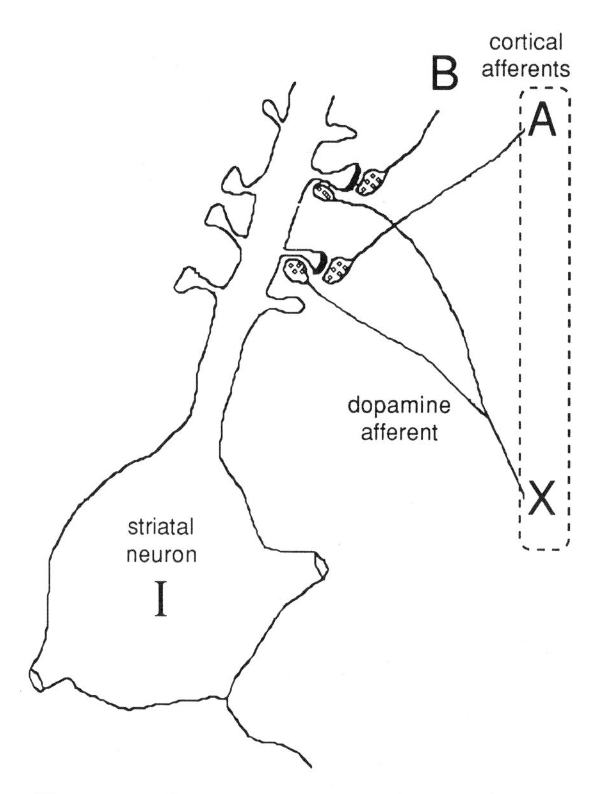

Figure 12.3 Synaptic arrangement of inputs from cortex and dopamine neurons to medium-sized spiny striatal neurons. The dendritic spine is contacted at its tip by a cortical axon and at its stem by a dopamine axon. In the basic design of hypothetical dopamine-dependent heterosynaptic plasticity induced by a reward-related event, cortical neurons A and B converge at the tip of dendritic spines on a single striatal neuron I. These connections might be modifiable by increased use, e.g., long-term facilitation. However, the modification occurs only when dopamine input X coming indiscriminately to the stems of the same dendritic spines is active at the same time. In the present example, cortical input A, but not B, is active at the same time as dopamine neuron X, as during the occurrence of a reward-related event. This leads to modification of the *A*→*I* transmission, but leaves the B→I transmission unaltered. (Anatomical drawing modified from Smith and Bolam, 1990.)

and X, the activity of neuron X may influence synaptic transmission between A and I but leave the transmission at the inactive $B \rightarrow I$ synapse unchanged. Thus, the message about a reward-related event coming from dopamine neuron X specifically influences the $A \rightarrow I$ neurotransmission. The key function of dopamine neuron X would be to signal the event (reward) that is particularly important for behavior and influence as a kind of gate the highly structured activity circulating in corticostriatal and limbic-striatal connections. The following discussion speculates on how this synaptic constellation could be used for influencing behavior.

Focusing

The immediate effect of dopamine release is probably a reduction in postsynaptic excitability. This results in a general reduction of corticostriatal processing, thus focusing striatal activity onto the strongest inputs, whereas weaker activity is lost (Toan and Schultz, 1985; Filion et al., 1988). In the model of figure 12.3, the focusing effect of dopamine input X would reduce all input activity of neurons A and B. This would only let the information from input A, as the strongest of the convergent inputs, pass beyond the impulse-generating mechanism at the cell body, whereas weaker activity would be lost. Figure 12.4 illustrates the focusing effect in a more complex network where several presynaptic (e.g., cortical) neurons converge in an ordered fashion onto postsynaptic (e.g., striatal) neurons, corresponding to convergence from somatosensory and motor cortex inputs to striatal regions (Flaherty and Graybiel, 1993). A contrast-enhancing nonlinear element would be the threshold phenomenon in the opening of voltage-dependent sodium channels at the excitable membrane. Thus, the phasic release of dopamine induced by a reward-related event would give higher impact to all information pertaining to the currently present reward event. The immediate result could be a reorientation of behavior towards salient appetitive stimuli.

Learning

Although the evidence from behavioral deficits arising from impaired dopamine neurotransmission does not unequivocally support a specific role of dopamine in learning and memory,it is tempting to suggest a hypothetical model according to which the reward-related responses of dopamine neurons could influence neuronal plasticity in the striatum and frontal cortex.

In our model of figure 12.3, let us make the synaptic weights of cortical inputs A and B to striatal neuron I hebbian-modifiable, and assume that this plasticity depends on dopamine input X. Neurons A and X would be simultaneously active when a reward-related stimulus occurs, whereas neuron B does not modify its activity. This would

No dopamine activity

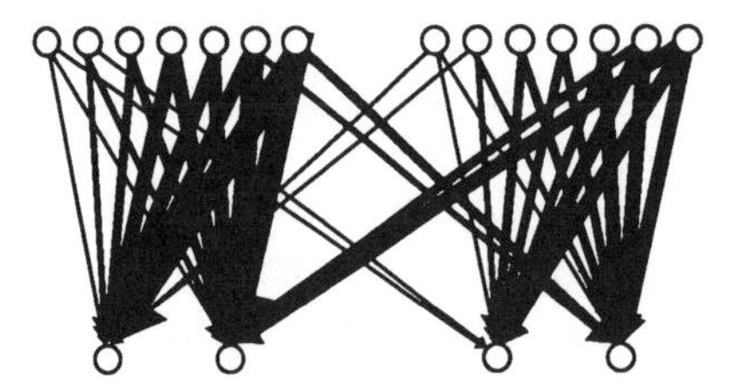

Dopamine-induced focussing

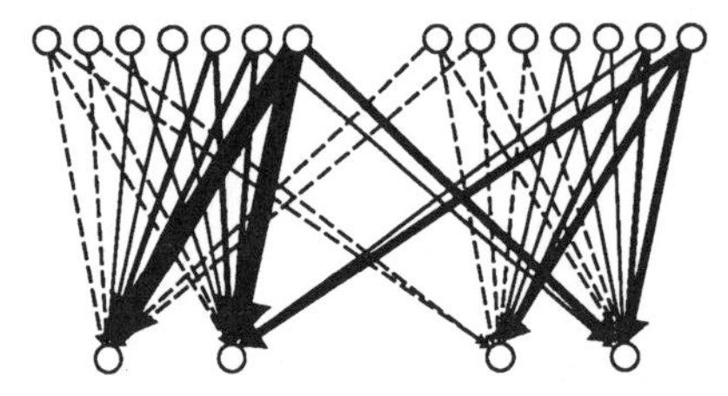

Dopamine-induced long term facilitation

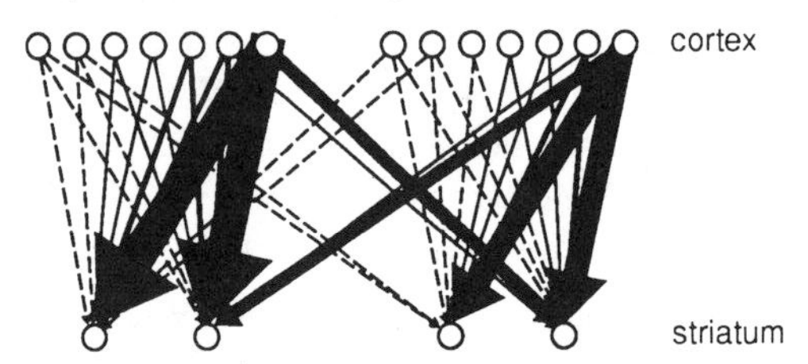

Figure 12.4 A possible influence of a phasic dopamine signal on the selection of striatal information processing based on convergent inputs. Suppose that inputs from different cortical origins converge in an ordered manner on single striatal neurons. The different strengths of these inputs reflect the differential activation of cortical neurons by the current behavioral situation. In the example, neurons could be activated by the physical stimuli emanating from the reward and its surroundings, by reaching movements of the arm toward the reward, and by event-related memory and expectation activity. (*Top*) In the absence of dopamine, cortical inputs would influence striatal neurons in a poorly contrasted manner. (*Middle*) Dopamine has an immediate focusing effect which nonlinearly enhances the strongest inputs occurring at the time of the dopamine signal relative to weaker inputs which are suppressed. (*Bottom*) In a hypothetical learning mechanism, dopamine facilitates long-term changes at hebbian-modifiable synapses. Arrow width represents the relative synaptic influences on postsynaptic impulse activity, consisting in a combination of presynaptic influence and synaptic strength.

increase the synaptic weight at the active synapse $A \rightarrow I$, but leave the weight at the inactive $B \rightarrow I$ synapse unchanged. If neuron A is subsequently activated again by a stimulus that shares some features with the stimulus having activated both neurons A and X before, but without necessarily also activating neuron X, the response in neuron I would be increased, whereas any input from neuron B leads to an unchanged postsynaptic response. Thus, neuron X need only be active in a subset of situations in which neuron A is active. In this model a reward-related event coming from dopamine neuron X would heterosynaptically facilitate plasticity of the $A \rightarrow I$ synapse. Or, the synaptic plasticity of $A \rightarrow I$ and $B \rightarrow I$ neurotransmission is conditional of X being conjointly active with A or B. In the more elaborate network of figure 12.4, the dopamine input would induce a long-term increase in transmission in those connections that were active when a reward-related event occurred. This selection of anatomically convergent inputs would in the future favor the processing of the inputs present at the striatal neurons when dopamine neuron X discharged, such as sensory responses from the rewarding event, short-term memory and expectation-related activity surrounding the event, and activity related to the particular movements leading to the event.

This model assumes that dopamine neurons may participate in learning by signaling an unpredicted reward. This signal might well code the most attracting aspects of the reward and be particularly important during the learning phase. However, dopamine neurons do not convey a message about the reinforcer once the reward contingencies are established and reward is predicted by phasic stimuli, although this reinforcer signal is important for the performance of learned behavior and for avoiding extinction. A neuronal network model using the dopamine signal for learning may take these characteristics into account and include a separate reinforcer signal for maintaining established behavior. Such a reinforcer signal during established task performance exists in striatal neurons (Apicella et al., 1991). Because of the difference against the reward response of dopamine neurons, the signal may enter the striatum from elsewhere, possibly from the amygdala. It would be interesting to see how the striatum and artificial networks would profit from different neuronal messages signaling the physically same reinforcing event during learning and performance, respectively. Taken together, this suggests that the dopamine signal could be an important substrate for reinforcement learning which appears to be appropriate for the basal ganglia, in contrast to postulated learning mechanisms in the cerebellum using error or deviation signals.

Recent biological findings appear to suggest a role of dopamine in neuronal plasticity. Dopamine neurotransmission may not operate in the same time range of milliseconds as cortically induced excitations. The association of dopamine receptors with G proteins, and particularly the involvement of a second messenger in the case of the D1 receptor, suggests onsets and durations of postsynaptic membrane effects in the range of seconds. Thus, the dopamine signal appears to influence postsynaptic neurons in a different manner than most other striatal and cortical neurotransmitters (for further details see chapter 4). In addition, inputs to the striatum and nucleus accumbens are subject to posttetanic potentiation and depression (Calabresi et al. 1992b; Pennartz et al. 1993), some of which depend on intact dopamine transmission (Calabresi et al., 1992a). The slower time course of dopamine membrane action may leave a trace of the reward event and influence all subsequent activity. The possibility of dopamine-dependent striatal plasticity and the central importance of unpredicted reward for learning suggest that the observed responses of dopamine neurons may contribute to changes in postsynaptic neuronal circuits underlying reward-directed learning. The reward-related input to dopamine neurons could be the decisive signal upon whose reception the striatal synapses concurrently activated by the events leading to the obtention of reward would be strengthened.

ACKNOWLEDGMENTS

The work was supported by the Swiss National Science Foundation, the Fyssen Foundation (Paris), the Foundation pour la Recherche Médicale (Paris), the United Parkinson Foundation (Chicago), the Roche Research Foundation (Basel), the National Institutes of Mental Health (Bethesda, Md.), and the British Council.

REFERENCES

Alexander, G. E., DeLong, M. R., and Strick, P. L. (1986) Parallel organization of functionally segregated circuits linking basal ganglia and cortex. *Annu. Rev. Neurosci.* 9:357–381.

Apicella, P., Ljungberg, T., Scarnati, E., and Schultz, W. (1991) Responses to reward in monkey dorsal and ventral striatum. *Exp. Brain Res.* 85:491–500.

Apicella, P., Scarnati, E., Ljungberg, T., and Schultz, W. (1992) Neuronal activity in monkey striatum related to the expectation of predictable environmental events. *J. Neurophysiol.* 68:945–960.

Calabresi, P., Maj, R., Mercuri, N. B., and Bernardi, G. (1992a) Coactivation of D1 and D2 dopamine receptors is required for long-term synaptic depression in the striatum. *Neurosci. Lett.* 142:95–99.

Calabresi, P., Pisani, A., Mercuri, N. B., and Bernardi, G. (1992b) Long-term potentiation in the striatum is unmasked by removing the voltage-dependent magnesium block of NMDA receptor channels. *Eur. J. Neurosci.* 4:929–935.

DeLong, M. R., Crutcher, M. D., and Georgopoulos, A. P.: Relations between movement and single cell discharge in the substantia nigra of the behaving monkey. *J. Neurosci.* 3:1599–1606.

Dickinson, A. (1980) *Contemporary Animal Learning Theory.* Cambridge: Cambridge University Press.

Dickinson, A., Hall, G., and Mackintosh, N. J. (1976) Surprise and the attenuation of blocking. *J. Exp. Psychol* [*Anim. Behav.*] 2:313–322.

Doucet, G., Descarries, L., and Garcia, S. (1986) Quantification of the dopamine innervation in adult rat neostriatum. *Neuroscience* 19:427–445.

Filion, M., Tremblay, L, and Bédard, P. J. (1988) Abnormal influences of passive limb movement on the activity of globus pallidus neurons in parkinsonian monkey. *Brain Res.* 444:165–176.

Flaherty, A. W., and Graybiel, A. (1993) Two input systems for body representations in the primate striatal matrix: Experimental evidence in the squirrel monkey. *J. Neurosci.* 13:1120–1137.

Freund, T. T., Powell, J. F., and Smith, A. D. (1984) Tyrosine hydroxylase–immunoreactive boutons in synaptic contact with identified stiatonigral neurons, with particular reference to dendritic spines. *Neuroscience* 13:1189–1215.

Goldman-Rakic, P. S., Leranth, C. Williams, M. S., Mons, N., and Geffard, M. (1989) Dopamine synaptic complex with pyramidal neurons in primate cerebral cortex. *Proc. Natl. Acad. Sci. U. S. A.* 86:9015–9019.

Ljungberg, T., Apicella, P., and Schultz, W. (1991) Responses of monkey midbrain dopamine neurons during delayed alternation performance. *Brain Res.* 586:337–341.

Ljungberg, T., Apicella, P., and Schultz, W. (1992) Responses of monkey dopamine neurons during learning of behavioral reactions. *J. Neurophysiol.* 67:145–163.

Mackintosh, N. J. (1975) A theory of attention: Variations in the associability of stimulus with reinforcement. *Psychol. Rev.* 82:276–298.

Miller, J. D., Sanghera, M. K., and German, D. C. (1981) Mesencephalic dopaminergic unit activity in the behaviorally conditioned rat. *Life Sci.* 29:1255–1263.

Niki, H., and Watanabe, M. (1979) Prefrontal and cingulate unit activity during timing behavior in the monkey. *Brain Res.* 171:213–224.

Nishijo, H., Ono, T., and Nishino, H. (1988) Single neuron responses in amygdala of alert monkey during complex sensory stimulation with affective significance. *J. Neurosci.* 8:3570–3583.

Pearce, J. M., and Hall, G. (1980) A model for Pavlovian conditioning: Variations in the effectiveness of conditioned but not of unconditioned stimuli. *Psychol. Rev.* 87:532–552.

Pennartz, C. M. A., Ameerun, F. F., Groenewegen, H. J., and Lopes da Silva, F. H. (1993) Synaptic plasticity in an in vitro slice preparation of the rat nucleus accumbens. *Eur. J. Neurosci.* 5:107–117.

Rescorla, R. A., and Wagner, A. R. (1972) A theory of Pavlovian conditioning: Variations in the effectiveness of reinforcement and nonreinforcement. In A. H. Black and W. F. Prokesy (eds), *Classical Conditioning* II: *Current Research and Theory.* New York: Appleton Century Crofts, pp. 64–99.

Robbins, T. W. and Everitt, B. J. (1992) Functions of dopamine in the dorsal and ventral striatum. *Semin. Neurosci.* 119–128.

Robinson, T. E. and Berridge, K. C. (1993) The neural basis for drug craving: an incentive-sensitization theory of addiction. *Brain Res. Rev.* 18(3):247–91

Romo, R. and Schultz, W. (1990) Dopamine neurons of the monkey midbrain: Contingencies of responses to active touch during self-initiated arm movements. *J. Neurophysiol.* 63:592–606.

Schultz, W. (1986) Responses of midbrain dopamine neurons to behavioral trigger stimuli in the monkey. *J. Neurophysiol.* 56:1439–1462.

Schultz, W. (1992) Activity of dopamine neurons in the behaving primate. *Semin. Neurosci.* 4:129–138.

Schultz, W., and Romo, R. (1990) Dopamine neurons of the monkey midbrain: Contingencies of responses to stimuli eliciting immediate behavioral reactions. *J. Neurophysiol.* 63:607–624.

Schultz, W., Ruffieux, A., and Aebischer, P. (1983) The activity of pars compacta neurons of the monkey substantia nigra in relation to motor activation. *Exp. Brain Res.* 51:377–387.

Schultz, W., Apicella, P., and Ljungberg, T. (1993) Responses of monkey dopamine neurons to reward and conditioned stimuli during successive steps of learning a delayed response task. *J. Neurosci.* 13:900–913.

Smith, A. D., and Bolam, J. P. (1990) The neural network of the basal ganglia as revealed by the study of synaptic connections of identified neurones. *Trends Neurosci.* 13:259–265.

Smith, Y., Bennett, B. D., Bolam, J. P., Parent, A., and Sadikot, A. F. (1993) Synaptic interactions between the dopaminergic afferents and the cortical or thalamic input at the single cell level in the striatum of monkey. *Soc. Neurosci. Abstr.* 19:977.

Steinfels, G. F., Heym, J., Strecker, R. E, and Jacobs, B. L. (1983) Behavioral correlates of dopaminergic unit activity in freely moving cats. *Brain Res.* 258:217–228.

Sutton, R. S., and Barto, A. G. (1981) Toward a modern theory of adaptive networks: Expectation and prediction. *Psychol. Rev.* 88:135–170.

Thorpe, S. J., Rolls, E. T., and Maddison, S. (1983) The orbitofrontal cortex: Neuronal activity in the behaving monkey. *Exp. Brain Res.* 49:93–115.

Toan, D. L., and Schultz, W. (1985) Responses of rat pallidum cells to cortex stimulation and effects of altered dopaminergic activity. *Neuroscience* 15:683–694.

Widrow, G., and Hoff, M. E. (1960) Adaptive switching circuits. In IRE Western Electronic Show and Convention, *Convention Record,* Part 4, pp. 96–104.

Wise, R. A., Spindler, J., de Wit, H., and Gerber, G. J. (1978) Neuroleptic-induced "anhedonia" in rats: Pimozide blocks reward quality of food. *Science* 201:262–264.

13 A Model of How the Basal Ganglia Generate and Use Neural Signals That Predict Reinforcement

James C. Houk, James L. Adams, and Andrew G. Barto

INTRODUCTION

Reinforcement resulting from reward or punishment is important in shaping human and animal behavior, and there is considerable evidence that the dopamine (DA) system of the basal ganglia is a crucial brain mechanism for mediating learning by reinforcement (Beninger, 1983; Wise and Rompré, 1989; Wickens, 1990) Reinforcement is also emerging as a powerful strategy for confronting difficult problems in engineering control (Barto et al., 1990; Werbos, 1992). Although learning by reinforcement has many advantageous properties, it has an important limitation, called the credit assignment problem, that can be critical in learning applications (Minsky, 1963; Barto et al., 1983) In biological terms, this is the problem of getting reinforcement signals to the right synapses (spatial credit assignment) at the right time (temporal credit assignment) for them to be effective in guiding the learning process. Here we present a neural model that addresses the temporal credit assignment problem through the detection of events that predict subsequent reinforcement. We introduce this model based on the anatomy and physiology of the striosome compartments of the striatum (Gerfen et al 1987; Graybiel, 1991); and on the signaling properties of DA neurons (see chapter 12). We then give it a conceptual foundation that is based on the theory of adaptive critics described by Barto in chapter 11. Preliminary versions of the model have been published previously (Houk, 1992).

DOPAMINE NEURONS

There is now substantial evidence that DA neurons, located in the pars compacta of the substantia nigra and in the ventral tegmental area, play an essential role in both the primary reinforcement of behavior and in guiding preparatory behavior on the basis of the likelihood that the animal will subsequently receive reinforcement (Wickens, 1990; Apicella et al., 1991). In chapter 12, Schultz and colleagues review their findings with microelectrode recordings from DA neurons in behaving monkeys showing that, at an early stage in learning a new behavioral task, these cells discharge in response to primary reinforcement. Later, as the animal learns the task, the cells begin to discharge in response to stimuli that regularly (or even probabilistically) precede the primary reinforcement and thus function as predictors of reinforcement. For example, Ljungberg et al. (1992) observed bursts of DA neuron discharge in response to any unexpected delivery of a drop of liquid reward in the initial phase of their experiment. Then the liquid reward was used as primary reinforcement in a task that required the monkey to reach and depress a lever when a small light was illuminated. As the monkey learned to respond to the light, bursts of DA discharge began to appear in response to the light, while responses to the liquid primary reinforcement progressively disappeared.

In another experiment probing the signaling properties of DA neurons, monkeys reached blindly into a box in search of a morsel of food, and DA bursts occurred whenever the fingers touched morsels of cookie or apple or a raisin (Romo and Schultz, 1990). In contrast, the neurons rarely responded to the touch of nonfood objects. Since this state of responsiveness served as the starting point for an experiment in which the animal was conditioned to an earlier event (the opening of the trap door to the food box) the authors treated contact with the food as primary reinforcement. However, these bursts are clearly in response to rather complex patterns of tactile input, as opposed to primary reinforcement produced by consuming the food. Here we postulate that the tactile responses to food objects are acquired, or secondary, reinforcers, because, during past experiences, they reliably predicted subsequent primary reinforcement obtained by food consumption and because they come to function as reinforcers themselves. In the next phase of the Romo and Schultz experiment, the door to the food box was left normally closed, and its abrupt opening signaled the availability of a morsel of food. During this phase the burst of DA discharge transferred from food contact to the opening of the food box that regularly preceded food contact. The noise and appearance of the trap door opening thus served as a predictor of food contact, which in turn predicted food consumption. These data suggest that DA discharge ratchets backward in time, in a sequence of familiar events, so as to respond to earlier and earlier predictors of reinforcement.

These intriguing results have both input and output implications that are each important to contemplate. In this chapter our main emphasis is on the input issues, i.e., what neural mechanisms enable DA neurons to fire in response to earlier and earlier predictors of reinforcement. We propose a model to explain how striosomal modules of the basal ganglia could predict future reinforcement in a recursive manner, and we show

how these signaling properties are analogous to those of an adaptive critic in the actor-critic architecture discussed by Barto in chapter 11. We also provide some discussion of the output issue, i.e., how might these signaling properties facilitate the control of the motor behaviors that ultimately secure the primary reinforcement. We will see how the system as a whole is potentially capable of addressing both the temporal and the spatial aspects of the credit assignment problem.

ORGANIZATION OF STRIOSOMAL MODULES

The input layer of the basal ganglia, the striatum, is divided into circumscribed regions called striosomes that are surrounded by matrix regions (Graybiel 1994; see also chapter 5). Both kinds of striatal region contain spiny neurons, so called because their processes are covered with dendritic spines that receive highly convergent input from the cerebral cortex and thalamus. However, the two regions differ in their chemical makeup and, most important, in the targets to which their neurons project (Gerfen et al., 1987; Graybiel, 1991). Spiny neurons in the striosomes project to DA neurons in the substantia nigra and ventral tegmental area, whereas spiny neurons in the matrix regions project to pallidal output neurons of the basal ganglia located in the internal division of the globus pallidus and in the pars reticulate of the substantia nigra. This chapter deals particularly with the striosomal spiny (SPs) neurons that project to DA neurons in the manner shown in figure 13.1.

The *solid black* arrow symbolizes the inhibitory GABAergic nature of the direct projection from SPs neurons to the DA neuron. The indirect projection involving a sideloop through the subthalamic (ST) nucleus instead has a net excitatory action. This ST sideloop is actually more complex than shown in figure 13.1, having a multisynaptic organization similar to the sideloops present in matrix circuits, as described by Graybiel and Kimura in chapter 5 and summarized by Houk in chapter 1 of this book. The simplified diagram will suffice for present purposes.

Spiny neurons in both striosome and matrix compartments have similar specializations in electrical properties (Kawaguchi et al., 1989) and receive organized, convergent input from widespread areas of the cerebral cortex (see chapter 5). The C's in figure 13.1 illustrate three of the thousands of cortical cells that send convergent input to the SPs neuron. As noted in chapter 1 and elsewhere, the neuronal architecture of the striatum is ideally suited for the recognition of complex patterns of cortical afference. We assume here that spiny neurons come to recognize complex contextual patterns through the reinforcing influence of the dopaminergic input to the striatum. In support of this hypothesis, there is growing evidence for dopamine-dependent plasticity in corti-

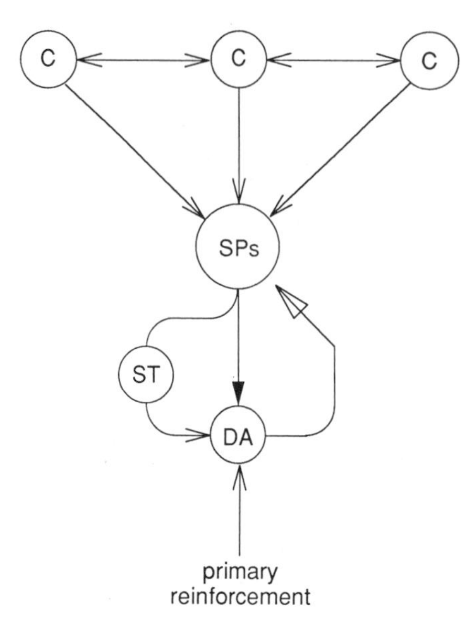

Figure 13.1 Striosomal module. See text for details. Open arrowheads signify net excitation, the black arrowhead in the direct SPs-to-DA projection signifies net inhibition, and the triangular arrowhead signifies neuromodulation. C, cerebral cortical columns; SPs, spiny neurons in striosomal compartments of the striatum; ST, subthalamic sideloop that includes neurons in the subthalamus; DA, dopamine neurons of the substantia nigra pars compacta.

costriatal synapses (see chapters 4 and 10 and below). The proposed specialization of SPs neurons for pattern recognition is assumed to allow them to detect particular contexts that are valid predictors of reinforcement. These responses would then be transmitted to DA neurons, which could explain how DA signaling predicts reinforcement. This hypothesis is elaborated below.

Another striking characteristic of the anatomy of striosomal modules is the projection that DA neurons make back to the same striatal zones that send them input (see figure 13.1). The reciprocal nature of projections between clusters of SPs and DA neurons has now been demonstrated in four laboratories (Gerfen et al., 1987; Jiménez-Castellanos and Graybiel, 1987; Selemon and Goldman-Rakic, 1990; Hedreen and DeLong, 1991); and can be considered a well-established aspect of striosomal modules. The model proposed here builds importantly on this characteristic, which we believe is responsible for the ability of DA neuron signaling to make progressively earlier predictions of reinforcement. As is explained more fully below, after an SPs neuron learns to fire in response to one context that predicts reinforcement, it can use this feedback pathway to reinforce itself for firing to an even earlier context that predicts reinforcement, which can then function as an antecedent secondary reinforcer. This anatomy gives rise to a recursive feature, analogous to the adaptive critic's predictions of its own predictions, and we propose it as being essential for the resolution of the temporal credit assignment problem discussed in the opening paragraph.

A third input to DA neurons shown in figure 13.1 is labeled primary reinforcement. This connection is based on indirect argument, as opposed to direct anatomical demonstration. The technique of microdialysis has played an important role in demonstrating the likelihood that DA neurons receive signals from the lateral hypothalamus that are related to primary reinforcements of an appetitive nature (Wise and Bozarth, 1984; Hoebel et al., 1989).

MECHANISM OF RESPONSIVENESS TO PREDICTORS OF REINFORCEMENT

The organization of the striosomal module shown in figure 13.1 calls attention to three sources of input to DA neurons that might interact in the generation of DA firing patterns. Figure 13.2 shows hypothetical time courses of these signals in response to a "predictor of reinforcement" presented as a cortical input pattern that regularly precedes primary reinforcement, the latter signal being sent from the lateral hypothalamus.

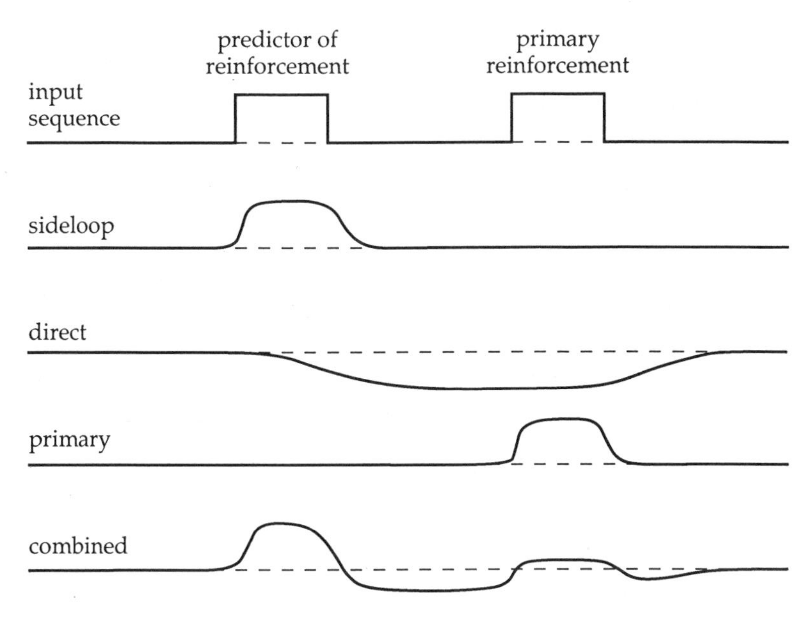

Figure 13.2 Hypothetical time course of signals generating dopamine neuron firing patterns. A pattern of excitatory signals which predicts reinforcement is delivered via projections from the cerebral cortical columns. The sideloop excitatory, direct inhibitory, and primary reinforcement signals converge on dopamine neurons, as shown in figure 13.1. The bottom trace shows the combination of those three signals.

Imagine, in analogy with the experiments of Schultz and colleagues, that the predictor of reinforcement is comprised of either the small light in the lever task, or the complex set of tactile stimuli that occur when the monkey contacts a piece of apple in the food box task. The primary reinforcement would be the liquid or food reward. We wish to explain why DA neurons respond to predictors of reinforcement and not to subsequent primary reinforcement, despite the latter being sent as excitatory input to DA neurons from the lateral hypothalamus.

In this section we assume that the SPs neuron has already learned, through a cellular mechanism that is described below, to fire a burst of discharge when the predicting context occurs. The traces in figure 13.2 labeled *sideloop* and *direct* respectively illustrate the postulated postsynaptic responses in the DA neuron produced by transmission of this burst through the excitatory ST sideloop and through the inhibitory direct projection. Our assumption that the excitation precedes the inhibition is in accord with electrophysiological observations showing that responses to electrical stimulation of the cortex or striatum evoke excitation followed by inhibition in substantia nigra (and globus pallidus) neurons (Kita and Kitai, 1991; Fujimoto and Kita 1992; Kita, 1992). The model, however, assumes a time course of inhibition slower than that observed in the electrophysiological experiments. These experiments thus far have demonstrated only a relatively rapid inhibition lasting about 25 ms and mediated by $GABA_A$ receptors. Although slow inhibition has not yet been demonstrated, its presence is anticipated based on the high density of $GABA_B$ receptors in the substantia nigra (Bowery et al., 1987; Martinelli et al., 1992). $GABA_A$ receptors act via G proteins to mediate slow inhibitory processes (lasting several hundred milliseconds) both in postsynaptic neurons and in presynaptic terminals (Isaacson et al., 1993).

The trace labeled *primary* in figure 13.2 shows an excitatory postsynaptic event postulated to occur in response to a primary reinforcement input from the lateral hypothalamus. If the latter were presented in isolation, the DA neuron would respond to it. However, if primary reinforcement is preceded by a predictive context that excites the SPs neuron, the complex sequence of postsynaptic events shown in the trace labeled *combined* would occur. The initial excitatory phase, owing to excitation transmitted through the sideloop, would fire the DA neuron. The subsequent inhibitory phase, transmitted through the direct pathway from the SPs neuron, might then largely cancel the excitatory potential produced by the primary reinforcement. The cancellation illustrated in figure 13.2 assumes that the inhibition is postsynaptic, as supported by monoclonal antibody staining (Martinelli et al., 1992). Cancellation might be even more complete if $GABA_B$ inhibition also occurred in the presynaptic terminals of the primary reinforcement input.

Thus far we have considered the relatively simple case of responding to a first-order predictor, i.e., a contextual event that is an immediate antecedent of primary reinforcement. In general, there may be a longer sequence of events and actions that ultimately leads to primary reinforcement. For example, the input sequence in figure 13.3 includes two contextual stimuli that function as successive predictors of reinforcement. Imagine that the later of the two predictors, C_a, represents the context that occurs when the monkey puts its hand into the food box and contacts a piece of apple. C_b represents an earlier, second-order, predictor, which might be the opening of the food box in the experiment of Schultz and colleagues. We assume the primary reinforcement, r, is the consumption of the apple.

The time plots below the input sequence illustrate the net postsynaptic activations in response to each event in the input sequence. (Note that this decomposition of input components is different from that used in figure 13.2.) The responses to C_a and to C_b each consist of initial excitatory phases followed by prolonged inhibitions, whereas the response to r is simply an excitation. The DA neuron would be expected to add together these components to generate its net output. It is apparent that the inhibitory phase of the response to C_b will cancel the excitatory phase of the response to C_a, and the inhibitory phase of the response to C_a will cancel the excitatory response to r. The bottom trace ignores the minor fluctuations of this summation process to illustrate that the net DA response, after all of these cancellations, would consist of a single excitatory phase produced by the earliest prediction of reinforcement, the context C_b. This is the pattern observed in recordings from DA neurons after DA signaling has transferred to the context associated with food box opening (see chapter 12).

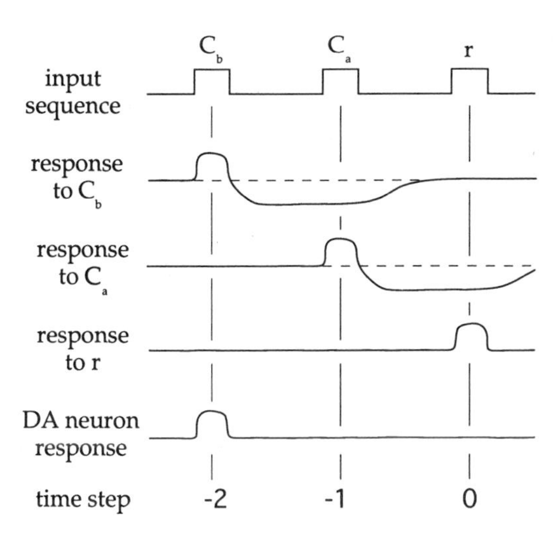

Figure 13.3 Dopamine (DA) neuron response to earlier predictors of reinforcement. See text for details. C_b and C_a, successive contextual stimuli; r, primary reinforcement.

CORRESPONDENCE WITH THE THEORY OF ADAPTIVE CRITICS

Barto in chapter 11 describes an adaptive critic as a device that learns to anticipate reinforcing events, pointing out that the effective reinforcement signals it generates can greatly enhance the performance of reinforcement learning systems. Influenced by the proposals of Klopf (1982), the adaptive critic was developed by Sutton (1988) and appeared as a neuronlike component of a control system (Barto et al., 1983) as well as a model of classical conditioning (Sutton and Barto 1981, 1990). Sutton (1988) used the term *temporal difference* (TD) algorithms for this general class of predictive mechanisms. There is a remarkable similarity between the discharge properties of DA neurons and the effective reinforcement signal generated by a TD algorithm under conditions representing the experimental situation of terminal primary reinforcement (see chapter 11.) The latter is the situation in which a sequence of contexts and associated actions must transpire before a primary reinforcing event ultimately occurs. Here we pursue this analogy by showing a particular correspondence between the signals that might by generated in striosomal modules and the terms in the TD equation for effective reinforcement.

Based on equation (10) from chapter (11), the effective reinforcement $\hat{r}_t$ at time step t is:

$$\hat{r}_{t-} = P_t - P_{t-1} + r_t \tag{1}$$

where P_t is the prediction at step t of future primary reinforcement, and P_{t-1} is the prediction at the previous time step. (We set the discount factor γ to 1 in order to simplify the discussion.) Equation (1) is a discrete-time version of the TD equation for effective reinforcement. To relate it to the activity of DA neurons, we suggest how these discrete time steps might relate to the continuous flow of real time by treating each discrete time step as the time of a salient event sensed by the animal. Various cellular events occur throughout the time intervals between these events.

Let us equate $\hat{r}_t$ with the discharge of the DA neuron in figure 13.1. Similarly, r can be equated with the primary reinforcement input to the DA neuron, which is zero except at the end of the requisite sequence of contexts and actions. At this point, it is natural to suggest that SPs neurons generate the predictions of reinforcement. P_t would then be transmitted by the excitatory sideloop and $-P_{t-1}$ by the direct, but slowly acting, inhibitory process. In essence, the slow inhibitory process functions as a kind of short-term memory of a negative image of the reinforcement that was predicted at the previous time step.

The time plots in figure 13.3 can be used to analyze $\hat{r}$ (DA neuron response) at each time step in the behavioral sequence from box opening (C_b), to food contact (C_a), to the food consumption that results in a positive value for the primary reinforcement signal r_t at $t=0$. The earliest predictor C_b occurs at $t=-2$ and evokes a positive postsynaptic response that would represent the prediction P_{-2}. This is the only term that contributes to $\hat{r}_{-2}$, since there was no response of the SPs neurons at the previous time step $t=-3$. The response to C_b then goes through its negative phase, providing a negative trace of P_{-2} for the computation at $t=-1$. The predictor C_a then evokes a positive response for P_{-1}, but this is canceled by the negative trace of P_{-2}. Finally, at $t=0$, the primary reinforcement r_0 evokes a positive response that is canceled by the negative trace of P_{-1}.

The interpretation suggested by this comparison is that the burst discharges of SPs neurons might be thought of as predictions of subsequent reinforcement. In order for the model to conform with the specifics of the TD equation [equation (1)], these bursts would have to be generated at each stage of a behavioral sequence (though not necessarily by the same SPs neuron). This is one of the predictions of the present model that needs to be tested experimentally.

LEARNING TO PREDICT PRIMARY REINFORCEMENT

Earlier we described a mechanism capable of explaining the predictive responses of DA neurons that are observed after monkeys have learned a new behavioral task. This explanation was based on the assumption that, during the learning phase, SPs neurons projecting to the recorded DA neuron acquire an ability to respond to contextual inputs that are predictive of reinforcement. This is not a trivial learning task, since primary reinforcement is typically delayed, occurring a substantial time interval after the predictive contexts that need to be reinforced. Although synaptic plasticity has been demonstrated in striatal neurons (Calabresi et al., 1992), the issue of delayed reinforcement has not been addressed. In this section we present a cellular model of delayed DA-sensitive synaptic plasticity to explain how SPs neurons might learn to recognize contexts that are antecedent to DA reinforcement.

Lisman (1989) proposed a model of synaptic plasticity that is founded on the unique properties of a complex protein called calcium-calmodulin–dependent protein kinase II (CaM PK II). This molecule is a major component of the postsynaptic density (Kennedy et al., 1983), suggesting a general role in plasticity, and it is present in rather high concentrations in the striatum (Newman-Gage and Graybiel, 1988), suggesting a particular role in spiny neurons. Calmodulin (CaM) binding activates CaM PK II and subsequent autophosphorylation greatly prolongs the active state, presumably because it traps CaM in molecular pockets (Meyer et al., 1992; Schulman and Hanson, 1993). The prolonged active state then potentiates glutamate receptors so as to produce long-term potentiation (LTP) (McGlade-McCulloh et al., 1993). The model proposed here incorporates this CaM trapping mechanism and also includes interactions between CaM PK II and the DA-stimulated intracellular signals studied extensively by Greengard and collaborators (Hemmings et al., 1987).

The flow diagram of figure 13.4 outlines the anticipated interactions between several intracellular signals in the spines of SPs neurons, and figure 13.5 illustrates how the dynamics of these interactions might increment the synaptic responsiveness to a context C_a that precedes reinforcement r. At the upper left in figure 13.4, glutamate (released at the terminals of cortical afferents in response to C_a) is shown to initiate a cascade of intracellular signaling. The three types of receptor for glutamate produce a mixture of membrane depolarization and an increase of intracellular Ca^{2+} (Bliss and Collingridge, 1993). The latter effect mediates plasticity, since it activates CaM, which activates CaM PK II, which then potentiates glutamate receptors. Bound CaM is nec-

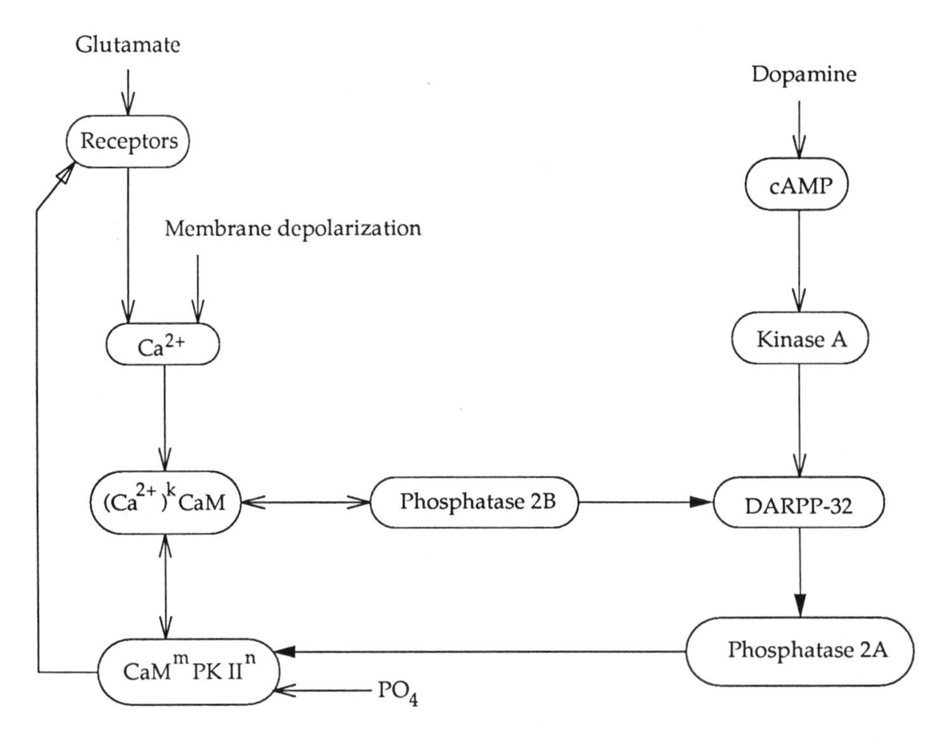

Figure 13.4 Cellular model of interactions which support learning earlier predictors of reinforcement. See description in text. Open arrowheads signify excitation or activation, black arrowheads signify inhibition or deactivation, and the triangular arrowhead signifies a possible potentiation or depression of receptors. Ca^{2+}, free calcium ions; $(Ca^{2+})^k CaM$, calcium-activated calmodulin with k (up to four) bound calcium ions; $CaM^m PK\ II^n$, activated Ca^{2+}-calmodulin–dependent protein kinase II with m pockets occupied by activated calmodulin molecules and n sites phosphorylated; PO_4, phosphate group; cAMP, the second messenger cylic adenosine monophosphate; DARPP-32, dopamine- and cAMP-regulated phosphoprotein.

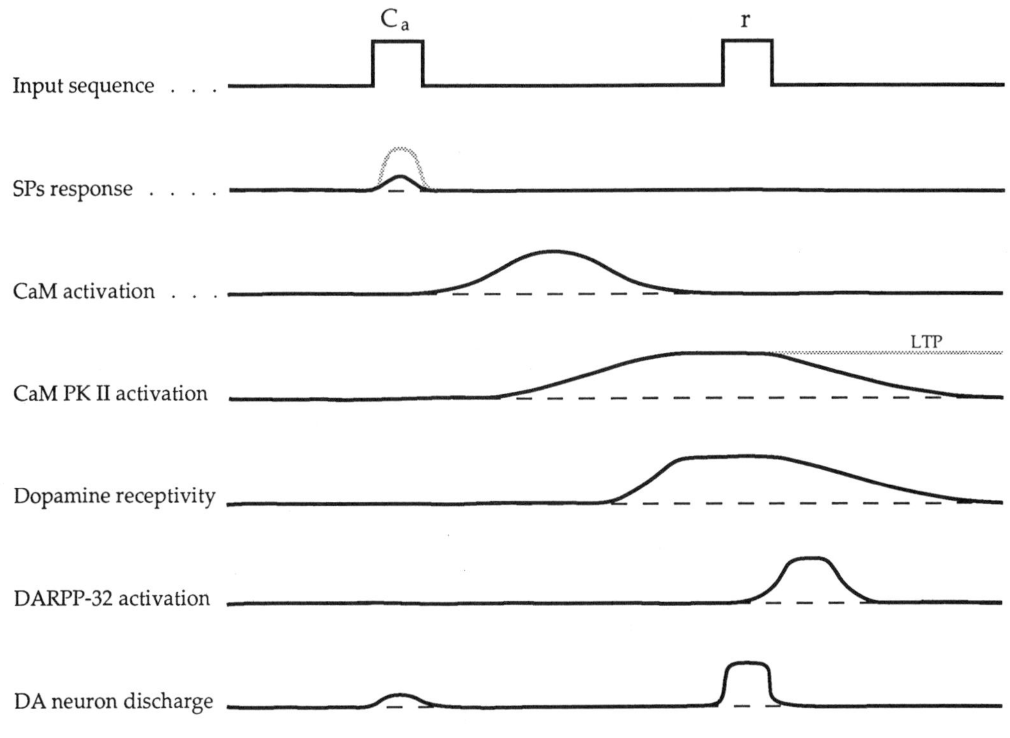

Figure 13.5 The interplay of intra- and extracellular events in creating long-term potentiation (LTP) of a predictor of reinforcement. See text and previous figures for details and abbreviations.

essary to activate its substrates and to initiate LTP (Mody et al., 1984). In figure 13.4, k in the expression $(Ca^{2+})^k CaM$ is the number of calcium ions bound to CaM, up to four; the delayed buildup of activated CaM in figure 13.5 is postulated to result from this multiple binding requirement. Although CaM has a high affinity for many of its substrates, including the phosphatase 2B shown in figure 13.4, it has a much lower affinity for binding to protein kinase II (PK II). This low affinity would be expected to raise the threshold and delay the time course for the activation of CaM PK II, as is illustrated by the solid trace of CaM PK II activation in figure 13.5.

The production of LTP requires an additional reaction, the autophosphorylation of CaM PK II. Autophosphorylation drastically alters the affinity of PK II for CaM, thus keeping this molecule activated for minutes as opposed to a fraction of a second in the dephosphorylated state (Schulman and Hanson, 1993). The dashed trace of CaM PK II activation labeled *LTP* in figure 13.5 illustrates the prolongation of activation that would be produced by autophosphorylation. We assume that this autophosphorylation step is normally counteracted by the presence of a phosphatase that drives the reaction toward dephosphorylation. In figure 13.4 the enzyme that dephosphyhorylates PK II is assumed to be phosphatase 2A, because the usual enzyme, phosphatase 1, is apparently not present in striatal neurons (Nairn et al., 1988); furthermore, phosphatase 2A is more specific than, and has properties similar to, phosphatase 1 (Cohen, 1989). The several-minutes time course of CaM PK II activation (and LTP) produced by autophosphorylation could be extended to longer memories by protein synthesis mechanisms involving polyribosomes (Weiler and Greenough, 1993).

In our model, the conversion of CaM PK II into the autophosphorylated form, the critical step that initiates LTP, depends on the arrival of a properly timed DA reinforcement signal. DA works through DARPP-32, a *d*opamine- and-adenosine 3′:5′-cyclic monophosphate (*c*AMP)–*r*egulated *p*hospho*p*rotein extensively studied by Greengard's laboratory (Hemmings et al., 1987). This regulatory protein has its highest brain concentration in the striatum where it appears to be localized to the medium-sized spiny neurons. DA activates D1 receptors located on spines, which then activate cAMP, kinase A, and finally DARPP-32, as illustrated in figure 13.4. Activation of DARPP-32 inhibits phosphatase 2A, and thus can disinhibit the autophosphorylation of CaM PK II to initiate LTP, but this requires that the DA neuron fire during a brief period of DA receptivity.

The timing of DA neuron firing and DARPP-32 activation we assumed in the construction of figure 13.5 is appropriate to initiate LTP. Activated DARPP-32 inhibits phosphatase 2A, thus removing its inhibition of CaM PK II autophosphorylation. This, by itself, will not phosphorylate CaM PK II. Autophosphorylation occurs only under the condition that CaM PK II is already activated by CaM binding. Thus, the time course of CaM PK II activation shown in figure 13.5 is a permissive factor in defining the period of DA receptivity. Another condition for DA being effective relates to phosphatase 2B activity. The latter is rapidly activated and inactivated by transients in CaM activation, and, when activated, phosphatase 2B blocks the ability of DA to activate DARPP-32 (Halpain et al., 1990). Thus, the time course of CaM activity shown in figure 13.5 defines a prohibitive factor. The time course of DA receptivity shown in figure 13.5 reflects the combination of the permissive and prohibitive factors.

DA receptivity is analogous to the eligibility traces that have been invoked in certain computational models of reinforcement and classical conditioning (Klopt, 1982; Sutton and Barto, 1990; see also chapter 11). Like an eligibility trace, DA receptivity is a potential for reinforcement that becomes elevated after the occurrence of a synaptic input and decays slowly over time. However, unlike most eligibility traces used in network modeling, the elevation in DA receptivity does not begin immediately after a synaptic input; instead, it is postulated to be specifically delayed by the restrictive effects of phosphatase 2B activation.

As a consequence, spiny neuron synapses would tend to ignore contexts that precede reinforcement by very short time intervals. Instead, SPs neurons would preferentially learn antecedent contexts that precede reinforcement by a longer time interval, or, more precisely, by a range of longer time intervals as determined by the time course of the delayed DA receptivity.

While there are insufficient data on the time course of intracellular signaling to specify the preferential time interval discussed in the preceding paragraph, we postulate that it should be of the order of a few hundred milliseconds. This postulate is based on a requirement of the model, namely that the preferred interval for delayed reinforcement should not exceed the duration of the slow inhibitory event illustrated in figure 13.2. This is because an acquired response to a predictor needs to evoke an inhibition of sufficient duration to cancel a subsequent response to primary reinforcement, as was discussed earlier.

LEARNING EARLIER PREDICTORS OF REINFORCEMENT

The delayed receptivity of SPs neuron spines to DA reinforcement discussed in the previous section could provide an effective mechanism for learning a first-order predictor that occurs reliably at a fixed time interval prior to primary reinforcement. The predictive context would have to precede reinforcement by a few hundred milliseconds, which is a relatively short time interval on the scale of behavior. Learning to respond to higher-order predictors that may occur at longer and more variable delays is a more difficult problem. In this section, we explore the learning properties of an entire striosomal module when it contains an embedded SPs neuron with delayed DA receptivity. We show that these modules have an emergent property, a recursive capacity for learning to recognize a sequence of contextual events that are predictive of reinforcement. In the following section we discuss the similarity of this mechanism to the operations performed by an adaptive critic.

We have illustrated above how a context C_b, that reliably precedes a context C_a, that reliably precedes primary reinforcement r, could elicit discharge in the DA neurons of a striosomal module at the early time step when C_b occurs (see figure 13.3). There we assumed that the SPs neuron had already learned to fire in response to these contexts, whereas in this section we address the learning process itself. In particular, we describe how the DA released by a response to C_a could serve as a secondary reinforcer that trains the SPs neuron to recognize the context C_b that is an antecedent predictor of reinforcement. We then generalize this result to suggest how these modules could ultimately learn to recognize long chains of sequential events that lead to primary reinforcement.

The upper time plot in figure 13.6 shows the stimulus sequence, from C_b (the opening of the food box), which precedes the occurrence of C_a (the contact of the fingers with a piece of apple), which precedes the occurrence of the consumption of the piece of apple (the primary reinforcement *r*). We start with a state in which the SPs neuron has a strong response to C_a, owing to a previous history of primary reinforcement *r*, and an unreinforced, weak response to C_b, as shown by the second trace in figure 13.6. The weak response to C_b leaves in its wake a rising and falling wave of receptivity to DA in the spines that were excited by C_b. This is followed by a strong response to C_a, which gives rise to another wave of receptivity to DA, but this time in the spines that were excited by C_a. The DA response trace shows the net effect of these SPs neuron responses and the response to *r* on DA neuron discharge. Referring back to the examples of figures 13.2 and 13.3, one can mentally

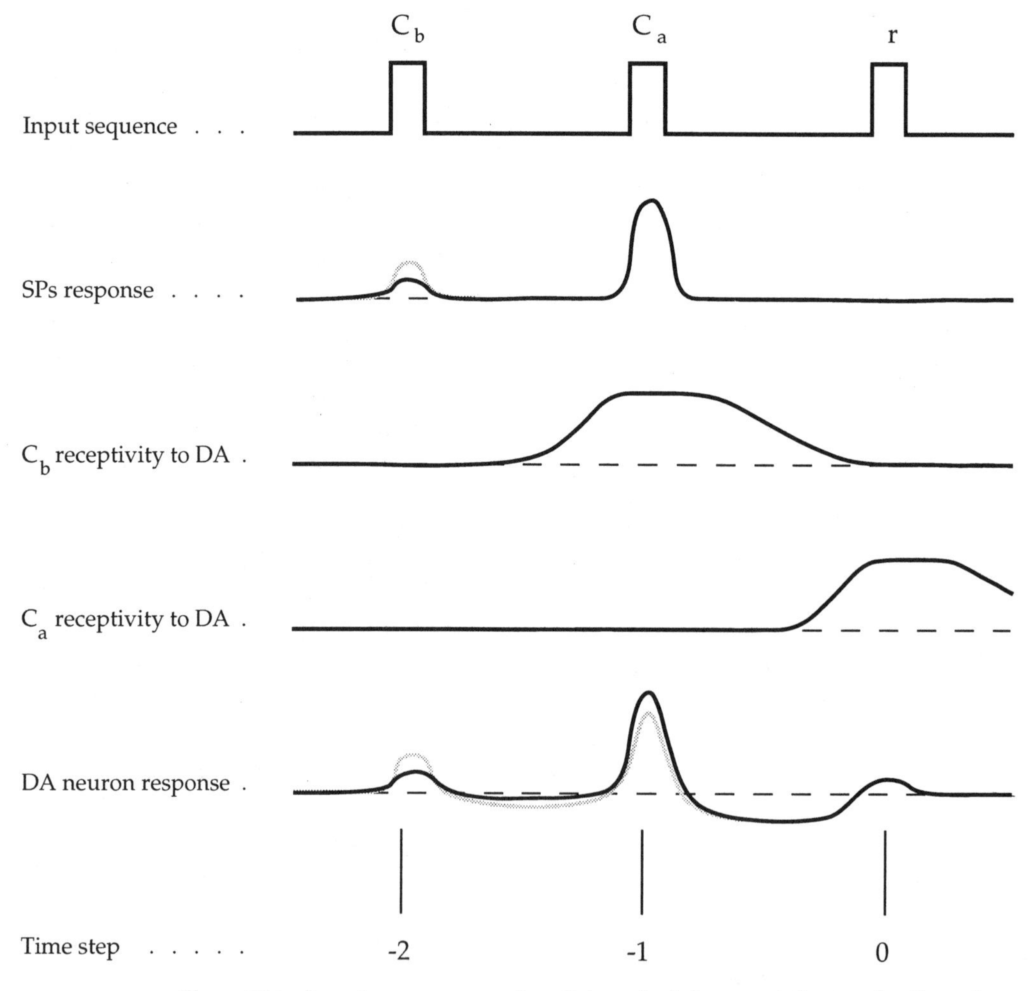

Figure 13.6 Learning a sequence of predictors of reinforcement. See text for discussion. Abbreviations as given for figures 13.1 and 13.3.

combine the different components of response to see how they would give rise to the DA response trace illustrated in figure 13.6. The salient feature of this trace, ignoring the smaller fluctuations, is the burst of discharge that occurs at the -1 time step.

Now we can contemplate how these several events that occur during a single trial would interact to produce learning. Based on the cellular model of plasticity discussed above, there would have to be a coincidence of DA release and of receptivity to DA in order to instantiate a change in synaptic strength. Figure 13.6 shows that the required coincidence occurs at $t=-1$ in those spines that responded to C_b at $t=-2$, and this instantiates an increase in synaptic strength in the C_b-responsive spines. The stippled traces in figure 13.6 show how this change in synaptic strength would affect responding in a subsequent trial with the same input sequence. The increased SPs neuron response to C_b is a direct result of enhanced transmission at the synapses responding to C_b. The response to C_a is unaltered in the SPs neuron, but is depressed in the DA neuron, as a consequence of the delayed inhibitory component of the enhanced response to C_b. The striosomal module as a whole thus acquires a stronger response to the earlier context and a weaker response to the later one. Over the course of many such trials, the context C_b would come to evoke an SPs neuron burst as strong as that shown in the original response to C_a. At this point, the burst of DA discharge, the earliest prediction of reinforcement, will have moved to the $t=-2$ time step, and the DA response to C_a will be completely inhibited. Note that responses to later contexts are not forgotten; instead, we predict that they are simply canceled by the delayed inhibitory components of responses to earlier contexts.

Having thus acquired this new antecedent of reinforcement, the striosomal module can then progress through yet another cycle of learning an even earlier predictor of reinforcement. Given the recursive nature of this process, striosomal modules might, through extended experiences, become capable of detecting contextual stimuli that occur at very long time periods in advance of a primary reinforcement. It is difficult to anticipate what the ultimate limit of this process might be. However, one might expect that the reliability of the prediction of reinforcement might progressively decrease at earlier times, thus diminishing the likelihood of finding a new context that predicts the DA response with sufficient regularity to become established as an antecedent predictor.

RELATION TO THE ACTOR-CRITIC ARCHITECTURE

The model of a striosomal module described in this chapter fulfills the main functions of the adaptive critic in the actor-critic architecture described by Barto in chapter 11. This architecture is an effective way of implementing a reinforcement learning system and is currently being studied by engineers and computer scientists as an approach to solving difficult nonlinear control problems (Barto et al., 1983, 1990). The basic idea is to let predictions of reinforcement, which are generated by an adaptive critic, serve as surrogate (or secondary) reinforcers for controlling an actor, which is the system that generates the command signals that control actions. In the present section, we use this architecture as a framework for exploring how two kinds of information-processing module in the basal ganglia might function like interacting critics and actors in the control of motor behavior.

Earlier we reviewed the contrasting connectivity of spiny neurons in striosome and matrix compartments of the basal ganglia. As illustrated in figure 13.7, SPs neurons project to DA neurons and matrix spiny (SPm) neurons instead project to the pallidal (PD) output stage of the basal ganglia. Figure 13.7 indicates how this and other anatomical features serve to define matrix modules, which are partly analogous to the striosomal modules discussed above. Like the SPs neurons, SPm neurons have separate pathways for transmitting excitatory and inhibitory inputs to their target neurons, in this case the PD neurons. The descending connections of PD neurons are omitted in figure 13.7 to focus on the more prominent ascending pathways to columns of frontal cortical neurons (*F*), via specific divisions of the thalamus (*T*). We assume

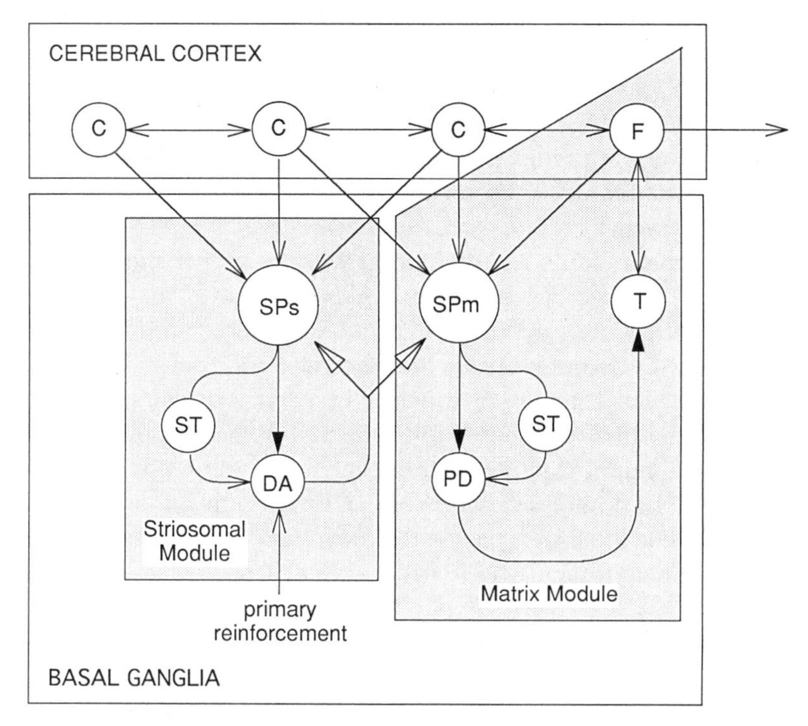

Figure 13.7 Modular organization of the basal ganglia including both striosomal and matrix modules. See figure 13.1 for the significance of the arrowheads and abbreviations for the striosomal module. Additional abbreviations: F, columns in frontal cortex; SPm, spiny neurons in the matrix compartment of the striatum; PD, pallidal neurons; T, thalamic neurons.

that the signals in F neurons function as the outputs of matrix modules. A specific hypothesis about the information-processing operations that go on in matrix modules is summarized in chapter 1 of this book.

Let us assume that striosomal molecules function as adaptive critics. In addition, we assume in figure 13.7 that the DA innervation of the striatum is sufficiently diffuse for the DA signal generated by a striosomal module to diverge to adjacent matrix modules. (This assumption is reconsidered in the next section.) Although only a single matrix module is shown, 95% of the spiny neurons are in matrix zones (Wilson, 1990), suggesting a ratio of about 20 matrix modules to each striosomal module. On this basis, one can justify divergence to the set of competitively interacting SPm units illustrated in figure 11.5 of chapter 11. Similarly, we postulate here that the groups of matrix modules innervated by a given striosomal module interact competitively with one another. It is then natural to assume that these groups of matrix modules function like the actor in the actor-critic architecture, thus generating signals that command various actions. (A more realistic assumption is discussed below.) Correspondingly, the branching of the DA fiber illustrated in figure 13.7 to both SPm and SPs neurons would provide the requisite coupling between a striosomal module functioning as an adaptive critic and surrounding matrix modules functioning as an actor. The predictions of reinforcement generated by the striosomal module would serve as secondary reinforcers for training the matrix modules. Under this training influence, the matrix modules would learn to use the convergent cortical input onto SPm neurons as a basis for generating action commands in sets of F neurons (see figure 13.7). The DA signals would tend to shape these commands into ones that regularly lead to the achievement of primary reinforcement.

Although the analogy proposed here between the basal ganglia and the actor-critic architecture is clearly a simplified abstraction of the actual anatomy and physiology of this part of the brain, it helps to translate into biological terms certain principles and concepts that may be important in basal ganglia function. As an example, we will use it to explain how striosomal modules might help to resolve the temporal credit assignment problem raised in the introductory paragraph. The temporal credit assignment problem arises because the process of adjusting a module's behavior on the basis of reinforcement feedback has to accommodate the fact that consequences are delayed and depend on the environmental conditions in which an action was made as well as on other actions executed before and after the action in question. As a result, both the nature of the consequences and their time courses are both delayed and highly variable. If an event occurs that has primary reinforcing significance, it is not easy to determine which elements of the preceding activity deserve the credit (or the blame in the case of an aversive event) for causing this event (the credit assignment problem alluded to above). The adaptive critic addresses this problem by learning to predict, or anticipate, primary reinforcement over long time periods of variable and uncertain duration. The critic's predictions contribute to secondary, or acquired, reinforcement signals that provide *immediate* feedback about *anticipated* consequences of current activity (where "immediate" means after a relatively small and relatively fixed delay). Under the influence of these acquired reinforcement signals, which we postulate to be the signals produced by DA neurons, the matrix modules can learn to exert actions and thus influence future sensory input in desirable ways using synaptic modification rules that do not need to accommodate the highly variable temporal relationship between matrix activity and primary reinforcement.

MORE REALISTIC ASSUMPTIONS

While the model described in the previous section is attractive for its simplicity, some of its assumptions ignore the known anatomy and physiology of the system. Here we call attention to a few of these discrepancies, and discuss potential consequences of considering a more realistic model.

In the previous section we discussed the role of the basal ganglia as if there were only one striosomal module, computing a single prediction of reinforcement, and broadcasting that prediction broadly to all of the matrix modules. This most certainly is a gross oversimplification. Instead, we know that there are many, spatially segregated striosomes, each projecting to a somewhat separate cluster of DA neurons in a relatively topographic manner (Gerfen et al., 1987; Jiménez-Castellanos and Graybiel, 1987; Selemon and Goldman-Rakic, 1990; Hedreen and DeLong, 1991). Considering that each cluster of DA neurons is likely to receive somewhat different primary reinforcement inputs, the different modules associated with each striosome should become involved in predicting different qualities of primary reinforcement. Furthermore, since each striosome contains a large number of SPs neurons, each receiving a somewhat different constellation of afference, one would expect that different modules belonging to a given striosome would find different bases for predicting their special quality of primary reinforcement.

These anatomical considerations suggest considerable opportunity for generating a large diversity of secondary reinforcers that would then be available for training different groups of matrix modules. Why then have Schultz and colleagues (see chapter 12) observed such a high degree of homogeneity in the responses of the DA neurons that they have sampled? Perhaps this results, at least in part, from the fact that their animals are engaged in learning just a single behavioral task over the period during which the population is being sampled. It would be interesting to train the animals to perform concurrently several behavioral tasks, to see if different DA neurons specialize to predict in different ways. Concurrent learning of several tasks, some with a common primary reinforcement and others with different ones, probably reflects more accurately the experiences of the animal in the wild. Certainly this situation presents a much more challenging, and more interesting, computational problem, appropriately suitable for the parallel architecture of the brain.

The discussion in the preceding section treated the outputs of matrix modules as if they were signals that command specific actions. In contrast, single-unit recordings from frontal cortical F neurons, the outputs of matrix modules, indicate that these signals are not immediate commands, but rather signify higher-order properties (Goldman-Rakic, 1987; Schultz and Romo, 1992). Looking at these properties from a motor perspective, they would appear to represent plans that might then organize other systems to generate the actual command signals for actions. If one instead looks at these signals from a sensory perspective, they appear to signal complex contexts that could indeed be useful in the formulation and implementation of plans and actions. Houk and Wise (1993, 1994) adopted the latter view of matrix module function and suggested mechanisms whereby salient contexts might be detected and registered into working memory for subsequent use by motor program generators in the cerebellum. Salient contexts were postulated to include the state of the organism, the desirability of the action, the actions planned in the near future, the location of targets of action, and sensory inputs that both select and trigger motor programs.

The linkages between F signals and actual motor commands is apparently quite flexible, which gives rise to an additional source of uncertainty with which the adaptive critics would have to contend. These concepts do not invalidate the actor-critic model of the basal ganglia suggested earlier. Instead, they cast this model within the perspective of more flexible, complex, and seemingly more powerful control options.

SUMMARY

The model presented here explains how dopamine (DA) neurons in the basal ganglia might acquire their ability to predict reinforcement and how outputs from these neurons might then be used to reinforce behaviors that lead to primary reinforcement. DA neurons are embedded in striosomal modules that include reciprocal connections with spiny neurons in the striatum. We propose a cellular learning rule whereby spiny neurons are trained by their DA input to detect contexts that precede reinforcement by a short time interval. Striosomal spiny neurons then use these acquired responses to control their own DA input. Through this recursive mechanism, DA neurons could learn to detect earlier and earlier predictors of reinforcement. DA signals also diverge to reinforce spiny neurons in matrix modules, training the latter to detect and register contexts that are useful in planning and controlling motor behavior. The proposed scheme has interesting parallels with an actor-critic architecture that has been used to solve difficult engineering control problems.

ACKNOWLEDGMENTS

The authors are grateful to Richard S. Sutton for helping us relate the anatomy of striosomal modules to the mathematical operations performed by the temporal differences algorithm. This work was supported by a contract from the Office of Naval Research (N00014-88-K-0339).

REFERENCES

Apicella, P., Ljungberg, T., Scarnati, E., and Schultz, W. (1991) Responses to reward in monkey dorsal and ventral striatum. *Exp. Brain Res.* 85:491–500.

Barto, A. G., Sutton, R. S., and Anderson, C. W. (1983) Neuronlike elements that can solve difficult learning control problems. *IEEE Trans. Syst. Man Cyber.* 13:835–846.

Barto, A. G., Sutton, R. S., Watkins, and C. J. C. H. (1990) Learning and sequential decision making. In M. Gabriel and J. Moore (eds.), *Learning and Computational Neuroscience: Foundations of Adaptive Networks.* Cambridge, Mass.: MIT Press, pp. 539–602.

Beninger, R. J. (1983) The role of dopamine in locomotor activity and learning. *Brain Res.* 287:173–196.

Bliss, T. V. P., and Collingridge, G. L. (1993) A synaptic model of memory: long-term potentiation in the hippocampus. *Nature* 361:31–39.

Bowery, N. G., Hudson, A. L., and Price, G. W. (1987) GABA-A and GABA-B receptor site distribution in the rat central nervous system. *Neuroscience* 20:365–383.

Calabresi, P., Pisani, A., Mercuri, N. B., and Bernardi, G. (1992) Long-term potentiation in the striatum is unmasked by removing the voltage-dependent magnesium block of NMDA receptor channels. *Eur. J. Neurosci* 4:929–935.

Cohen, P. (1989) The structure and regulation of protein phosphatases. *Annu. Rev. Biochem* 58:453–508.

Fujimoto, K., and Kita, H. (1992) Responses of rat substantia nigra pars reticulata units to cortical stimulation. *Neurosci. Lett.* 142:105–109.

Gerfen, C. R., Herkenham, M., and Thibault, J. (1987) The neostriatal mosaic. II. Patch- and matrix-directed mesostriatal dopaminergic and non-dopaminergic systems. *J. Neurosci* 7:3935–3944.

Goldman-Rakic, P. S. (1987) Circuitry of primate prefrontal cortex and regulation of behavior by representational memory. In F. Plum and V. B. Mountcastle (eds.), *Handbook of Physiology. The Nervous System V,* Part 1. Bethesda, Md., American Physiological Society, pp. 373–417.

Graybiel, A. M. (1991) Basal ganglia—input, neural activity, and relation to the cortex. *Curr. Opin. Neurobiol.* 1:644–651.

Halpain, S., Girault, J-A., and Greengard, P. (1990) Activation of NMDA receptors induces dephosphorylation of DARPP-32 in rat striatal slices. *Nature* 343:369–372.

Hedreen, J. C., and DeLong, M. R. (1991) Organization of striatopallidal, striatonigral, and nigrostriatal projections in the Macaque. *J. Comp. Neurol* 304:569–595.

Hemmings, H. C., Walaas, S. I., Ouimet, C. C., and Greengard, P. (1987) Dopamine regulation of protein phosphorylation in the striatum: DARPP-32. *Trends Neurosci.* 10:377–383

Hoebel, B. G., Hernandez, L., Schwartz, D. H., Mark, G. P., and Hunter, G. A. (1989) Microdialysis studies of brain norepinephrine, serotonin and dopamine release during ingestive behavior: Theoretical and clinical implications. *Ann. N. Y. Acad. Sci.* 575:171–191.

Houk, J. C. (1992) Learning in modular networks. In K. S. Narendra (ed.), *Proceedings of the Seventh Yale Workshop on Adaptive and Learning Systems.* New Haven, Conn.: Center for Systems Science, pp. 80–84.

Houk, J. C. (1992) *Learning in Modular Networks. NPB Technical Report 7,* Northwestern University University Medical School, Department of Physiology, Ward Building 5-342, 303 E. Chicago Ave., Chicago IL 60611-3008.

Houk, J. C., and Wise, S. P. (1993) Outline for a theory of motor behavior. In P. Rudomin, M. A. Arbib, and F. Cervantes-Perez (eds.), *Neuroscience: from Neural Networks to Artificial Intelligence, Research Notes in Neural Computing,* Vol. 4. Heidelberg: Springer-Verlag, pp. 452–470.

Houk, J. C., and Wise, S. P. (1994) Distributed modular architecture linking basal ganglia, cerebellum and cerebral cortex: Its role in planning and controlling action. *Cereb. Cortex* (submitted for publication)

Isaacson, J. S., Solis, J. M., and Nicoll, R. A. (1993) Local and diffuse synaptic actions of GABA in the hippocampus. *Neuron* 10:165–175.

Jiménez-Castellanos, J., and Graybiel, A. M. (1987.) Subdivisions of the dopamine-containing A8-A9-A10 complex identified by their differential mesostriatal innervation of striosomes and extrastriosomal matrix. *Neuroscience* 23:223–242.

Kawaguchi, Y., Wilson, C. J., and Emson, P. C. (1989) Intracellular recording of identified neostriatal patch and matrix spiny cells in a slice preparation preserving cortical inputs. *J. Neurophysiol* 62:1052–1068.

Kennedy, M. B., Bennett, M. K., and Erondu, N. E. (1983) Biochemical and immunochemical evidence that the "major" postsynaptic density protein is a subunit of a calmodulin-dependent protein kinase. *Proc. Natl. Acad. Sci. U. S. A.* 80:7357–7361.

Kita, H. (1992) Responses of globus pallidus neurons to cortical stimulation: intracellular study in the rat. *Brain Res.*

Kita, H., and Kitai, S. T. (1991) Intracellular study of rat globus pallidus neurons: Membrane properties and responses to neostriatal, subthalamic and nigral stimulation. *Brain Res.* 564:296–305.

Klopf, A. H. (1982) *The Hedonistic Neuron: A Theory of Memory, Learning and Intelligence.* New York: Hemispheres.

Lisman, J. (1989) A mechanism for the Hebb and the anti-Hebb processes underlying learning and memory. *Proc. Natl. Acad. Sci. U. S. A.* 86:9574–9578.

Ljungberg, T., Apicella, P., and Schultz, W. (1992) Responses of monkey dopamine neurons during learning of behavioral reactions. *J. Neurophysiol.* 67:145–163.

Martinelli, G. P., Holstein, G. R., Pasik, P., and Cohen, B. (1992) Monoclonal antibodies for ultrastructural visualization of L-baclofen-sensitive GABA-B receptor sites. *Neuroscience* 46:23–33.

McGlade-McCulloh, E., Yamamoto, H., Tan, S.-E., Brickey, D. A., and Soderling, T. R. (1993) Phosphorylation and regulation of glutamate receptors by calcium/calmodulin-dependent protein kinase II. *Nature* 362:640–642.

Meyer, T., Hanson, P. I., Stryer, L., and Schulman, H. (1992) Calmodulin trapping by calcium-calmodulin–dependent protein kinase. *Science* 256:1199–1202

Minsky, M. L. (1963) Steps toward artificial intelligence. In E. A. Feigenbaum and J. Feldman (eds.), *Computers and Thought*. New York: McGraw-Hill, pp. 406–450.

Mody, I., Baimbridge, K. G., and Miller, J. J. (1984) Blockade of tetanic- and calcium-induced long-term potentiation in the hippocampal slice preparation by neuroleptics. *Neuropharmacology* 23:625–631.

Nairn, A. C., Hemmings, H. C., Jr., Walaas, S. I., and Greengard, P. (1988) DARPP-32 and phosphatase inhibitor-1, two structurally related inhibitors of protein phosphatase-1, are both present in striatonigral neurons. *J. Neurochem.* 50:257–262.

Newman-Gage, H., and Graybiel, A. M. (1988) Expression of calcium/calmodulin-dependent protein kinase in relation to dopamine islands and synaptic maturation in the cat striatum. *J. Neurosci.* 8:3360–3375.

Romo, R., and Schultz, W. (1990) Dopamine neurons of the monkey midbrain: Contingencies of responses to active touch during self-initiated arm movements. *J. Neurophysiol.* 63:592–605.

Schulman, H., and Hanson, P. I. (1993) Multifunctional Ca^{2+}/calmodulin-dependent protein kinase. *Neurochem. Res.* 18:65–77.

Schultz, W., and Romo, R. (1992) Role of primate basal ganglia and frontal cortex in the internal generation of movements. I. Preparatory activity in the anterior striatum. *Exp. Brain Res.* 91:363–384.

Selemon, L. D., and Goldman-Rakic, P. S. (1990) Topographic intermingling of striatonigral and striatopallidal neurons in the Rhesus monkey. *J. Comp. Neurol.* 297:359–376.

Sutton, R. S. (1988) Learning to predict by the method of temporal differences. *Machine Learning* 3:9–44.

Sutton, R. S., and Barto, A. G. (1981) Toward a modern theory of adaptive networks: Expectation and prediction. *Psychol. Rev.* 88:135–170.

Sutton, R. S., and Barto, A. G. (1990) Time-derivative models of pavlovian reinforcement. In M. Gabriel and J. Moore (eds.), *Learning and Computational Neuroscience: Foundations of Adaptive Networks*. Cambridge, Mass.: MIT Press, pp. 497–537.

Weiler, I. J., and Greenough, W. T. (1993) Metabotropic glutamate receptors trigger postsynaptic protein synthesis. *Proc. Natl. Acad. Sci. U. S. A.* 90:7168–7171.

Werbos, P. J. (1992) Neurocontrol and supervised learning: An overview and evaluation. In D. White and D. Sofge (eds.), *Handbook of Intelligent Control*. New York: Van Nostrand Reinhold

Wickens, J. R. (1990) Striatal dopamine in motor activation and reward-mediated learning: Steps towards a unifying model. *J. Neural Transm.* 80:9–31.

Wilson, C. J. (1990) Basal Ganglia. In Shepherd, G. M. (ed.), *The Synaptic Organization of the Brain*. Oxford: Oxford University Press, pp. 279–316.

Wise, R. A., and Bozarth, M. A. (1984) Brain reward circuitry: Four circuit elements "wired" in apparent series. *Brain Res. Bull.* 297:265–273.

Wise, R. A., and Rompré, P.-P. (1989) Brain dopamine and reward. *Annu. Rev. Psychol.* 40:191–225.

Editors' Commentary on Part III

There is now considerable evidence that dopamine functions as a reinforcement signal that guides the learning of diverse behavioral responses. In chapter 10, Wickens and Kötter concisely summarize this evidence and relate it to the pattern recognition ideas that were discussed in earlier chapters and commentaries. Dopamine (DA) is released by neurons that project to the basal ganglia and frontal cortex. The heaviest projections of these DA neurons are to the striatum, and the authors of chapters 10 through 13 uniformly adopt the working hypothesis that a key function of the DA system is to adjust the synaptic strengths of the cortical inputs to striatal spiny (SP) neurons through reinforcement. In this manner, SP neurons could learn to recognize contextual states and events that are of behavioral significance to the animal. The four chapters contained in part III discuss these reward mechanisms from several perspectives.

CRITICS, ACTORS, AND REWARDING BEHAVIOR

Chapter 11 provides a conceptual model for considering the learning problems discussed in the other chapters of part III. According to this model, we may think of the basal ganglia as performing two complementary functions. Units that Barto calls *actors* perform the function of generating commands that control behavior. The actors are trained to generate their commands by an *adaptive critic* that learns to predict the likelihood that a reward will be obtained. The adaptive critic uses these predictions to generate an *effective reinforcement* signal that it uses to train the actors. Barto's effective reinforcement signals are strikingly similar to the DA neuron signals that Schultz and colleagues describe in chapter 12, lending credence to the model. An adaptive critic may be analogous to a striosomal module as discussed later in this commentary (and in chapter 13). An actor, instead, may be analogous to a spiny neuron (SPm) in the matrix zone of the striatum. Alternatively, an actor may be likened to an entire cortical-basal ganglionic module (chapters 1 and 13) that would both recognize a context, with its SPm neuron, and register the occurrence of that context in the frontal cortex as a working memory. This working memory signal could then be used as a motor command or as a motor plan, as was discussed in the Commentary to part II.

While it seems likely that dopamine provides reinforcement to spiny neurons, the specific cellular effects on synaptic strength remain controversial. Chapter 10 provides an overview of this issue and suggests a theoretical framework for interpreting the results of several synaptic plasticity studies. This framework, a three-factor cellular learning rule with clear correlates to learning at the behavioral level, states that synapses are strengthened in response to a coincidence of presynaptic ("situation"), postsynaptic ("action"), and dopaminergic ("reinforcement") activity. This learning rule allows presynaptic contexts representing a "situation" to be associated only with active spiny neurons that are firing to request an "action." For example, before learning has occurred, a context might activate several groups of SPm neurons with equal probability. If by chance the context engages a spiny neuron (and thus a module) associated with a behavioral action that leads to dopamine "reinforcement," the strength of the association between the context and the module's activation would increase, while associations with the other inactive modules would weaken. This would gradually narrow the field of potential responses.

It is interesting to contemplate the consequences of the trans-striatal loops discussed in chapter 6 and in the part II Commentary. Because of this loop, the "action" term in the three-factor learning rule would be represented both pre-and postsynaptically by the activity within the trans-striatal loop. Indeed, assuming that each SPm neuron is part of one trans-striatal loop with the frontal cortex, it receives inputs that represent both "situation" (inputs from nonfrontal cortex) and "action" (input from frontal cortex). Accordingly, as dopamine reinforcement increases synaptic strength, the gain of an individual trans-striatal loop would be increased as well as the strength of its association with a particular context. Strengthening the cortical synapses of a trans-striatal loop would effectively increase the loop's response to a given context. Effectively, this would bias striatal activity in favor of the reinforced module and its corresponding motor plan or command. This idea is supported by evidence cited in chapter 10 that the striatal application of dopamine produces repetition of a limited set of stereotyped behaviors.

DEVELOPING PREDICTIONS OF DELAYED REINFORCEMENT

A reward always follows the actions used to obtain it. Thus, if learning is to be guided by reinforcement, mechanisms are necessary for looking backward in time, to determine which neural events contributed to the particular actions that led to the reward. The single-unit results summarized by Schultz in chapters 2 and 11 demonstrate that both striatal spiny and DA neurons are capable of signaling predictions of reinforcement. DA neurons always fire with a brief burst discharge that is time locked to a phasic environmental event, or to the onset of a primary reward. Spiny neurons, on the other hand, can show either burst or sustained discharge. Chapter 13 suggested that the capacity for prediction originates in striatal spiny neurons in each of these cases. Spiny neurons in striosomal compartments (SPs neurons) were postulated to burst in response to contextual events predictive of reinforcement, and their projections to DA neurons could then explain the ability of DA neurons to predict reinforcement. Spiny neurons in matrix compartments (SPm neurons) might respond in bursts like SPs neurons, or they might respond with sustained discharge due to feedback through the trans-striatal loops via the frontal cortex, as discussed in the part II Commentary. Whereas SPs neurons might simply predict the likelihood of obtaining a reward, SPm neurons would respond to predictive contexts that are also useful in planning the actions that culminate in a reward.

The above ideas are based on the assumption that both SPs and SPm neurons learn to respond in a predictive manner as a consequence of dopamine reinforcement controlled by DA neuron discharge. The cellular learning rule discussed earlier might explain how DA neurons teach spiny neurons to recognize contexts occurring *simultaneously* with the delivery of primary reinforcement. Returning to the coffee cup example in the Commentary to part II, we will assume that the taste of coffee constitutes a reward. DA activity in response to the taste of coffee reinforces concurrent sipping behavior and creates an association between sipping from a coffee cup and the coffee reward. However, this learning rule cannot account for the ability to *predict* reinforcement. Prediction is required to attach value to earlier contextual events, for example, to the sight of a full coffee cup, in order that its sensation can be used to control the various movements that eventually lead to sipping the coffee. The adaptive critic model presented in chapters 11 and 13 provides a mechanism for bridging the gap between prediction and reinforcement, based on two mechanisms that are biologically feasible. The first involves traces of synaptic activity left by intracellular chemical signals (also discussed in chapter 10) and the second involves reciprocal connectivity between SPs and DA neurons.

The key to learning first-order predictions, that is, predictions based on contexts occurring just prior to primary reinforcement, may be chemical signals that convey delayed dopamine receptivity in spiny neurons. In chapters 10 and 13, delayed dopamine receptivity was related to the dynamics of protein kinases in spiny neurons. This could provide a mechanism for reinforcing SPs and SPm synapses whose activity falls within a time window that precedes the primary reinforcing event. In addition, the delayed character of this receptivity (chapter 13) creates a situation where contexts occurring immediately before reinforcement are ignored. In our coffee cup example, the context of raising the cup to our lips would be reinforced in the SPs neurons while the context of sipping the coffee would be ignored.

In the adaptive critic model, first-order predictions by SPs neurons are then used recursively to control their own dopamine input. Two parallel pathways serve distinct roles within this recursive mechanism. First, a fast excitatory connection via the subthalamic nucleus allows DA neurons to respond to rewarding contexts recognized by SPs neurons. Given the SPs neuron's potential for pattern recognition, these contexts could be quite specific. This fits well with Schultz's observations during a blind reach task, when he found that unexpected contact with food rewards, which elicited a strong DA response, were quite discriminatory and could not be elicited by similarly shaped nonfood items. The second pathway, a slow inhibitory connection, provides a means for canceling the response of the DA neurons to the primary reinforcement as is consistently observed. Furthermore, inhibition unopposed by excitation could explain the observation in chapter 12 that when rewards are not delivered following the presentation of a predictive stimulus, DA activity is depressed at the time when the reward would have arrived. These postulated mechanisms thus provide a potential explanation for several of the salient features of DA neuron discharge and dopamine reinforced learning.

IV Cognitive and Memory Operations

14 Contribution of the Basal Ganglia to Skill Learning and Working Memory in Humans

John Gabrieli

It has long been appreciated that diseases primarily affecting the basal ganglia, such as Parkinson's disease (PD) and Huntington's disease (HD), lead to profound motor disorders. In Parkinson's disease, patients exhibit tremor, rigidity, and akinesia as a consequence of cell death in the substantia nigra (around an 80% loss) and a resultant depletion of dopamine (DA) in the striatum (also about an 80% reduction). HD patients, with primary lesions in the caudate and putamen, exhibit athetosis and chorea. More recently, there is direct evidence that Tourette's syndrome (TS), a chronic neurological disorder characterized by involuntary motor and phonic tics, is also a disease of the basal ganglia. TS patients show an abnormally large number of DA receptors in postmortem striatum (Singer et al., 1991), and abnormal volumes of caudate, putamen, and globus pallidus (Peterson et al., 1993; Singer et al., 1993).

Patients with diseases of the basal ganglia offer an opportunity to explore the contribution of the striatum and related structures to human information processing, including learning and memory. There is a growing body of evidence from studies with PD, HD, and TS patients indicating that the basal ganglia play an important role in two forms of learning and memory-skill learning and working memory. Such evaluations, however, must consider the cognitive abilities and disabilities shown by these patients. In broad terms, although some PD patients exhibit depression or dementia, pervasive changes in intellectual ability do not appear to be necessary consequence of the disease (although I present data below that specific changes in working memory ability may be a necessary consequence). In contrast, HD patients always develop a progressive dementia which results in pervasive deficits in cognition. Early in the disease, however, some forms of learning and memory appear relatively spared, whereas other forms appear to be greatly compromised. Little is known about learning and memory, or cognitive abilities altogether, in TS patients. Some TS patients exhibit attention disorders as children, but not all TS patients do so. There have been few systematic studies of cognition in adult TS patients, and none has examined forms of learning and memory linked to the basal ganglia via studies of PD or HD patients.

SKILL LEARNING IN PATIENTS WITH DISEASES OF THE BASAL GANGLIA

In skill-learning paradigms, subjects perform a challenging task on repeated trials in one or more sessions. The (indirect) measure of learning is the improvement in speed or accuracy achieved by a subject across trials and sessions. The neural basis of skill learning has been dissociated from that of declarative or explicit memory (the direct recall or recognition of events or facts learned in those events) through the performance of patients with global amnesia (who typically have bilateral lesions of medial temporal lobe or diencephalic structures). Amnesic patients have shown normal or near-normal acquisition and retention of *sensorimotor skills* on such tasks as mirror tracing (Milner, 1962; Gabrieli et al. 1993), rotary pursuit (Corkin, 1968; Cermak et al., 1973: Brooks and Baddeley, 1976) and serial reaction time (Nissan and Bullemer, 1987). They have also learned some *perceptual skills,* such as reading mirror-reversed text (Cohen and Squire, 1980; Martone et al., 1984), and some *problem-solving skills,* such as solving the Tower-of-Hanoi problem (Cohen and Corkin, 1981; Saint-Cyr et al., 1988; but see Butters et al., 1985) and learning an algorithm to square two-digit numbers mentally (Charness et al., 1988; Milberg et al., 1988). Amnesic patients gain and maintain mastery on these difficult tasks despite being unable to recall the episodes in which they learned their skill, unable to recognize the materials they encountered in those episodes, and unaware of their newly acquired prowess.

Mishkin and his collaborators have called skills *habits,* and proposed that the basal ganglia and cerebellum may mediate critical aspects of habit learning (Mishkin and Petri, 1984; Mishkin et al., 1984). Studies of HD and PD patients have provided support for that idea in humans. HD patients have shown impaired skill learning on rotary pursuit (Heindel et al., 1988, 1989), serial reaction time (Knopman and Nissen, 1991; Willingham and Koroshetz, 1993), and mirror reading (Martone et al., 1984). PD patients have shown impaired skill learning on rotary pursuit (Heindel et al 1989; Harrington et al., 1991), serial reaction time (Ferraro et al., 1993), and Tower of Hanoi (Saint-Cyr et al., 1988) tasks. We recently examined skill learning for the first time in TS patients, and found a mild but significant deficit on rotary pursuit learning (Stebbins et al., 1992). Patients with basal ganglia diseases have shown impairments on skill-learning tasks despite having better recall for their skill-learning experience and better recognition of the testing materials than the amnesic patients who learned these tasks normally. Thus, the skill-learning deficit in HD, PD, and TS patients cannot be accounted for by a generalized memory problem. Rather, these findings point toward a specific contribution of the basal ganglia to skill learning in humans.

We have tried to specify the nature of the contribution of the basal ganglia to skill learning by attempting to find out what sort of skills can be learned by early-stage HD patients. We had HD patients perform two skill-learning tasks: rotary pursuit and mirror tracing. In the rotary pursuit task, subjects try to maintain contact between a handheld stylus and a target metal disk, the size if a nickel, on a revolving turntable. With practice, control subjects demonstrate skill learning by increasing the time per trial that they are able to maintain contact with the disk. Examining learning on this task by patients with diseases of motor control poses a problem, because patients have a substantial performance deficit. Comparing learning between patients and controls performing at quite different levels of competence raises the difficult question of how learning measures ought to be compared. One attempt to deal with this problem for the rotary pursuit task is to equate performance by setting a speed of rotation for each subject such that initial performance for all subjects consists of about 5 seconds of contact for each 20 seconds. Thus, any learning deficit cannot be attributed simply to a performance deficit (the disadvantage of this method is that different levels of task difficulty could recruit different motor mechanisms in control and patient groups).

We examined the performance of HD patients on the rotary pursuit task on three series of eight 20-second trials. An initial block of practice trials was used to determine the speed of rotation for each subject such that the score (time on target) was close to 5 seconds. Normal subjects showed skill learning by increasing time-on-target across trials (figure 14.1). HD patients, by contrast, showed almost no learning at all, despite their initially equivalent performance. These results are similar to prior findings with HD patients (Heindel et al., 1988, 1989). We were concerned whether matching the initial level of performance at 5 seconds was potentially misleading; perhaps HD patients were performing at some sort of ceiling level at 5 seconds. To address this issue, we tested three additional HD patients, and set their initial performance at a higher level than that of controls and at the midrange of performance, about 10 seconds time-on-target. These patients (HD+) also showed no skill learning on the rotary pursuit task (figure 14.2).

On the mirror-tracing task, the subject sits facing a mirror, mounted on the far side of a metal board that obscures a six-pointed star consisting of nonconducting tape arranged on an aluminum plate. The subject holds a stylus in his or her preferred hand, and is told to trace the pattern, from a starting mark, and to proceed as quickly and as accurately as possible. The subject is able to see his or her hand and the pattern only in the mirror, so that many overlearned associations between vision and movement are reversed. A ground wire between the aluminum plate and a scoring box completes a low-voltage circuit

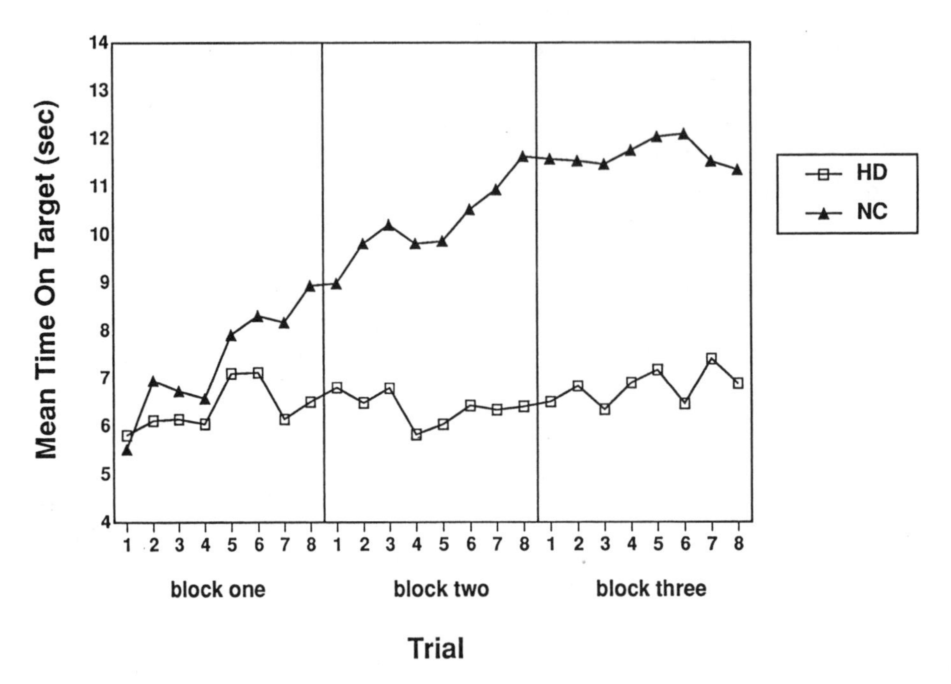

Figure 14.1 Performance on the rotary-pursuit task of patients with Huntington's disease (HD) and normal control (NC) subjects.

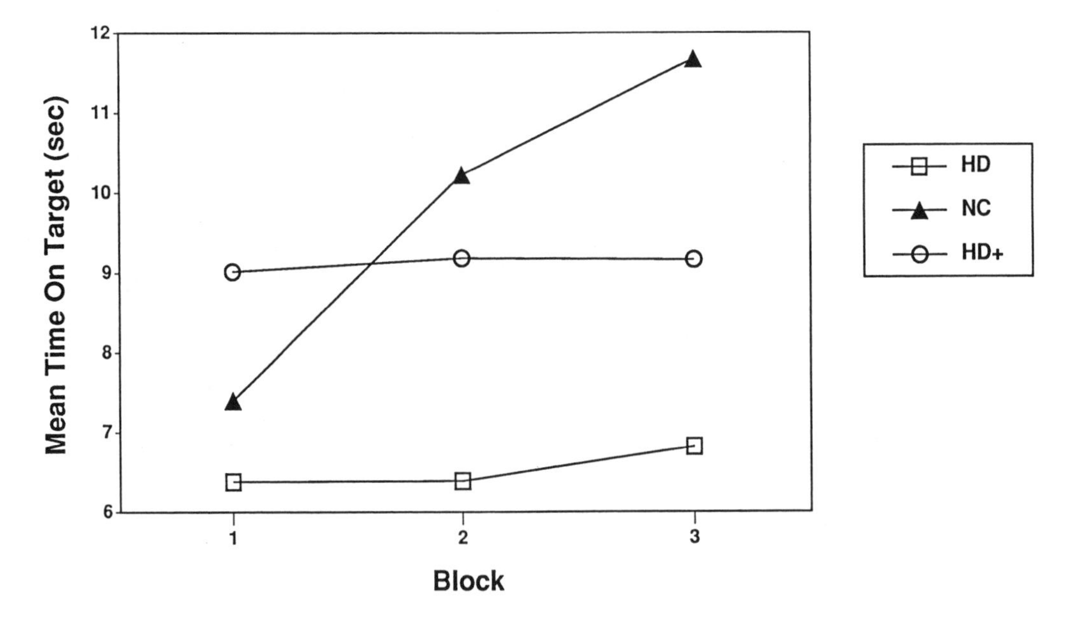

Figure 14.2 Performance on the rotary-pursuit task of patients with Huntington's disease (HD), normal control (NC) subjects, and additional patients wtih Huntington's disease (HD+) whose initial performance was made superior to that of control subjects by slowing the speed of rotation.

whenever the stylus touches the nonconducting tape. The scoring box records the number of times the subject goes off the pattern (errors), and time to trace around the pattern (completion time) is measured by a stopwatch. Subjects performed two blocks of trials with five trials per block. Normal control (NC) subjects initially performed poorly, tracing the star slowly and often departing from the star pattern. With practice, NC subjects traced the star more quickly and made fewer errors (figures 14.3 and 14.4). HD patients seemed to show robust learning by both skill-learning measures; their learning was similar to that of NC subjects. Their somewhat (although not statistically reliable) worse overall performance complicates interpretation of the results. It appears, however, that the same HD patients who showed no skill learning at all on the rotary pursuit task exhibited relatively normal learning on the mirror-tracing task.

ROLE OF THE BASAL GANGLIA IN SKILL LEARNING

The apparent dissociation in HD patients between impaired rotary pursuit and intact mirror-tracing skill learning can be interpreted in many ways, because the two tasks differ in many ways. One possibility is that the basal ganglia are critical for perceptual-motor sequencing, the rapid selection of perceptual, cognitive, and motor operations that allows for good performance and skill learning (see also Willingham, 1992). Such sequencing may be a critical aspect of rotary pursuit skill

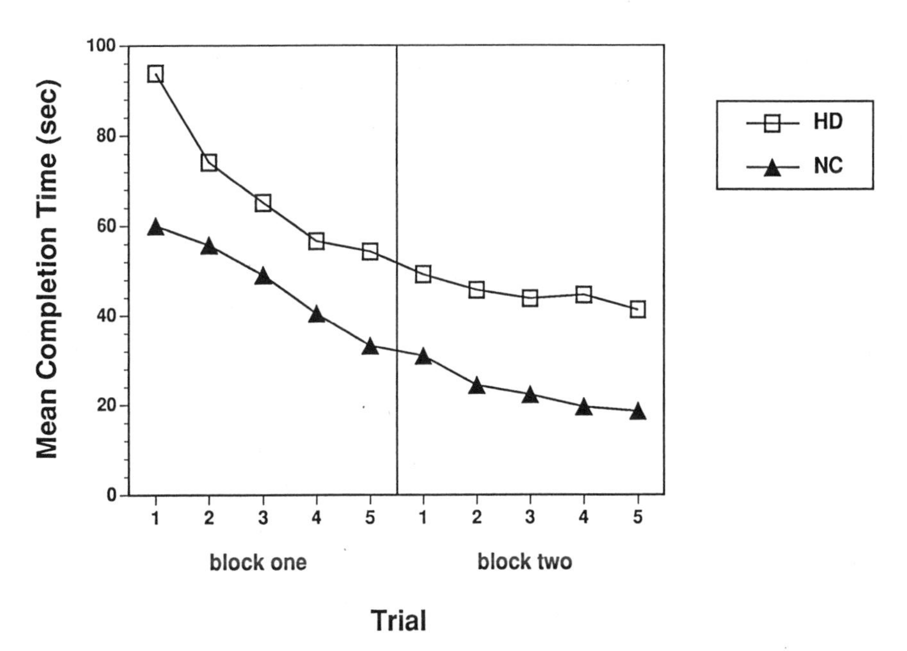

Figure 14.3 Performance on the mirror-tracing task of patients with Huntington's disease (HD) and normal control (NC) subjects.

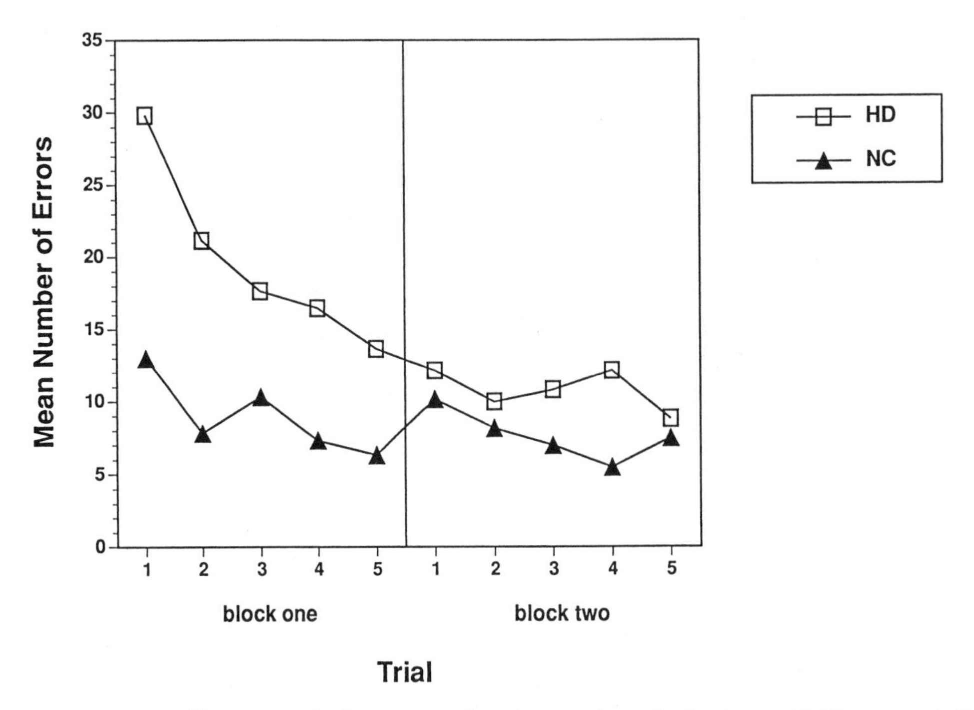

Figure 14.4 Performance on the mirror-tracing task of patients with Huntington's disease (HD) and normal control (NC) subjects.

learning. Perhaps subjects gain rotary pursuit skill as they learn to execute smoothly a series of movements that maintains contact between the stylus and the revolving target. Skill learning on the mirror-tracing task may be less dependent on sequencing, and more dependent on the learning of novel perceptual-motor mappings. Perhaps this form of learning is more dependent on cerebellar circuits (Sanes et al., 1990). The idea that the basal ganglia are critical for skill learning dependent on perceptual-motor sequencing but not for skill learning dependent on perceptual-motor mapping is supported by findings with HD patients on the serial reaction time task (Willingham and Koroshetz, 1993).

A recent positron emission tomography (PET) study is relevant because it examined which neural brain regions were activated, as measured by increased relative cerebral blood flow, while normal subjects performed rotary pursuit across a number of trials (Grafton et al., 1992). Motor execution (the difference between rotary pursuit performance and a baseline condition) was associated with cortical, nigrostriatal, and cerebellar activation. Activation in the last two regions is consistent with the evidence from the patient studies reviewed above. Skill learning, however, was linked with increasing activation in primary motor cortex, the supplementary motor area, and the pulvinar thalamus contralateral to the hand used to do the task. One difficulty in interpreting these results is that the actual motor performance of a skilled subject differs in many ways from that of a less skilled subject. For example, unskilled subjects make many ballistic movements after they lose contact with the disk and attempt to regain contact. In contrast, skilled subjects perform in a smooth fashion, maintaining a circular movement that sustains contact with the disk. Thus, the skill-learning comparison includes a mixture of acquired knowledge underlying newfound skill and changes in motor performance that are the expression of that skill. The most simple interpretation of these results, however, is that contributions from the cerebellum and basal ganglia may be important in the formation and perhaps early expression of newly acquired skills, but that the long-term memory underlying skilled performance is located cortically. In that case, the cerebellum and basal ganglia could play roles in skill learning that are somewhat analogous to the role of the hippocampus in declarative memory. The hippocampus appears to be necessary for the acquisition and early expression of declarative memories, but the final locus of well-established declarative information appears to be neocortical (Zola-Morgan and Squire, 1990; Squire and Zola-Morgan, 1991). The opposite interpretation is also plausible at present, that the increases in motor cortex activation reflect changes in performance, and that the absence of increasing striatal and cerebellar activation reflects increased efficiency due to knowledge acquired in those structures.

WORKING MEMORY IN PATIENTS WITH DISEASES OF THE BASAL GANGLIA

Working memory is a multicomponent psychological system that supports the temporary storage, manipulation, and transformation of information needed to perform cognitive tasks. Working memory conducts interactions between perception of the outside world, long-term knowledge, and actions in the service of intelligent goals and plans. The three central features of this computational arena for processing and storing information are that (1) it has sharply limited resources; (2) it plays an increasingly vital role as a cognitive task becomes increasingly complex; and (3) it is used for the full range of high-level cognitive performance. Goldman-Rakic (1987) has provided considerable and compelling evidence for the participation of frontal cortical regions in working memory. Further, Goldman-Rakic and her collaborators have shown that the neurotransmitter DA plays a critical role in the working memory aspect of performance on delayed-response tasks as shown by the local depletion of DA (Brozoski et al., 1979) and the injection of receptor-specific D1 antagonists (Sawaguchi and Goldman-Rakic, 1991).

The importance of DA for working memory in infrahuman primates raises the question of its role in human working memory. Patients with PD provide an opportunity to examine this issue, because PD primarily affects DA systems. PD patients, in fact, do perform poorly on a range of problem-solving (e.g., Tower of London, Wisconsin Card Sorting) and other tasks demanding the flexible shifting of strategy, especially when that shift must overcome habitual responses and when the strategy must be guided by internal plans rather than external cues (e.g., Brown and Mardsden, 1988). Two studies that manipulated DA administration to PD patients lend support to the view that DA plays a critical role in human working memory. In one study, levodopa treatment was temporarily withdrawn from PD patients, and the patients performed poorly on a test of planning (Tower of London; Lange et al., 1992; but see Gotham et al., 1988, for a more complex consequence of levodopa withdrawal). In a second study, previously untreated PD patients were examined before and after beginning levodopa treatment. Patients receiving levodopa demonstrated not only the expected improvement in motor performance, but also improved performance on a digit-ordering task thought to be a measure of working memory. The same patients did not show similar improvements on a broad range of other cognitive abilities (Cooper, et al., 1992).

The idea that working memory is important for problem-solving and digit-ordering tasks seems reasonable, but research in the past decade has produced more direct measures of the construct of working memory. Daneman and Carpenter (1980) created a direct measure of working capacity or span that has correlated with a broad range of intellectual abilities in normal subjects. We used a version of their task developed by Salthouse and Babcock (1991) for studies of norman aging. Subjects were instructed to listen to a sentence (e.g., "The boy ran with the dog"), to immediately afterward select from among three choices the correct answer to a question about the sentence (e.g., Who ran? — boy;—man; —girl), and to remember the last word from each sentence for later recall. The number of sentences presented on each trial increased successively from one to seven, with three trials presented at each series length. After selecting the answers for the questions about each sentence in a trial, subjects were asked to turn to write down the last word of each sentence in the order that they had been presented. The span score was the maximum number of sentences for which a subject recalled all final words in the correct order for at least two of three trials. Thus, this measure operationalizes working memory directly by making subjects simultaneously (a) process sentences one after another in order to answer a question about each sentence, and (b) hold in memory the final words of each sentence for later recall.

We tested a group of early-stage PD patients who were untreated, not demented, not depressed, and did as well on a vocabulary test as their age- and education-matched normal control group (NC). The PD patients had half the working memory capacity or span of the control group (figure 14.5). Furthermore, their working memory impairment

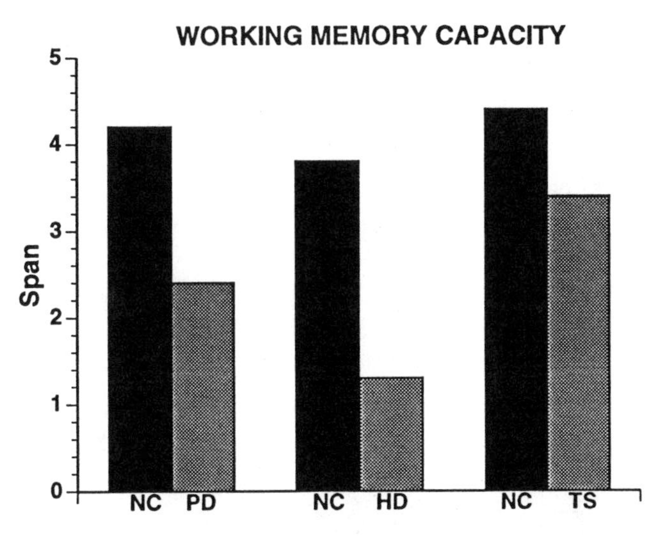

Figure 14.5 Working memory capacity of normal control (NC) subjects and patients with Parkinson's disease (PD), Huntington's disease (HD), and Tourette's syndrome (TS).

was manifest across knowledge domains: the PD patients had a similar impairment on a test of working memory for arithmetic and numbers. Indeed, substantially reduced working memory capacity appears to be the rule in diseases of the basal ganglia: patients with early-stage HD and younger adults with TS also demonstrated substantially reduced working memory spans with the same measure (see figure 14.5).

Working memory may be an important resource for performance on a class of strategic memory tasks that are often failed by patients with dorsolateral frontal lobe lesions who do not show overt memory problems. These tasks include memory for the temporal order of events, frequency of events, the source (as opposed to the content) of recently encountered information, and self-ordered pointing to an array of stimuli. These tasks share the property that optimal performance requires a strategy that may be said to work with or work on memory for previous experience in order to accomplish a novel goal. The distinction between memory tasks that demand strategy, or considerable working memory, and tasks that demand relatively little strategy may be seen in the contrast between recall and recognition of words lists. We presented subjects with two lists of 24 words each. After one list, and a short distractor period, subjects were asked to recall the words. After the other list, subjects were shown pairs of words in a two-alternative forced-choice recognition memory test. Recall is more demanding of working memory because it requires a self-initiated retrieval strategy, whereas subjects can use the presented words to guide their retrieval for recognition. PD, HD, and TS patients showed a pattern of performance that may be a signature of impaired frontostriatal function: good recognition and poor recall (figure 14.6).

MEMORY FOR WORDS

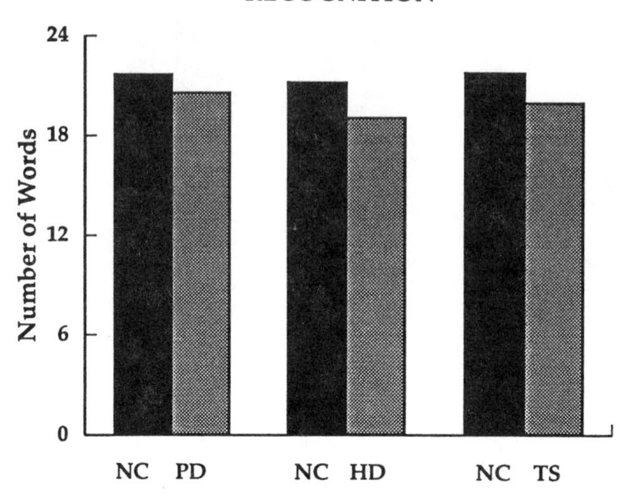

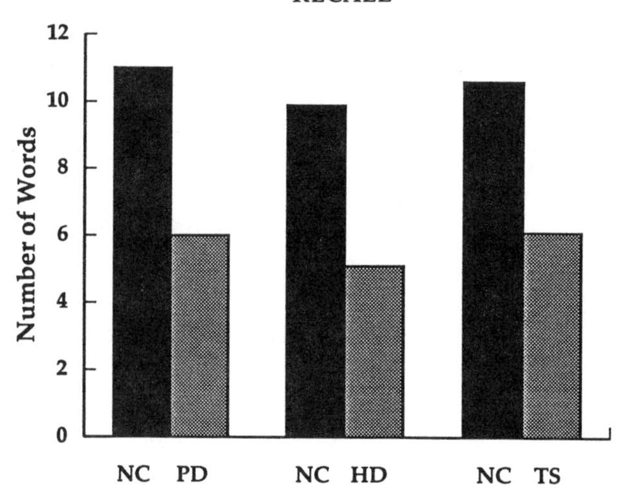

Figure 14.6 Memory for words in normal control (NC) subjects and patients wtih Parkinson's disease (PD), Huntington's disease (HD), and Tourette's syndrome (TS). (*Top*) Mean number of words recognized correctly (maximum 24). (*Bottom*) Mean number of words recalled correctly (maximum 24).

Patients with basal ganglia lesions thus have striking deficits in working memory and in strategic memory tasks that appear to rely on working memory. Patients with focal dorsolateral frontal lobe lesions show a strikingly similar pattern of strategic memory failure despite relatively intact recognition memory (table 14.1). The similarity may arise for several reasons. One possibility is that many or all PD, HD, and perhaps TS patients have some degree of frontal lobe damage. For example, there is evidence that in addition to the nigrostriatal damage in PD, there is also damage to the mesocortical dopamingeric projection from the ventral tegmental area to the frontal lobes (Javoy-Agid and Agid, 1980; Scatton et al., 1982). A second possibility is that there is minimal direct damage to frontal cortex, but a functional deafferentation in corticostriatal loops that results in frontal lobe dysfunction. PET studies have reported frontal lobe hypoactivity in early stage HD (Martin et al., 1992) and PD (Bes et al., 1983; Wolfson et al., 1985). Therefore, it seems appropriate to conceptualize the working and strategic memory failures consequent to frontal or basal ganglia lesions as evidence for the importance of frontostriatal circuits or loops (Alexander et al., 1986) for those sorts of memory. The notion that damage to any point in the loop can lead to similar consequences in memory performance is similar to what is known in amnesia, where damage to anatomically disparate medial temporal lobe, diencephalic, or basal forebrain structures can all lead to similar failures in declarative or explicit memory.

One final intriguing link between DA, frontostriatal loops, and working and strategic memory comes from studies of normal aging. Studies comparing memory performance between older and younger subjects have often found relatively selective age-associated declines in strategic

Table 14.1 Memory impairments common to frontal lobe and Parkinson's disease (PD) patients and older subjects

	Study		
Memory impairment	Frontal Lobe Patients	PD Patients	Older Subjects
Temporal order	Milner, 1971	Sagar et al., 1988	Naveh-Benjamin, 1990
Source	Janowsky et al., 1989	Taylor et al., 1990	McIntyre and Craik, 1987
Conditional associative learning	Petrides, 1985	Gotham et al., 1988*	
Self-ordered pointing	Petrides and Milner, 1982	Gotham et al., 1988*	Shimamura and Jurica, in press
Recall relative to recognition	Jetter et al., 1986	Taylor et al., 1986	Craik and McDowd, 1987

*On levadopa only.

memory tasks (see table 14.1). Using the test of working memory described above, Salthouse and Babcock (1991) found a steady, linear decrease in working memory span from age 20 through 70 years. Like the linear life span decrease in working memory, a number of investigators examining postmortem cell counts in the substantia nigra, postmortem indices of striatal DA function, and in vivo PET indices of striatal DA function have reported linear life span decreases in the nigrostriatal DA system (table 14.2). Thus, normal aging features a correlative triad that is similar, but less severe psychologically and biologically, to PD: reduced nigrostriatal DA activity, reduced working memory capacity, and reduced strategic memory performance. Perhaps a nonclinical decline in DA-striatal activity accounts for a substantial portion of normal age-associated declines in working and strategic memory.

ROLE OF THE BASAL GANGLIA IN WORKING MEMORY

What is the specific role of the basal ganglia in working memory? As discussed above, it is difficult at present to disentangle the specific contributions of the basal ganglia vs. frontal cortex to working memory, because the two brain regions may operate in such close functional concert. One interesting lead, which may or may not differentiate basal

Table 14.2 Linear age-dependent reduction in striatal dopaminergic function

Measure*	N	Reduction/Decade	Reference
Post mortem			
No. of pigmented neurons in substantia nigra	36	4.7%	Fearnley and Lees, 1991
Cell count in substantia nigra	28	6%	McGeer et al., 1977
Caudate dopamine	18	13%	Riederer and Wuketich, 1976
Alpha-dihydrotetrabenazine binding in caudate	49	10%	Scherman et al., 1989
Dopaminergic uptake sites in putamen	32	6%	De Keyser et al., 1990
Dopaminergic uptake sites in caudate	44	7.5%	Severson et al., 1982
[^{3}H]Spiroperidol binding in substantia nigra and putamen	78	6.4%	Rinne, 1987
In vivo			
PET 6-fluoro-L-dopa binding in striatum	10	9%	Martin et al., 1989
PET [^{11}C] NMSP binding in caudate and putamen	44	9%	Wong et al., 1984

*PET, positron emission tomography; NMSP, *N*-methylspiperone.

ganglia and frontal cortical contributions to working memory, comes from two correlations we examined between simple measures of psychomotor speed and the working memory span measure that had been administered to PD patients. The Symbol Digit Modalities Test (SDMT) is a test of psychomotor speed in which subjects are presented with a "key" consisting of nine simple geometric designs (e.g., a plus sign, a circle, an inverted T); each design is matched with a unique digit from 1 to 9. The test form consists of a series of the nine designs, and the subject's task is to write down next to each design the corresponding digit from the key. Subjects are asked to perform quickly and accurately, and their score is the number of correctly completed targets in a 90-second trial. We correlated performance on the SDMT with the two span scores (verbal and arithmetic) that the nine PD patients received. Despite the small number of subjects, there were substantial correlations between SDMT scores and both verbal ($r=.65$, $p=.08$) and arithmetic ($r=.73, p<.05$) working memory spans (such that lower SDMT scores correlated with lower span scores). Patients also performed the Purdue Pegboard, a test of manual speed and dexterity in which subjects are asked to place grooved pegs into a series of vertically placed holes in a formboard. The subject's score is the total number of pegs placed in the holes in three 30-second trials. There were no correlations between pegboard performance and working memory span scores, indicating that it is the interaction between perception and motor execution rather than the motoric process per se that underlies the working memory deficit. In a second set of 25 PD patients who represent a broader clinical spectrum than the untreated original sample, we found further evidence for a link between psychomotor slowing and working memory impairment. There was a strong correlation between SDMT scores and verbal working memory span ($r=.61, p<.001$) and free recall performance ($r=.53, p<.01$), but not between SDMT scores and recognition performance.

These results suggest that the critical contribution of the basal ganglia to working memory performance is mediated through psychomotor speed. Further, the lack of a correlation between a more purely motor measure (pegboard) and working memory performance may reflect the separation of corticostriatothalamic loops, particularly the distinction between a motor loop (critical for pegboard performance) and a dorsolateral prefrontal loop (critical for SDMT and working memory span).

CONCLUSION

Skill learning and working memory are consistently impaired in patients with diseases of the basal ganglia. Both deficits are selective. For skill learning, HD patients showed intact mirror-tracing skill learning, but impaired rotary pursuit skill learning. On memory tests, PD, HD, and

TS patients had intact performance on a recognition test, but impaired performance on a recall test more demanding of working memory. In broad terms, both the skill-learning and working memory deficits may reflect the importance of the basal ganglia for psychomotor sequencing performance, with different loops supporting sequencing in different domains (motor, cognitive). One may speculate that an essential contribution of the basal ganglia to human learning and memory is to support the speeded execution of component processes of a multistep cognitive or motor action. When that support is lost due to a basal ganglia disease, components are executed too slowly to accomplish either the smooth sequence of movements that characterizes perceptual-motor skill or the rapid sequence of thoughts that characterizes flexible working memory capacities. The idea that speed of execution can be vital for effective skill learning and efficient working memory can seem computationally mundane, but computers themselves have shown dramatically how processing speed alone can account for success or failure in accomplishing the goals of computations.

ACKNOWLEDGMENTS

The above research was done in collaboration with Glenn Stebbins, Jaswinder Singh, Chrostopher Goetz, and Daniel Willingham. The writing of this chapter and the original research reported were supported by a grant from Office of Naval Research (grant N00014-92-J-184).

REFERENCES

Alexander, G. E., DeLong, M. R., and Strick, P. L. (1986) Parallel organization of functionally segregated curcuits linking basal ganglia and cortex. *Annu. Rev. Neurosci.* 9:357–381.

Bes, A., Guell, A., Fabre, N., Dupui, Ph., Victor, G., and Geraud, G. (1983) Cerebral blood flow studied by xenon-133 inhalation technique in Parkinsonism: Loss of hyperfrontal pattern. *J. Cereb. Blood Flow Metab.* 3:33–37.

Brooks, D. N., and Baddeley, A. (1976) What can amnesic patients learn? *Neuropsychologia* 14:111–122.

Brown, R. G., and Marsden, C. D. (1988) Internal versus external cues and the control of attention in Parkinson's disease. *Brain* 111:323–345.

Brozoski, T., Brown, R. M., Rosvold, H. E., and Goldman-Rakic, P. S. (1979) Cognitive deficit caused by regional depletion of dopamine in prefrontal cortex of rhesus monkey. *Science* 205:929–932.

Butters, N., Wolfe, J., Martone, M., Granholm, E. and Cermak, L. S. (1985) Memory disorders associated with Huntington's disease: Verbal recall, verbal recognition, and procedural memory. *Neuropsychologia* 6:729–744.

Cermak, L. S., Lewis, R., Butters, N., and Goodglass, H. (1973) Role of verbal mediation in performance of motor tasks by Korsakoff patients. *Percept. Mot. Skills* 37:259–263.

Charness, N., Milberg, W., and Alexander, M. P. (1988) Teaching an amnesic a complex cognitive skill. *Brain Cogn.* 8:253–272.

Cohen, N. J., and Corkin, S. (1981) The amnesic patient H. M.: Learning and retention of a cognitive skill. *Soc. Neurosci. Abstr.* 7:517.

Cohen, N. J., and Squire, L. R. (1980) Preserved learning and retention of pattern-analyzing skill in amnesia: Dissociation of knowing how and knowing that. *Science* 210:207–210.

Cooper, J. A., Sagar, H. J., Doherty, S. M., Jordan, N., Tidswell, P., and Sullivan, E. V. (1992) Different effect of dopaminergic and anticholinergic therapies on cognitive and motor function in Parkinson's disease. *Brain* 115:1701–1725.

Corkin, S. (1968) Acquisition of motor skill after bilateral medial temporal-lobe excision. *Neuropsychologia* 6:255–265.

Craik, F. I. M., and McDowd, J. M. (1987) Age differences in recall and recognition. *J. Exp. Psychol. [Learn. Mem. Cogn.]* 13:474–479.

Daneman, M., and Carpenter, P. (1980) Individual differences in working memory and reading. *J. Verbal Learn. Verbal Behav.* 19:450–466.

DeKeyser, J., Ebinger, G., and Vauquelin, G. (1990) Age-related changes in the human nigrostriatal dopaminergic system. *Ann. Neurol.* 27:157–161.

Fearnley, J. M., and Lees, A. J. (1991) Ageing and Parkinson's disease: Substantia nigra regional selectivity. *Brain* 114:2283–2301

Ferraro, R. F., Balota, D. A., and Connor, L. T. (1993) Implicit memory and the formation of new associations in nondemented Parkinson's disease individuals and individuals with senile dementia of the Alzheimer type: A serial reaction time (SRT) investigation. *Brain Cogn.* 21:163–180.

Gabrieli, J. D., Corkin, S., Mickel, S. F., and Growdon, J. H. (1993) Intact acquisition and long-term retention of mirror tracing skill in Alzheimer's disease and in global amnesia. *Behav. Neurosci.* 107:899–910.

Goldman-Rakic, P. S. (1987) Circuitry of primate prefrontal cortex and regulation of behavior by representational memory. In F. Plum (ed.), *Handbook of Physiology—The Nervous System V.* New York: Oxford University Press, pp. 373–417.

Gotham, A. M., Brown, R. G., and Marsden, C. D. (1988) "Frontal" cognitive function in patients with Parkinson's disease "on" and "off" levodopa. *Brain* 111:299–321.

Grafton, S. T., Mazziotta, J. C., Presty, S., Friston, K. J., Frackowiak, R. S. J. and Phelps, M. E. (1992) Functional anatomy of human procedural learning determined with regional cerebral blood flow and PET. *J. Neurosci.* 12:2542–2548.

Harrington, D. L., Haaland, K. Y., Yeo, R. A., and Marder, E. (1991) Procedural memory in Parkinson's disease: Impaired motor but not visuoperceptual learning. *J. Clin. Exp. Neuropsychol.* 12:323–339.

Heindel, W. C., Butters, N., and Salmon, D. P. (1988) Impaired learning of a motor skill in patients with Huntington's disease. *Behav. Neurosci.* 102:141–147.

Heindel, W. C., Salmon, D. P., Shults, C. W., Walicke, P. A., and Butters, N. (1989) Neuropsychological evidence for multiple implicit memory systems: A comparison of Alzheimer's, Huntington's, and Parkinson's disease patients. *J. Neurosci.* 9:582–587.

Janowsky, J. S., Shimamura, A. P., and Squire, L. R. (1989) Memory and metamemory: Comparisons between patients with frontal lobe lesions and amnesic patients. *Psychobiology* 17:3–11.

Javoy-Agid, F., and Agid, Y. (1980) Is the mesocortical dopaminergic system involved in Parkinson's disease? *Neurology* 30:1326–1330.

Jetter, W., Poser, U., Freeman, R. B. J., and Markowitsch, H. J. (1986) A verbal long term memory deficit in frontal lobe damaged patients. *Cortex* 22:229–242.

Knopman, D., and Nissen, M. J. (1991) Procedural learning is impaired in Huntington's disease: Evidence from the serial reaction time task. *Neuropsychologia* 29:245–254.

Lange, K. W., Robbins, T. W., Marsden, C. D., James, M., Owen, A. M., and Paul, G. M. (1992) L-Dopa withdrawal in Parkinson's disease selectively impairs cognitive performance in tests sensitive to frontal lobe dysfunction. *Psychopharmacology* 107:394–404.

Martin, W. R. W., Palmer, M. R., Patlak, C. S., and Caine, D. B. (1989) Nigrostriatal function in humans studied with positron emission tomography. *Ann. Neurol.* 26:535–542.

Martin, W. R. W., Clark, C., Ammann, W., Stoessl, A. J., Shtybel, W., and Hayden, M. R. (1992) cortical glucose metabolism in Huntington's disease. *Neurology* 42:223–229.

Martone, M., Butters, N., Payne, M., Becker, J. T., and Sax, D. S. (1984) Dissociations between skill learning and verbal recognition in amnesia and dementia. *Arch. Neurol.* 41:965–970.

McGeer, P. L., McGeer, E. G., and Suzuki, J. S. (1977) Aging and extrapyramidal function. *Arch. Neurol.* 34:33–35.

McIntyre, J. S., and Craik, F. I. M. (1987) Age differences in memory for item and source information *Can. J. Psychol.* 41:175–192.

Milberg, W., Alexander, M. P., Charness, N., McGlinchey-Berroth, R., and Barrett, A. (1988) Learning of a complex arithmetic skill in amnesia: Evidence for a dissociation between compilation and production. *Brain Cogni.* 8:91–104.

Milner, B. (1962) Les troubles de la mémoire accompagnant des lésions hippocampiques bilatérales. In *Psychologie de l'hippocampe*. Paris: Centre National de la Recherche Scientifique.

Milner, B. (1971) Interhemispheric differences in the localization of psychological processes in man. *Bri. Med. J.* 27:272–277.

Mishkin, M., and Petri, H. L. (1984) Memory and habits: Some implications for the analysis of learning and retention. In L. R. Squire and N. Butters (eds.), *Neuropsychology of Memory*. New York: Guilford Press, pp. 287–296.

Mishkin, M., Malamut, B., and Bachevalier, J. (1984) Memories and habits: Two neural systems. In G. Lynch, J. L. McGaugh and N. M. Weinberger (eds.), *Neurobiology of Learning and Memory*. New York: Guilford Press, pp. 287–296.

Neveh-Benjamin, M. (1990) Coding of temporal order information: An automatic process? *J. Exp. Psychol.* [*Learn. Mem. Cogn.*] 16: 117–126.

Nissen, M. J., and Bullemer, P. (1987) Attentional requirements of learning: Evidence from performance measures. *Cogn. Psychol.* 19:1 32.

Peterson, B., Riddle, M. A., Cohen, D. J., Katz, L. D., Smith, J. C., Hardin, M. T., and Leckman, J. F. (1993) Reduce basal ganglia volumes in Tourrette's syndrome using three-dimensional reconstruction techniques from magnetic resonance images. *Neurology* 43:941–949.

Petrides, M. (1985) Deficits on conditional associative-learning tasks after frontal- and temporal-lobe lesions in man. *Neuropsychologia* 20:249–262.

Petrides, M., and Milner, B. (1982) Deficits on subject-ordered tasks after frontal and temporal lobe lesions in man. *Neuropsychologia* 20:601–614.

Riederer, P., and Wuketich, S. (1976) Time course of nigrostriatal degeneration in Parkinson's disease. *J. Neural Transm.* 38:277–301.

Rinne, J. O. (1987) Muscarinic and dopaminergic receptors in the aging human brain. *Brain Res.* 404:162–168.

Sagar, H. J., Sullivan, E. V., Gabrieli, J. D. E., Corkin, S., and Growdon, J. H. (1988) Temporal ordering and short-term memory deficits in Parkinson's disease. *Brain* 111:525–539.

Saint-Cyr, J. A., Taylor, A. E., and Lang, A. E. (1988) Procedural learning and neostriatal function in man. *Brain* 111:941–959.

Salthouse, T. A., and Babcock, R. L. (1991) Decomposing adult age differences in working memory. *Dev. Psychol.* 27:763–776.

Sanes, J. N., Dimitrov, B., and Hallet, M. (1990) Motor learning inpatients with cerebellar dysfunction. *Brain* 113:103–120.

Sawaguchi, T., and Goldman-Rakic, P. S. (1991) D1 Dopamine receptors in prefrontal cortex: Involvement in working memory. *Science* 251:947–950.

Scatton, B., Rouquier, L., Javoy-Agid, F., and Agid, Y. (1982). Dopamine deficiency in the cerebral cortex in Parkinson disease. *Neurology* 32:1039–1040.

Scherman, D., Desnos, C., Darchen, F., Pollack, P., Javoy-Agid, F., and Agid, Y. (1989) Striatal dopamine deficiency in Parkinson's disease: Role of aging. *Ann. Neurol.* 26: 551–557.

Severson, J. A., Marcusso, J., Winblad, B., and Finch, C. E. (1982). Age-correlated loss of dopaminergic binding sites in human basal ganglia. *J. Neurochem.* 39:1623–1631.

Shimamura, A. P., and Jurica, P. J. (in press). Memory interference effects and aging: Findings from a test of frontal lobe function. *Neuropsychology.*

Singer, H. S., Hahn, I.-H., and Moran, T. H. (1991). Abnormal dopamine uptake sites in postmortem striatum from patients with Tourette's syndrome. *Ann. Neurol.* 30:558–562.

Singer, H. S., Reiss, A. L., Brown, J. E., Aylward, E. H., Shih, B., Chee, E., Harris, E. L., Reader, M. J., Chase, G. A., Bryan, R. N., and Denckla, M. B. (1993) Volumetric MRI changes in basal ganglia of children with Tourette's syndrome. *Neurology* 43:950–956.

Squire, L. R., and Zola-Morgan, S. (1991) The medial temporal lobe memory system. *Science* 253:1380–1386.

Stebbins, G. T., Singh, J., Gabrieli, J. D. E., Commella, C. L., and Goetz, C. G. (1992) Impaired working memory in unmedicated adults with Giles de la Tourette's syndrome. *Soc. Neurosci. Abstr.* 18:1213.

Taylor, A. E., Saint-Cyr, J. A., and Lang, A. E. (1986) Frontal lobe dysfunction in Parkinson's disease: The cortical focus of neostriatal outflow. *Brain* 109:845–883.

Taylor, A. E., Saint-Cyr, J. A., and Lang, A. E. (1990) Memory and learning in early Parkinson's disease: Evidence for a "frontal lobe syndrome." *Brain Cogn.* 13:211–232.

Willingham, D. B. (1992) Systems of motor skill. In L. R. Squire and N. Butters (eds), *Neuropsychology of Memory.* New York: Guilford Press, pp. 166–178.

Willingham, D. B., and Koroshetz, W. J. (1993) Evidence for dissociable motor skills in Huntington's disease patients. *Psychobiology* 21:173–182.

Wolfson, L. I., Klaus, L. L., Brown, L. L., and Jones, T. (1985) Alterations of regional cerebral blood flow and oxygen metabolism in Parkinson's disease. *Neurology* 35:1399–1405.

Wong, D. F., Wagner, H. N., Dannals, R. F., Links, J. M., Frost, J. J., Ravert, H. T., Wilson, A. A., Rosenbaum, A. E., Gjedde, A., Douglass, K. H., Petronis, J. D., Folstein, M. F., Thomas Toung, J. K., Burns, D., Kuhar, M. J. (1984) Effects of age on dopamine and serotonin receptors measured by positron tomography in the living human brain. *Science* 226:1393–1396.

Zola-Morgan, S. M., and Squire, L. R. (1990) The primate hippocampal formation: Evidence for a time-limited role in memory storage. *Science* 250:288–290.

15 Memory Limits in Sensorimotor Tasks

Dana H. Ballard, Mary M. Hayhoe, and Jeff Pelz

The very limited capacity of short-term memory is one of the most prominent features of human cognition and has been much studied (Miller 1956; Simon and Chase 1973; Nakayama 1990; Allport 1989). However, most of the studies have stressed delimiting the upper bounds of this memory and few studies have posited a role for short-term memory in cognition. In this chapter we suggest a computational role for a limited short-term memory, and in the context of this book, go on to indicate how a structure with some of the properties revealed by the basal ganglia might play a crucial part. Such a role may be understood from research in robot models of cognitive processes. Recently a range of complex tasks have been modeled efficiently using very limited memory representations. The key aspect of these models is that complex internal representations are avoided by allowing frequent access to the sensory input during the problem-solving process (Brooks 1986; 1991; Agre and Chapman 1987; Ballard 1989; 1991). The models use deictic primitives, which dynamically refer to points in the world with respect to their crucial describing features (e.g., color or shape). The word *deictic* means "pointing out" or "showing" and was first used in this context by Agre and Chapman (1987), building on work by Ullman (1984). It means that aspects of the scene can be referred to by denoting that part of the scene with a special referent or pointer.

DEICTIC STRATEGIES

We hypothesize that the human visuomotor system works by composing hand and eye behaviors from discrete behavioral components and that these components exploit the momentary binding between the perceptual-motor system and the world. This binding can be achieved by actively looking directly at, or fixating, an environmental point. This ability allows the use of a frame of reference centered at the fixation point. As shown in figure 15.1, the "fixation frame" is viewer-oriented, but not viewer-centered. The fixation frame allows for closed-loop behavioral strategies that do not require very precise three-dimensional

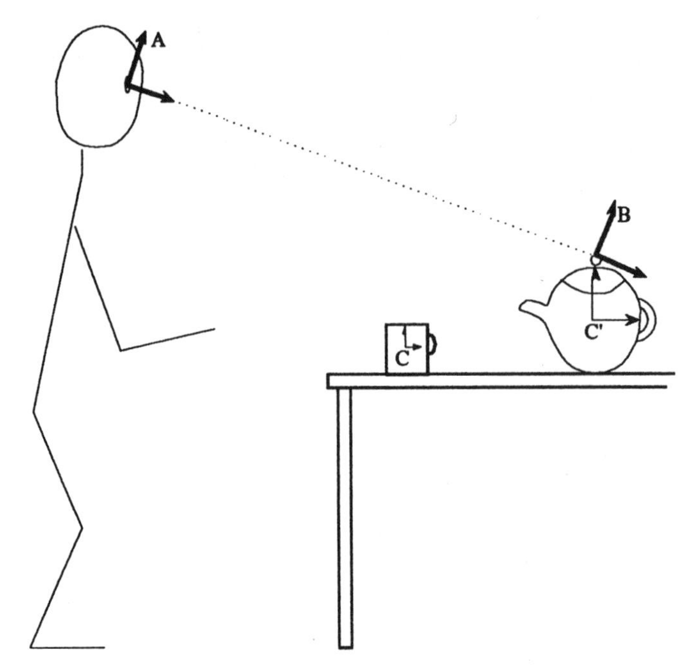

Figure 15.1 Much previous work in computational vision assumed that the vision system is passive and that computations are performed in a viewer-centered frame (*A*). Instead, biological and psychophysical data argue for the use of the fixation frame (*B*). This frame is selected by the observer to suit information-gathering goals and is centered at the fixation point. The origin of the frame is at the point of intersection of the two optical axes. To orient this frame, one axis can be parallel to the line joining the two camera centers and another chosen as the optical axis of the dominant eye. In changing gaze, the observer attaches the fixation point to object-centered frames (*C*).

information. For example, an object can be grasped by first looking at it and then directing the hand to the center of the fixation coordinate frame. In depth the hand can be servoed relative to the horopter by using binocular cues. This behavior, which we are calling a "fixation routine," is an example of a deictic strategy (Agre and Chapman 1987). Successions of deictic primitives can succinctly create complex behaviors as each primitive implicitly defines the context for its successor.

The deictic strategy of using the perceptual system to actively control the point of action in the world has precisely the right kind of invariance for a large amount of behaviors. No complex geometry is necessary, and thus the resultant descriptions transfer well to similar situations. As an example, consider the problem of picking up a green block that has another block stacked on top of it, as shown in figure 15.2.

This task can be accomplished by the following program:

Fixate(green)
Fixate(Top-of-stack)
PickUp
Fixate(Somewhere-on-the-table)

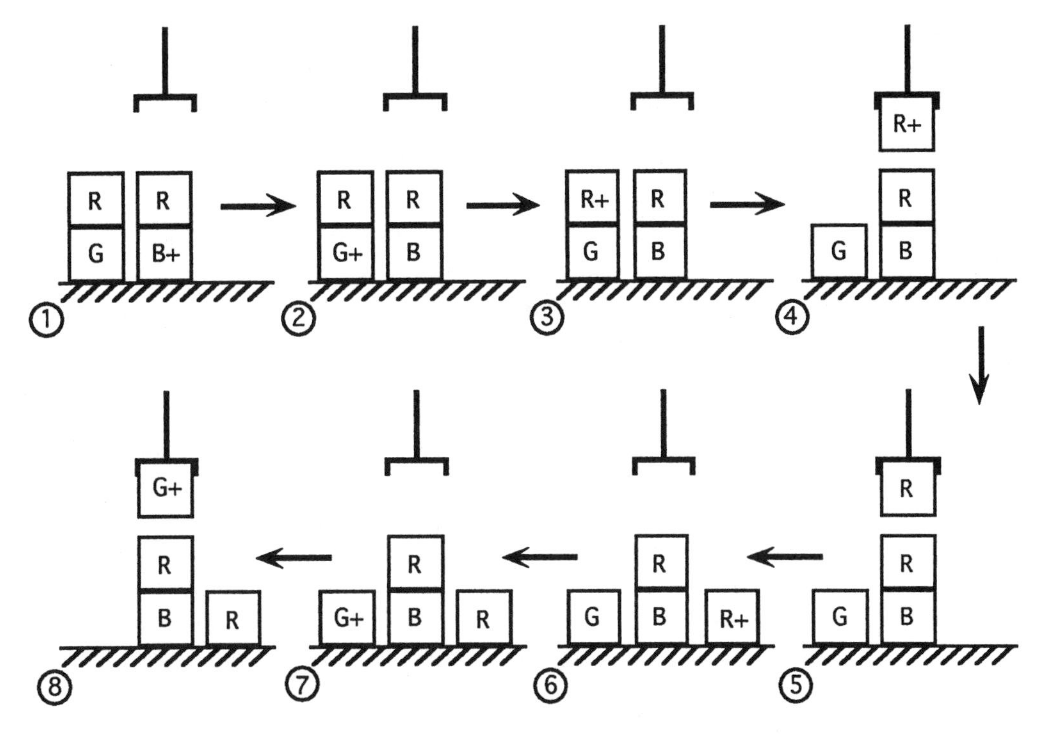

Figure 15.2 A graphical display from the output of a program that has learned the "Pick-up-the-green-block" task. The steps in the program use deictic references, and do not require geometrical coordinates. For each stage in the solution, the plus symbol shows the location of the fixation point. (From Whitehead and Ballard, 1991.)

PutDown

Fixate(green)

PickUp

In this program it is assumed that the instruction Fixate(image_feature) will orient the center of gaze to point to a place in the image with that feature. These actions are context-sensitive. For example, Fixate (Top-of-stack) will transfer the gaze to the top of the stack currently fixated. PickUp and PutDown are assumed to act at the center of the fixation frame. Whitehead and Ballard (1990) have shown that this task can be learned using a rather general set of primitives similar to the above. This includes instructions for moving a focus of attention. A focus of attention may be thought of as an electronic fovea, in terms of its ability to select target locations.

The simplifications of deictic strategies may be understood in terms of the computational complexity of the general problem of relating internal models to objects in the world. One way to interpret the need for sequential, problem-dependent eye movements is as a suggestion that the general problem of associating many models to many parts of the image simultaneously is too hard. In order to make it computationally tractable, it has to be simplified into a series of two kinds of tasks. These tasks either find information about location (one internal model) or identification (one world object). Table 15.1 summarizes this view. This partition is suggested from the organization of the human eye. The eye is distinguished from current commercial electronic cameras by virtue of having much better resolution near the optical axis. It has a high-resolution fovea where over a 1-degree range the resolution is better by an order of magnitude than that in the periphery. One feature of this design is the simultaneous representation of a large field of view and high acuity in the fovea. Because of this, we model the visual organization more crudely into a center and a surround. The center is "where-I'm-looking" and the surround is a source of new gaze points. A location task is to find the image coordinates of a single model in the presence of many alternatives. In this task the image periphery must be searched; one can assume that the model has been chosen a priori. An identification task is to associate the foveated part of the image with one of many possible models. In this task one can assume that the location of the material to be established is at the fixation point. Our work using color cues has shown that this simplification leads to dramatically faster algorithms for each of the specialized tasks. Thus we think of the eye movement as solving a succession of location and identification subtasks in the process of meeting some larger cognitive goal.

EVIDENCE FOR DEICTIC STRATEGIES

This different view of pointers as variables in a cognitive program raises a different set of questions for study. One can characterize the complexity of cognitive programs in terms of the average amount of state they require at any instant. In other words, programs that require the maximum number of variables would be complex and programs that

Table 15.1 The organization of visual computation into WHAT/WHERE modules may have a basis in complexity. Trying to match a large number of image segments to a large number of models may be too difficult.

		Models	
		One	Many
Image Parts	One	Manipulation: trying to do something with an object whose identity and location are known	Identification: trying to identify an object whose location can be fixated
	Many	Location: trying to find a known object that may not be in view	Too difficult?

require few or no variables would be simple. Do subjects prefer complex or simple programs? Within this context we designed a series of experiments to test the use of short-term memory in the course of a natural hand-eye task where subjects have the freedom to choose their own task parameters. Surprisingly, subjects choose not to operate at the maximum capacity of short-term memory but instead seek to minimize its use. In particular, reducing the instantaneous memory required to perform the task can be done by serializing the task with eye movements. These eye movements allow subjects to postpone the gathering of task-relevant information until just before it is required. The reluctance to use short-term memory to capacity can be explained if such memory is expensive to use with respect to the cost of the serializing strategy. Our experiments support this hypothesis.

The hand-eye task was to copy a pattern of colored blocks. This task was chosen to reflect basic sensory, cognitive, and motor operations involved in a wide range of human performance. A display of colored blocks was divided up into three areas: *model, source,* and *workspace.* The model area contains the block configuration to be copied; the source contains the blocks to be used; and the workspace is the area where the copy is assembled.

Subjects used the cursor driven by a mouse to "pick up" and "place" blocks on the screen. Picking up a block is accomplished by moving the cursor over the block and depressing a button attached to the mouse. Place the block is accomplished by moving the block to the desired location and releasing the button. For this reason the block-copying task used a set of coarse-grained, discrete locations for the blocks. Thus releasing the mouse button placed the block at the nearest discrete grid location. This obviated the need for very precise positioning and made the task easier to perform. Block sizes varied from 0.5 to 2.0 degrees. Using 1.7- $\times$ 1.3-degree blocks, the resultant grid was a 10 $\times$ 10 array, as can be inferred from figure 15.3, which shows the initial configuration for such an example. Displays were random configurations of eight blocks of four colors: red, green, yellow, and blue. Both the eye and cursor movements were monitored throughout the task. The eye movements were monitored using a Dual-Purkinje Image eye tracker, sampling the eye movements and hand movements every 20 ms. The head was held fixed throughout the experiment, using a bite-bar. At the outset of each set of trials for an individual subject, the subject's gaze was calibrated by measuring the recording signal over a grid of 25 positions that spanned the display screen. The accuracy of the tracker is considerably better than 1 degree so that fixations of individual blocks could be detected with high confidence.

One key feature of the experimental setup is that the task is sufficiently explicit so that the strategy adopted to solve it can be observed in detail. In particular, the colored blocks are difficult to group into larger gestalts and must be handled individually, allowing the separa-

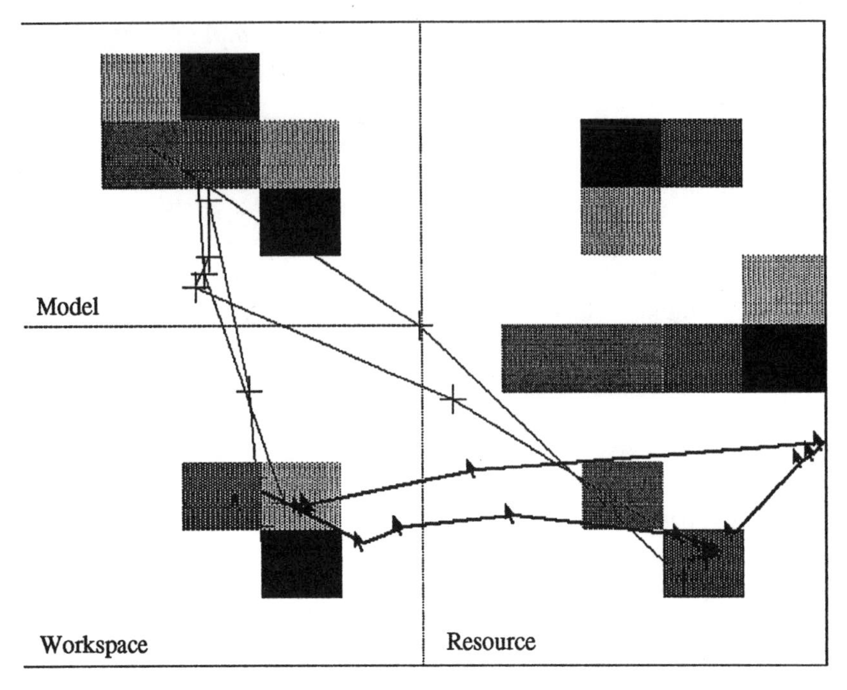

Figure 15.3 Annotated display used in the hand-eye coordination experiments. The subject's instructions are to build a copy of the model in the workspace area using blocks from the source area. Blocks are moved using a cursor controlled by a Macintosh Mouse. The display subtended 17 × 13 degrees of visual angle; individual blocks, 1.7 × 1.3 degrees. The right eye is tracked, and the eye position trace shown by the cross and the thin line. The cursor trace is shown by the arrow and dark line. A single cycle is represented, from dropping off block 2 to dropping off block 3. (In the experimental trial the blocks appear colored.) Immediately after dropping off block 2 (light gray) the fixation point is transferred to the model. Simultaneously the cursor is moved to the source area at the extreme right of the screen. Subsequently the fixation point is transferred to the source area at the location of block 3 (dark gray) and used to direct the hand for a PickUp action. Then the eye goes back to the model and the cursor is moved to the drop-off location. The eye moves to the drop-off location to facilitate the release of the block. This display is accomplished by a "replay" program that retraces the experimental course from saved data. In the experiment the block is erased immediately after it has been picked up, but here it has been left visible to mark its location; thus the moved block appears twice.

tion of perceptual and motor components of the task. Another key feature is that the detailed solution of the task is not specified; the choice of coordination of hand and eye movements is left entirely to the subject.

The Basic Task

When the task is performed while the model is visible throughout the trial, observations of individual eye movements suggest that information is acquired incrementally during the task and not in toto at the beginning of the task. For example, the trace for the third block used by subject K is depicted in figure 15.3. Initially the mouse movement and fixation point movements are in different directions, with the cursor being transferred to the source and the eye directed toward the model. The fixation point then moves to the source area at the location of block 3 (black) and is used to direct the hand for a PickUp action. Then the eye *goes back* to the model while the cursor is moved to the workspace. The eye moves to the drop-off location to facilitate the release of the block.

The fact that fixation is used for picking up and dropping off each block would have been expected from data on single hand-eye movements (Milner and Goodale 1991). However, the extent to which the eyes were used to check the model was unanticipated. It seems likely that humans use their ability to fixate to simplify the task in two ways. First, the fixation frame allows the use of deictic primitives. For example, an object is picked up by first looking at it and then directing the hand to the center of the fixation coordinate frame. The alternative requires programming a command in a world- or ego-based coordinate representation, with much greater demands on the fidelity of the representation. Second, fixation is used to acquire information en route at the point at which it is required. Consider the color of the third block. If this is memorized at the outset along with several other colors, then a corresponding number of memory locations would be required. However, a single location that encodes *the-color-of-the-next-block* can be used if the loading of that location is performed at the appropriate moment in the task.

Analysis

The basic cycle from the point just after a block is dropped off to the point where the next block is dropped off provides a convenient way of breaking up the task into component subtasks of single block moves. This allows us to explore the different sequences of primitive movements made in putting the blocks into place. A way of coding these subtasks is to summarize where the eyes go during a particular subtask. Thus the sequence in figure 15.2 can be encoded as "model-pickup-model-drop" (MPMD) with the understanding that the PickUp occurs in the source area and the drop in the workspace area. Four principal sequences of eye movements can be identified, as shown in figure 15.4, *a*.

In this task the crucial information is the color and relative location of each block. The observed sequences (in the main experiment) can be understood in terms of whether the subject has remembered either the color or location, or both, of the block currently needed. The necessary assumption is that the information is most conveniently obtained by explicitly fixating the appropriate locations in the model and that the

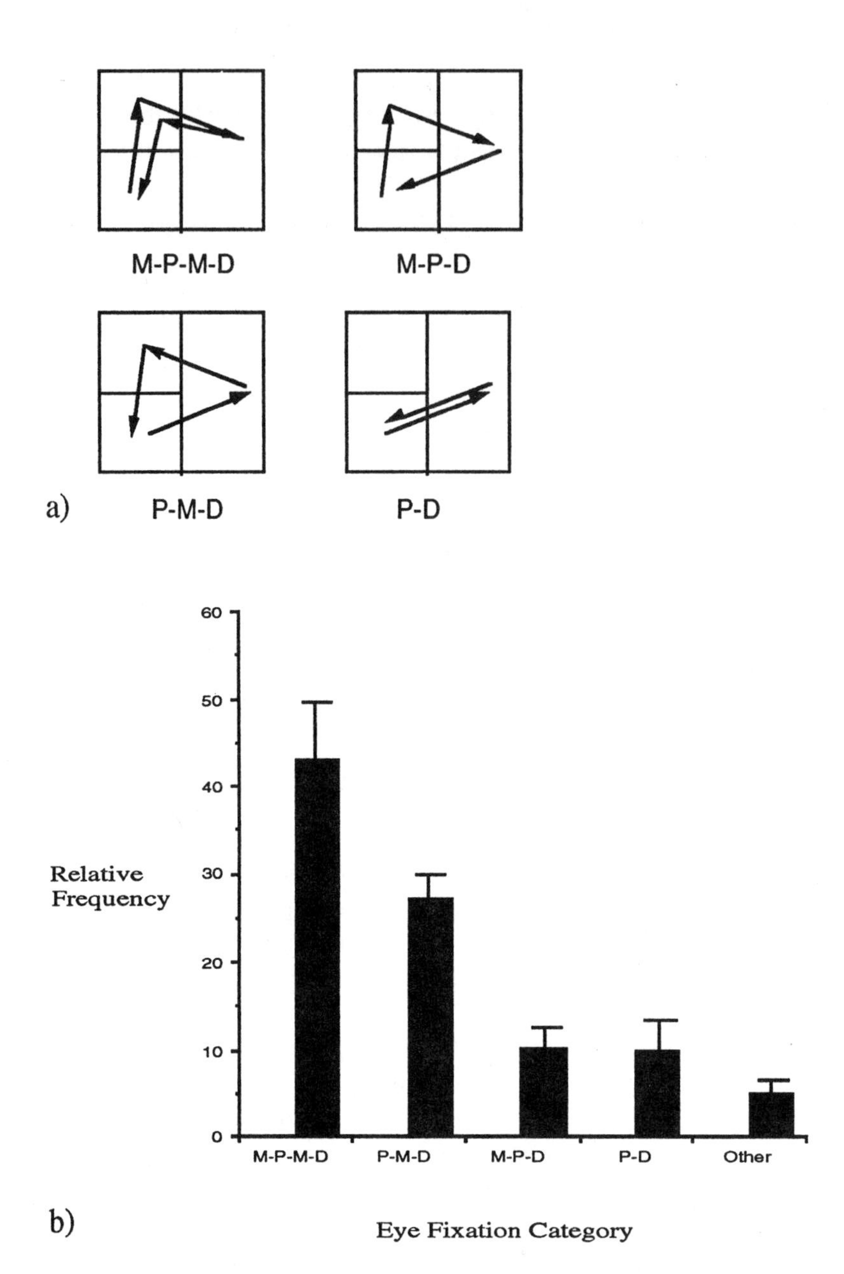

Figure 15.4 (*a*) The different categories of eye movements used in the task. *M* means that the eyes are directed to the model. *P* and *D* mean that the eyes and mouse are coincident at the PickUp point and PutDown point, respectively. Thus for the P-M-D strategy, the eye goes directly to the source for PickUp, then to the model area, and then to the source for drop-off. (*b*) Summary data for three subjects.

main preference is to acquire color or location information just before it is required. If both color and location are needed—if they have not been previously remembered—an MPMD sequence should result. If the color is known, a PMD sequence should result; if the location is known, an MPD sequence should result; and if both are known, then PD. In the data the PD sequences were invariably the last one or sometimes two blocks in the sequence. Thus MPMD sequences are memoryless and MPD, PMD, and PD sequences can be explained if the subjects are sometimes able to remember an extra location or color, or both, when they fixate the model area.

Summary data for the three subjects is shown as the dark bars in figure 15.4, *b*. The memoryless MPMD strategy is the most frequently used by all the subjects, far outweighing the others. Note that if subjects were able to complete the task from memory, then a sequence composed exclusively of PDs could have been observed, but the PD strategy is used only at the end of the construction. We take this frequent access to the model area during the construction of the copy as direct evidence of the incremental access to information in the world during the task.

Performing the Task from Memory

One necessary control is to establish how much memory can be used in this task. In a previous experiment, subjects were given 10 seconds to inspect the model, which was then removed from view. Performance was good up to four items, but degraded rapidly above this (Ballard et al., 1992). When subjects were allowed to inspect the model for a variable duration before it was removed from view, performance asymptoted by around 10 seconds at about 60% correct for models of eight blocks. On this basis we might have expected that subjects would use memory more extensively in the main experiment, but they clearly use only minimal memory when they are free to do so.

Computer Simulation

The observed sequences where subjects postpone the gathering of both the color of the next block and separately its location until just before it is needed represents a low memory approach to the task. But how much memory are they using? To answer this, we developed a simple computer model that allows color and location memory to have a variable capacity. If either of these locations are empty the eyes are drawn to the model area. To explain this in more detail, consider the model control program shown in figure 15.5.

In figure 15.5 actions are triggered by activating conditions obviating the need for precise sequential control. For example, if the pattern is not copied, then short-term memory receives a new color of the next

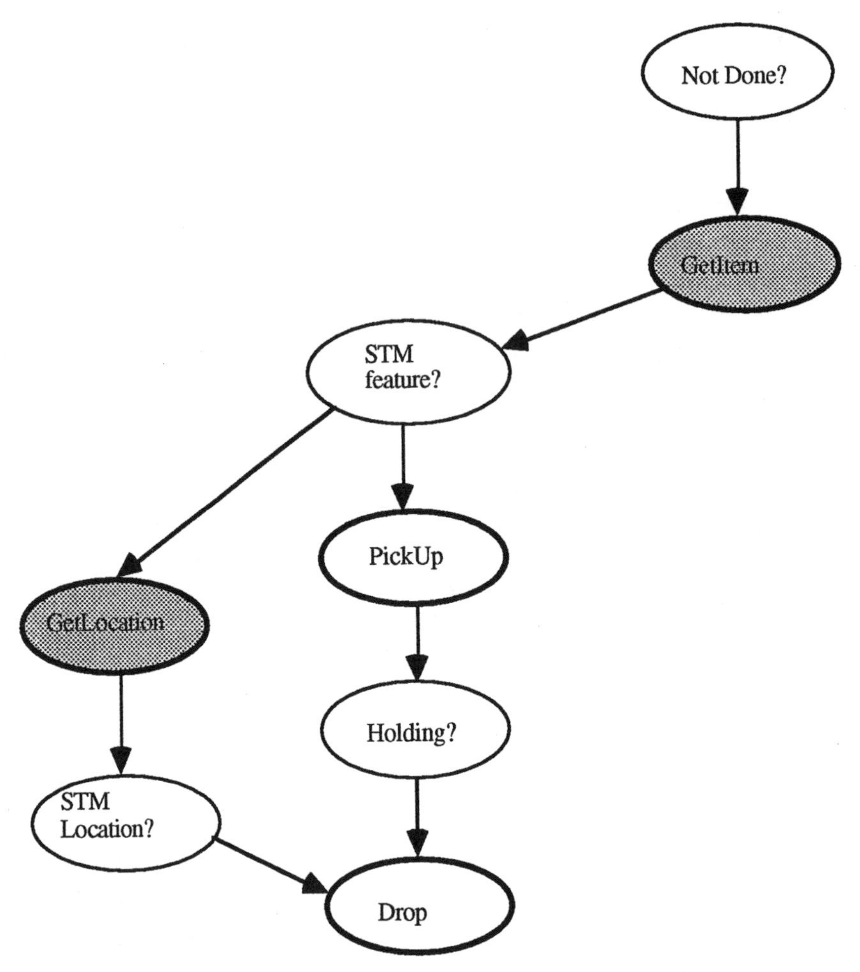

Figure 15.5 Asynchronous control program for the blocks task. The *bold ellipses* represent actions that are triggered by activating conditions. If the activating conditions are satisfied for a given action, then that action is carried out.

block; this in turn triggers PickUp. The PutDown action requires both that an object be held and that it has a target location. Actions have effects. For example, the perceptual actions GetItem and GetLocation act as producers of color and location information, respectively. Similarly, the manipulation actions PickUp and PutDown act as consumers of color and location information; i.e., they remove the corresponding item from short-term memory.

If there was never any color or location memory, each of the observed sequences would be of the form MPMD. To explain the observed sequences, we only have to allow the model fixations to probabilistically produce an extra color or location, or both, where now GetItem and GetLocation, in addition to determining a single color or location, are each allowed to determine the subsequent color with probability P_c and the subsequent location with probability P_1. The only remaining modification is that at the penultimate block, subjects invariably memorize the color and location of the last block. When this feature is added to the model as a special case the observed data can be modeled very closely, as shown by the gray bars in figure 15.6. To generate figure 15.6, $P_1 = 0.1$ and $P_c = 0.2$ were used. In other words, on each trip to the model to get a location there is a 10% chance that the location for the subsequent block move will be remembered. For color, the figure is 20%. Thus the task is performed very close to the memoryless condition.

Monitoring Progress

If in fact subjects are not using extensive memory for this task, then they should have to cope with an additional problem: If they have no extensive internal model, how do they keep track of what has been copied and what is left to do? The data suggest that humans do have

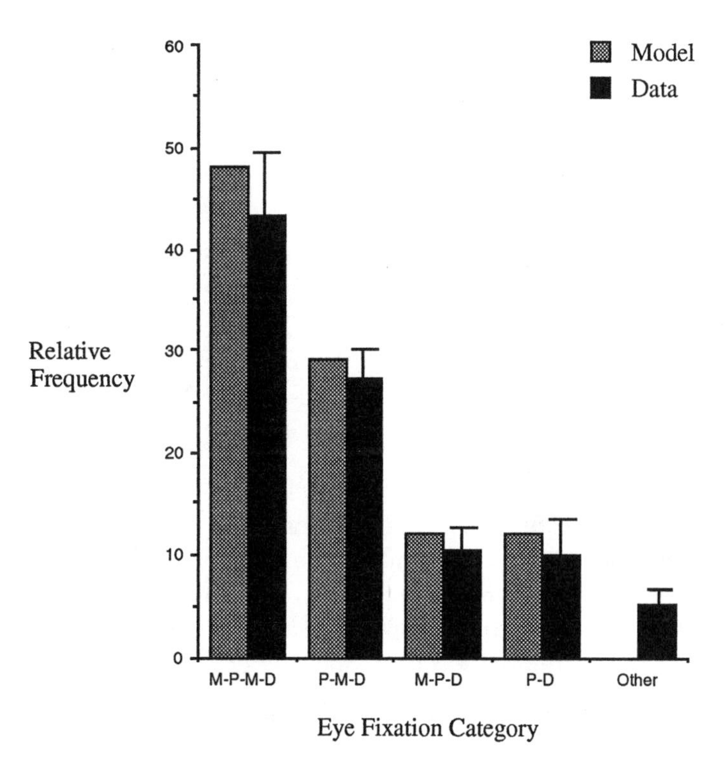

Figure 15.6 Model compared with the data of figure 15.4 showing frequency of category use of a sample of 100 block moves from four observers. Treating the addition of a block to the figure being built in the workspace allows the comparison of different strategies. A strategy that memorized the model configuration at the outset could then consist entirely of PickUp and PutDown operations. Instead, the data summary shows a number of different programs. Model used $P_c = 0.21$ and $P_1 = 0.11$, relatively low probabilities.

this problem and additionally are using their eye movements to keep track of their progress in the task incrementally. This can be understood by examining locations of fixation points in the model and workspace. When subjects fixated the model and workspace successively, they returned to the same point in the pattern. This finding is surprising as the model pattern and workspace pattern do not *appear* the same until the latter is complete. Despite their different appearances, corresponding points in the two patterns are selected for fixation. Our hypothesis is that the accuracy of the fixations is designed to help the cognitive program deal with a difficulty that is introduced by the sequential strategy. Since the pattern is being copied incrementally, memory must be used to keep track of what has been copied and what remains to be copied.

Here again the deictic strategy results in a simplification of the task. By looking at corresponding points in the model and copy, gaze could establish a common frame of reference. In looking at the workspace, the partially completed pattern is remembered as a unit and then compared to the model pattern once they are in registration. The comparison can be almost literal, thus revealing the uncopied part of the pattern. This method has the advantage of not having to encode the details of the pattern or the history of the copying procedure. In contrast, consider the bookkeeping problem of doing this in a viewer-centered frame of reference. In this case the nervous system has to handle the additional problem of building and comparing object-centered representations. Additional support for the comparison hypothesis comes from the fact that subjects do not fixate individual blocks even though they have the ability to do so, but instead choose several central reference points. This is even more obvious in the monochrome experiments, when some sequences of blocks are copied without reference to the model.

IMPLEMENTING ROLES

To do successful visual computations, the primitive instructions must be appropriate for the brain's way of doing things. For example, owing to these time constraints, a parallel operation carried out over the entire retinotopic array could easily be done, whereas a serial instruction that propagated information from one side of the array to the other would be less plausible. It is immediately seen that the instructions used in the simple computer model—*GetColor, GetLocation, PickUp,* and *PutDown*—are not revealing, as they place no specific constraints on the visual system. They must be broken down into a more concrete set of instructions that fits onto brain architecture. Let us illustrate some ways this could be done.

A crucial ability is to define appropriate targets for fixation. At one point in the blocks task subjects need a block of an appropriate color, and at another point they need to know where to put that block. For both of these subtasks a way of defining locations is required. The traditional way of thinking about this is to have vision build a three-dimensional model of the scene containing the blocks as components with detailed location and shape information. But the transient nature of vision motivates searching for a more economical way that requires a minimum of model-building.

One way of defining locations is by color. This is particularly useful in the blocks experiments since individual blocks have distinct colors (although there are several blocks with the same color). Reconsider the task of identifying the next block to use and further suppose that you are fixating in the model area. One way to define the location of the blue block would be to be able to ask: Where are the locations of blue on the retinotopic array at this moment? It is easy to see that this constraint alone is insufficient. An additional constraint must denote the resource area. Note that defining this area is difficult as it is a location in space that will occupy a different part of the visual field depending on the location of the eyes. One way this might be done is to compare the specific color "blue" to each of the retinospatial locations. Next the constraint that the resource area is to the left of the fixation point can be added to determine the candidates in the resource area. Finally, one of these must be chosen for a saccadic eye movement. This set of instructions uses just one color, but Swain (1990) and Swain and Ballard (1990, 1991) have shown in a number of experiments that these methods work even better with natural objects of many different colors.

The use of color points to important principles, namely that the crucial defining features are problem-dependent and have to be computed just before they are needed. The skeptic, however, will be thinking that colors work nicely in this instance but that in general the world cannot be depended on to use colors in this way. Happily, there are ways of determining locations that do not depend on color. Another way of defining a location is to directly record what it looks like in terms of its photometric intensity values. If a literal template of all the values near a point is remembered, then this template can be matched against the current image to recover the location as the point of best match. Template matching had lost support in computational vision owing to a number of reasons: it is vulnerable to lighting changes; it is vulnerable to transformations such as scale and rotation; and its storage requirements are prohibitive. However, recent developments (Jones and Malik, 1992; Manjunath et al., 1992) have ameliorated these disadvantages. If the template is created dynamically and has a limited lifetime, then the first objection is less important, as lighting conditions are likely to be constant over the duration of the task. As for the second and third objections, special compact codes for templates can overcome these difficulties. These codes express the photometric values near a point in terms of a sum of a fixed set of about 45 templates, termed *basis templates.* By fixating at a point, the responses of the basis templates can be recorded. For practical purposes, these responses provide a unique "ZIP code" that identifies that particular point. Thus when the gaze is not at the marked point, the location of the point relative to the current fixation can be determined by matching the remembered ZIP code with the ZIP codes at all current locations, as illustrated in figure 15.7. The best match defines the target point.

Figure 15.7 shows how this might be done. When the eyes are fixating a given point, the ZIP code is remembered in short-term memory. Later, at the moment the eyes are required to return to that point, the remembered responses are compared to the current responses on the retina. The best match defines a motor target that can be fixated.

We now turn to a possible implementation of such a mechanism. the basal ganglia have long been almost exclusively associated with motor behavior, but recent studies have shown that they are also linked to short-term memory. Huntington's and Parkinson's disease patients, who have specific deficits in their basal ganglia, have shown striking reductions in short-term memory capacity (see chapter 14). In turn, the role of short-term memory in motor programming has not been examined. Thus the relationship of the basal ganglia and short-term memory raises a new set of issues regarding the structure of motor programs. In parallel with this study, anatomical knowledge of the circuitry has steadily improved (see chapter 5). It is now appreciated that the basal ganglia have prominent connections to the parts of the thalamus involved in vision. Combining these observations with the observation on short-term memory, a possible implication is that the basal ganglia may control the loading and use of short-term visual memory. This means that the circuitry that controls perception and action in the brain may be much more tightly linked than has previously been appreciated.

Suppose that the processes described above were controlled by the action of the basal ganglia, whose function is to keep track of when it is appropriate to load short-term memory. In the task there are many different fixations with different uses; so there has to be some mechanism that uses the task context to pick the appropriate one for a given segment. Note also that the basal ganglia, via their pathway to the thalamus, have the necessary connections to send this command. This putative function is depicted schematically in figure 15.8. The figure shows highly schematized pathways to suggest the overall control pathways. The retinotopically indexed neurons of thalamus and cortex are summarized in terms of a single element that depicts the enhanced foveal representation. The blocks task exhibits model fixations to identical spatial locations where only the task context is different. In this case we posit that the basal ganglia motor program might provide this

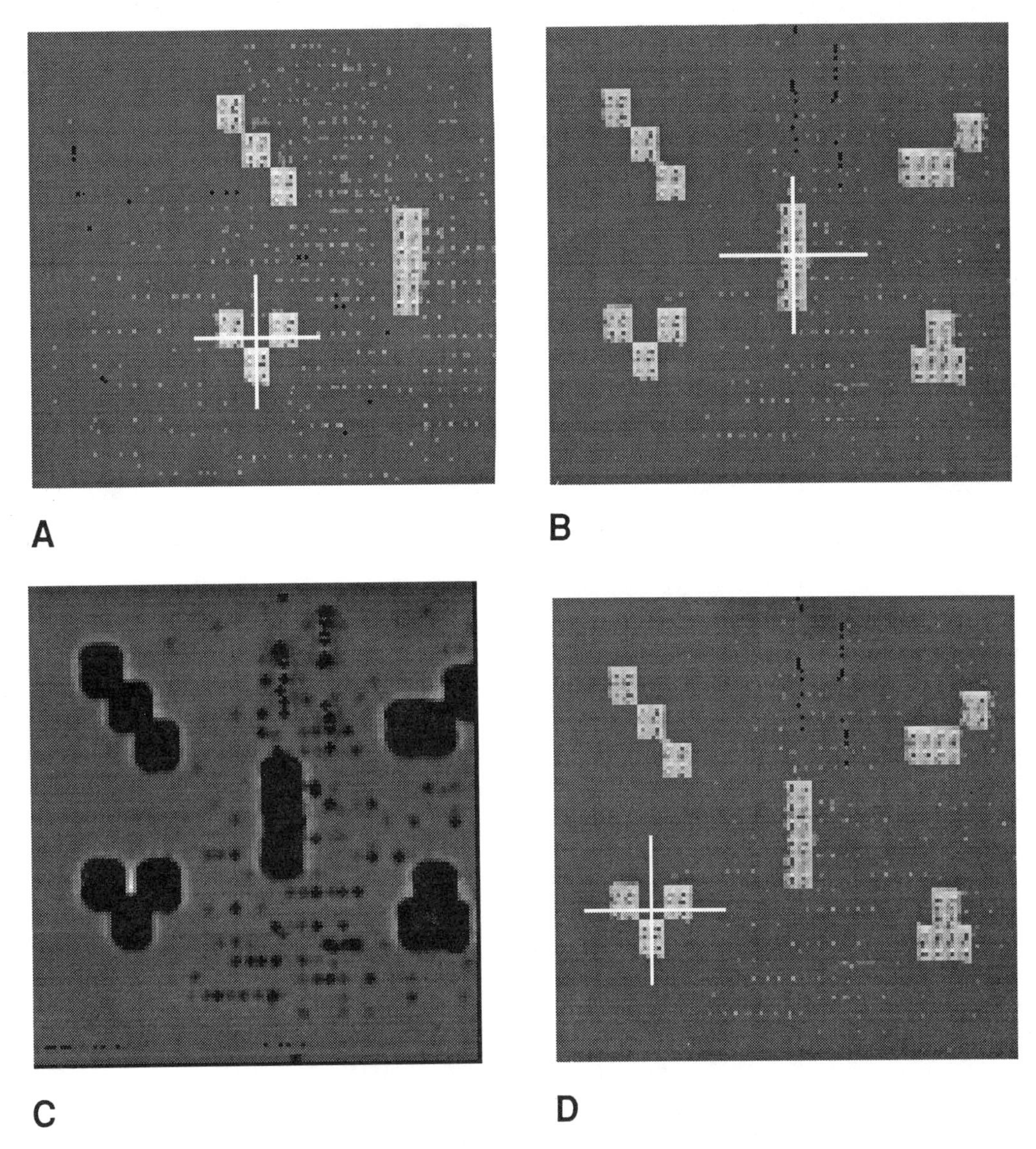

Figure 15.7 Using intensity information to define locations. (*A*) Original image and gaze points. (*B*) New gaze point. (*C*) Correlation image for the marked point. (*D*) Getting back to the original point.

context to the visual pathway so that appropriately different kinds of visual processing can be done.

Now consider the converse problem, where the task context requires that gaze be turned to a remembered location. This could be done in principle by matching the features of the desired location with features currently on the retina. Note two things. First, this requires the circuitry of visual cortex to be used in a different way. Now the features are communicated via feedback connections to a point of comparison with

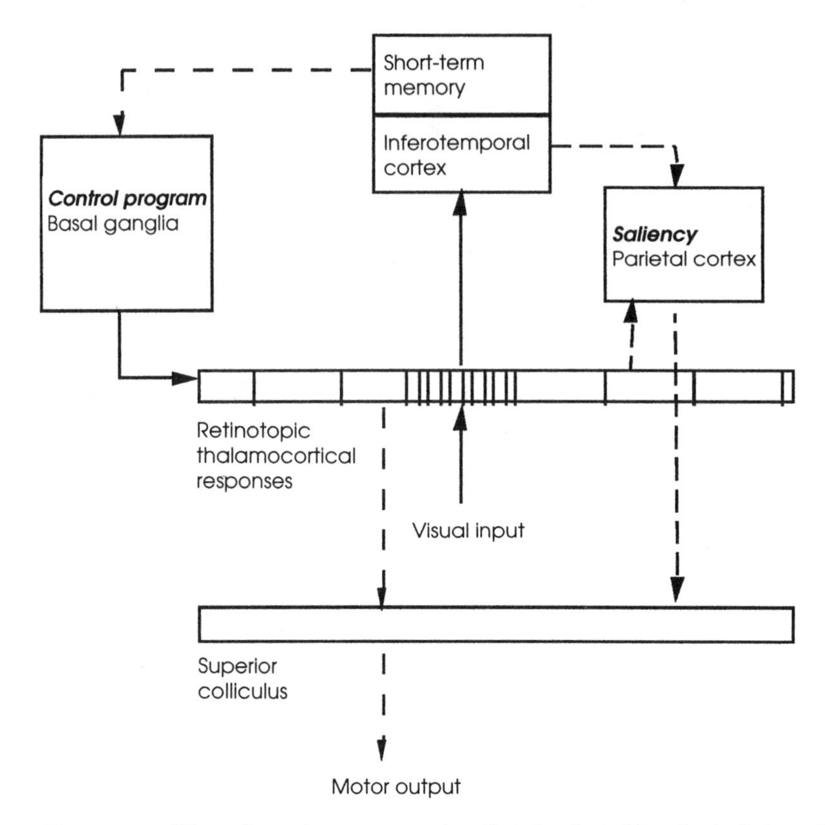

Figure 15.8 When the task context requires that the fixated location's features be remembered, a structure such as the basal ganglia could cue the thalamic input pathway through explicit connections (solid arrows).

their counterparts on the retinotopic cortex. Second, this requires the task context to specify how the visual cortical circuitry should be used. The basal ganglia are a plausible contender for such a role. It is not clear where this comparison might take place. Previously remembered features would be compared with retinotopically indexed features from all current locations. The best match would be the candidate current location of that feature.

CONCLUSIONS

New information about the basal ganglia has suggested that they may have a much more central role in the control of motor programs. Rather than just initiating movements, the basal ganglia may be involved in the momentary acquisition and use of perceptual information used in those movements. This has a number of ramifications that we have developed.

Making the connection between the computational model and the psychological model allows us to relate deictic pointers to short-term memory items. A pointer allows access to the contents of an item of short-term memory. In other words, we can think of the capacity of short-term memory as having a small number of pointers. The pointer notation raises the issue of binding, or setting the contents of a pointer. This is because pointers are general variables that can be reused for other computations. When are pointers bound? For a visual pointer one possible indication that it is being set could be fixation. Looking directly at a part of the scene provides special access to the features immediate to the fixation point, and these could be bound to a pointer during the fixation period. In all likelihood, binding can take place faster than this, say by using an attentional process, but using fixation as an upper bound would allow us to bind three pointers per second with a maximum capacity of having approximately seven pointers bound at any instant.

A structure such as the basal ganglia might send crucial timing information to the perceptual system. As the blocks task showed, information is acquired just prior to its use. The alternative strategy of memorizing the configuration to be copied in its entirety before moving blocks is never used. It is never used even though technically it could be: our memory experiments show that up to four blocks can be copied from memory without error. Instead, the observations point to the use of a small amount of prediction that only extends to the color and location of the block after the current one. Furthermore, this information seems to be acquired on separate fixations of the same area of the visual world. If this is true, then only the context of the task determines how identical retinotopic images are processed, and a potential keeper of such context is the basal ganglia. Such a structure must somehow send that context to the perceptual circuitry.

Information about the anatomy of the basal ganglia suggests that the circuitry is available to conduct this function. The basal ganglia are connected to the lateral geniculate nucleus of the thalamus via the substantia nigra. The substantia nigra operates through "pause cells" that normally have an inhibitory function but can be turned off via basal ganglia cells. This would be a fast way of gating perceptual functions.

The functions that such a structure must enable may have very different characteristics. In our model of how short-term memory is loaded, an explicit signal allows the thalamic input to propagate through the cortex via feedforward connections. In our model of how memory is used to recover a location, a different explicit signal allows extrastriatal features to be compared with striatal features via feedback connections that ultimately terminate in the superior colliculus. Implicating the basal ganglia in this model is admittedly speculative, but the model described in figures 15.8 and 15.9 clarifies the general kind of circuitry that might be involved.

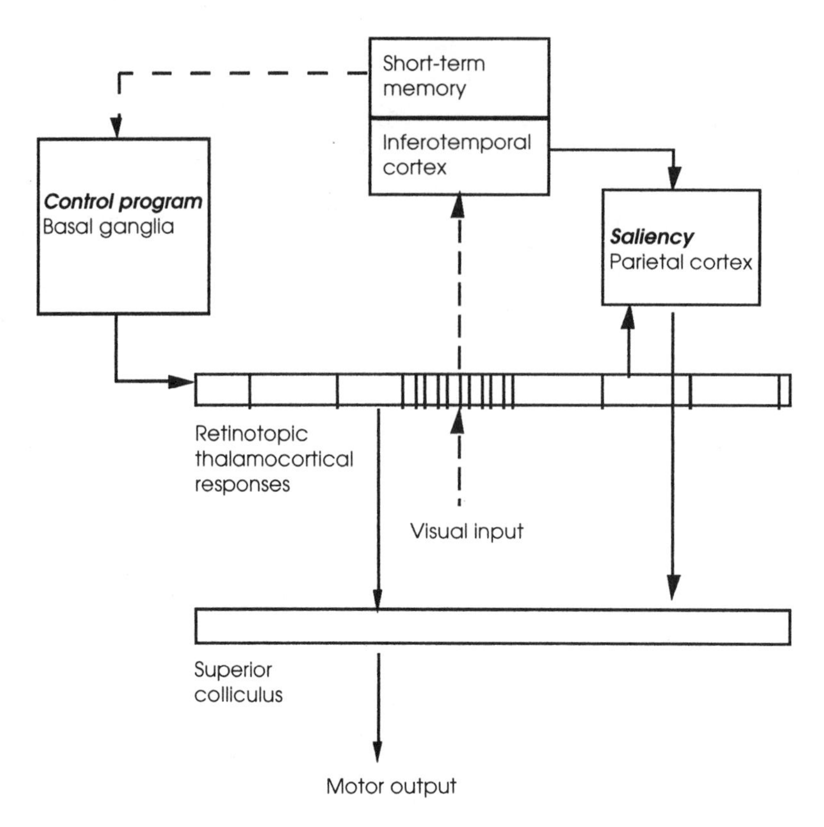

Figure 15.9 When the task context requires that the location of remembered features be recovered, the basal ganglia could cue the superior colliculus output pathway indirectly through connections through parietal cortex (solid arrows).

REFERENCES

Agre, P. E., and Chapman, D. (1987) Pengi: An implementation of a theory of activity. *Proc. A.A.A.I.* 87: 268–272.

Allport, A. (1989) Selective attention. In M. Posner (ed.), *Foundations of Cognitive Science* Cambridge, Mass.: MIT Press, pp. 631–682.

Ballard, D. H. (1989) Behavioral constraints on animate vision. *Image Vision Computing* 7: 3–9.

Ballard, D. H. (1991) Animate vision. *Artif. Intell. J.* 48: 57–86.

Ballard, D. H., Hayhoe, M. M., Li, F., and Whitehead, S. D. (1992) Hand-eye coordination during sequential tasks. *Philos. Trans. R. Soc. Lond.* [*Biol.*] 337: 331–339.

Brooks, R. A. (1986) A robust layered control system for a mobile robot. *I.E.E.E. J. Robotics Automation* 2: 14–22.

Brooks, R. A. (1991) Intelligence without reason, AI Memo 1293. Cambridge, Mass.: Massachusetts Institute of Technology, AI Laboratory.

Jones, D. G., and Malik, J. (1992) A computational framework for determining stereo correspondence from a set of linear spatial filters. *Image Vision Computing* 10: 699–708; also in *European Conference on Computer Vision,* Portofino, Italy, pp. 395–410.

Manjunath, B. S., Chellappa, R., and von der Malsburg, C. (1992) A feature-based approach to face recognition. In *I.E.E.E. Conference on Computer Vision and Pattern Recognition,* Champaign, Ill.: pp. 373–378.

Miller, G. (1956) The magic number seven plus or minus two: Some limits on your capacity for processing information. *Psychol. Rev.* 63: 81–96.

Milner, A. D., and Goodale, M. A. (1991) Visual pathways to perception and action, COGMEM 62. University of Western Ontario, Center for Cognitive Science.

Nakayama, K. (1990) The iconic bottleneck and the tenuous link between early visual processing and perception. In C. Blakemore (ed.), *Vision: Coding and Efficiency.* Cambridge: Cambridge University Press, pp. 411–422.

Swain, M. J. (1990) Color indexing, Ph.D. Thesis and TR 360. Rochester, N.Y.: University of Rochester, Computer Science Department.

Swain, M. J., and Ballard, D. H. (1990) Indexing via color histograms. Presented at *International Conference on Computer Vision* (ICCV-90), Kyoto, Japan, November.

Swain, M. J., and Ballard, D. H. (1991) Color indexing. *Int. J. Comput. Vision* 7: 11–32.

Ullman, S. (1984) Visual routines. *Cognition,* 18: 97–157; also in S. Pinker (ed.) (1984), *Visual Cognition.* Cambridge, Mass.: Bradford Books.

Whitehead, S. D., and Ballard, D. H. (1990) Active perception and reinforcement learning. *Neural Computation* 2: 409–419.

Whitehead, S. D., and Ballard, D. H. (1991) Learning to perceive and act by trial and error. *Machine Learning* 7: 45–83.

16 Neostriatal Circuitry as a Scalar Memory: Modeling and Ensemble Neuron Recording

Donald J. Woodward, Alexandre B. Kirillov, Christopher D. Myre, and Steven F. Sawyer

The interactions between neurons within the corticostriatothalamic loops (Alexander et al., 1990) clearly must encompass many of the processes of sensorimotor integration and the cognitive substrates of consciousness. This must be so if for no other reason than that much of the mass of the brain is included in these interactively nested looping circuits. Clarifying the principles of function within a representative circuit in this system is critical to an understanding of the role of the basal ganglia in integrating activity from all regions of cortex into a coherent action. Interconnections of the many interrelated physiological networks in the corticostriatal system can be expected to involve extensive computation of signals propagating through layers of neurons, or the creation of stable modes of distributed activity within parallel feedback systems. Adaptive changes may occur at many synaptic sites within the corticostriatothalamic loop systems.

If one examines a simplified schematic diagram of the basal ganglia "motor circuit" (figure 16.1), the problem of specifying realistic neural circuits for study through modeling becomes more acutely evident. The cerebral cortex contains numerous complex circuits interconnected as modules. Multiple cerebral cortical modules are known to produce oscillations, transients, and steady signals in their efferents that are assembled on convergent targets within the neostriatum. Felleman and Van Essen (1991), in their analysis of interconnections of the visual areas of the cortex, note that at least 19 separate visual cortical regions are interconnected by a minimum of 319 currently identified pathways. They commented that all of these cortical regions have efferent projections that compose subcomponents of the convergent signals to the caudate and putamen. A complexity of connectivity equivalent to that between cortical modules is likely to exist in the cortical connections with neostriatum. Our conclusion is that a comprehensive point-by-point mapping of this massive connectional array will need to continue for some time before a comprehensive view of the design of the networks begins to emerge. If our aims are to eventually clarify function within the multiple corticostriatothalamic loops, it certainly will be nec-

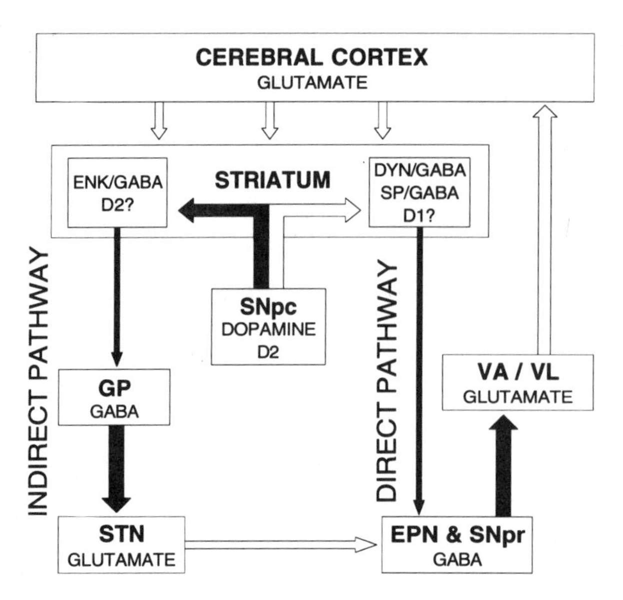

Figure 16.1 Schematic diagram of corticostriatothalamic system illustrating the "indirect" output to globus pallidus (GP; GP external of primates) and "direct" output to the entopeduncular nucleus (EPN; GP internal of primates) and substantia nigra pars reticulata (SNpr). Inhibitory and excitatory pathways are shown as solid black and open white arrows. The modulation of dopamine on neostriatal circuits is presented as "net inhibitory" and "net excitatory" effects. SNpc, substantia nigra pars compacta; STN, subthalamic nucleus; VA/VL, ventral anterior/ventral lateral thalamic nuclei; ENK, enkephalin; GABA, γ-aminobutyric acid; DYN, dynorphin; SP, substance P; D1 and D2, dopamine receptor subtypes

essary to clarify with some percision the computational processes occurring at each station in the loop. A question of immediate concern here has been the nature of the circuit computation thought to occur within the neostriatum.

A number of studies have revealed the global topography of corticostriatal compartmentation. The frontal cortex projects to the head and body of the caudate, and the sensory and motor cortices project to the lateral putamen in the primate (Alexander and Crutcher, 1990a,b) and the dorsolateral neostriatum in the rat (McGeorge and Faull, 1989). Massive interconnections between distant zones of neostriatum do not exist, which implies that each cortical efferent zone in neostriatum operates independently to some extent. Output axons of about half of the GABAergic inhibitory medium spiny neurons project as a *direct pathway* to the entopeduncular nucleus (EPN; globus pallidus internal segment of primates) or the substantia nigra pars reticulata (SNr). A largely separate population of GABAergic medium spiny neurons projects to the globus pallidus (GP; external segment of primates) as an *indirect pathway,* which has an inhibitory GABAergic projection to the subthalamic nucleus, which in turn emits excitatory influences to the EPN and SNr (see figure 16.1) (Gerfen, 1992).

A major component of neostriatal local circuitry is the extensive network of recurrent axon collateral connections between the mass of medium spiny inhibitory neurons, which make up about 90% to 95% of the neostriatal neuron population. Numerically smaller populations of interneurons exist, including the cholinergic large aspiny neuron and GABAergic aspiny neurons. However, the precise circuit connections of these interneurons are not known. The dominant local circuit within neostriatum appears to consist of arrays of local clusters of medium spiny neurons interconnected by a system of inhibitory collaterals driven by excitatory cortical and thalamic afferents (Wilson and Groves, 1980; Groves, 1983). There is no evidence concerning whether recurrent collaterals of medium spiny neurons exhibit anatomical specificity with respect to postsynaptic targets that give rise to direct and indirect pathways. But given that it is known that neurons in these two pathways are interspersed within the neostriatum (Flaherty and Graybiel, 1993), we assume that collateral networks of direct and indirect pathways are interconnected, as shown schematically in figure 16.2. The inhibitory interconnections tend to remain within local domains so that the neostriatum consists of repeated modules of such circuits (Penny et al., 1988; Kawaguchi et al., 1990).

The physiological role of the extensive collateral system of medium spiny neurons has remained somewhat of an enigma. Electrical stimulation of cortex can evoke an excitation-inhibition spike train sequence in neostriatum. This would seem to represent the simple consequence of cortical afferent excitation followed by collateral inhibition. However, ablation or cooling of cortex removes most of the inhibition, thereby suggesting that disfacilitation of cortical excitation may contribute to the suppression of spiking (Wilson et al., 1983). Stimulation of cortical afferents in a neostriatal slice preparation has been shown to evoke inhibitory postsynaptic potentials (IPSPs); however, antidromic invasion of axons of medium spiny neurons by stimulation of pallidum fails to evoke any overt signs of collateral inhibition (Wilson et al., 1989). This result was taken as evidence for inhibition produced by neostriatal interneurons, with a weak or absent inhibition by axon collaterals of the GABAergic medium spiny neurons. Other investigators (Nisenbaum and Berger, 1992) have interpreted IPSPs or inhibition after stimulation of cortical afferents as possibly due to collateral activation. The failure to evoke IPSPs after antidromic invasion might be interpreted as an indication of a strong downregulation of GABA release or receptor function under conditions of the study. Physiological regulation of GABA synapses by the noradrenergic system has been reported in the case of cerebellar Purkinje neurons (Woodward et al., 1991). A similar investigation could conceivably reveal GABA regulation by endogenous mechanisms in neostriatum. Overall, there is a potentially highly sig-

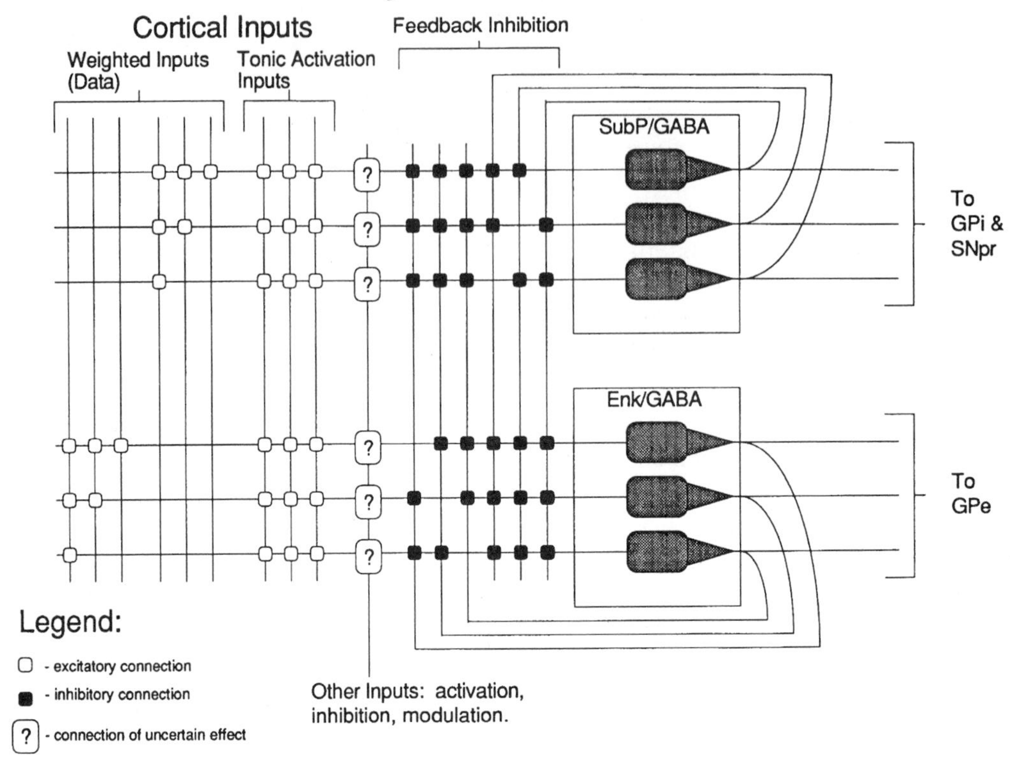

Figure 16.2 Schematic of a framework of local neostriatal circuits showing widespread cortical input to dendrite of medium spiny neurons. The direct and indirect outputs originate from neighboring neurons interconnected by inhibitory collaterals. For abbreviations, see legend to figure 16.1.

nificant regulation of both the strength and connectional organization of the collateral system that has yet to be clarified.

Many questions arise regarding these connections. Why should direct and indirect pathways exist, and what different forms of information are employed within this dual population of medium spiny neurons? If many medium spiny neurons converge on pallidal or SNr neurons, how does activity within local modules of neostriatum exert significant effects? What combination of cortical afferents carrying what kinds of information converge on the different neostriatal neurons that give rise to the direct and indirect pathways? What kinds of computations are mediated by the interconnections of the recurrent collateral system? How can computational modeling clarify these issues?

FUNCTION OF LOCAL NEOSTRIATAL CIRCUITS—HYPOTHESES

The local network of connections between medium spiny neurons would seem to be the simplest major local circuit in the corticostriatothalamic system and most amenable to study by computer simulation. We have considered three hypotheses concerning the function of the local collaterals in neostriatal circuits. The *first hypothesis* is that the *local collaterals* between medium spiny neurons function to *dampen the output of the neostriatum* produced by monosynaptic drive from cerebral cortex. The array of medium *spiny neurons may function a simple computing layer,* and may perform the simple discriminations of a "perceptron" acting on the input converging from cortex. According to this concept, the local transfer of a wave of input through the neostriatum would be dampened by the local feedback via the collateral inhibition.

A *second hypothesis* is that local focused heterogeneities exist in the input to clusters of medium spiny neurons, and when local excitation occurs, the inhibitory collaterals suppress activity in neighboring medium spiny neurons. This hypothesis postulates the existence of an intricate organization of lateral inhibition that functions as an *"either/or" logic circuit.* Throughout the neostriatum, strong excitatory inputs would suppress the influence of weaker adjacent inputs. This local logic circuit could be quite complex depending on the precise arrangement of local interconnections. Detailed knowledge of the microanatomy of local collateral interconnections and of the specificity of cortical afferent connections to single neurons would be needed to explore this concept. A basic characteristic implied in both this and the initial hypothesis is that the time course of computation would be restricted to the time envelope of afferent information arriving from the multiple cortical inputs.

A *third hypothesis* is that the parallel feedback arrangement of collaterals allows the *local circuits to mediate a form of local associative memory.* This property arises naturally from the computational properties inherent in the structure of recurrent parallel feedback circuits (Grossberg, 1973) that have been popularized by Hopfield (1984) and others in the past decade. We have recently begun to explore computational models inspired by the design of neostriatal circuits in order to comprehend more clearly the possible "memory" functions implied by the design of the circuit feedback interconnections. Spiking neuronal models have been created to approximate the membrane conductances of medium spiny neurons and the inhibitory spread of synaptic action on surrounding neurons. Simulations have included a range of connectional architectures, with different values for physiological parameters, to investigate the characteristics of a possible memory function.

SIMULATIONS OF NEOSTRIATAL CIRCUIT MODELS

A robust finding under many conditions has been that an excitatory influence broadcast across the population will evoke reciprocal ON/OFF states of inhibition or sustained firing of medium spiny neurons. The local neostriatal circuitry functions, in effect, as a biological N-flop (Myre et al., 1989; Myre and Woodward, 1990; Kirillov et al., 1991a,b, 1992) allowing many possible combinations of ON-OFF activity. The switch-like property of the inhibitory system requires a sustained excitatory influence to be expressed. The steady inhibitory influence of active neurons then prevents an otherwise identical excitatory drive from causing spiking in some of the neighbors. Excitation can arise from a number of conditions including a pacemaker property, as with a slowly decaying K current or a steady depolarization due to the action of a cholinergic excitation, or a random noisy excitatory input. In these models, the spontaneous firing rate can increase in proportion to the excitatory influence.

Switching from OFF to ON, or ON to OFF, by modeled neurons emerges when the net summed inhibition is strong enough to counteract the excitation on single neurons. A detailed investigation in a two-neuron version of the model has demonstrated that neurons fired stably in antiphase until the strength of inhibition reached a critical level at which two stable ON/OFF states appeared (Kirillov et al., 1993). ON/OFF states have been shown to appear within modeled circuits of two neurons, five totally interconnected neurons, and a 1000-neuron (10 × 10 × 10) array in which each neuron was connected locally with up to 32 nearest neighbors (Myre et al., 1989; Myre and Woodward, 1990; Kirillov et al., 1991a,b, 1992, 1993). ON/OFF states arise spontaneously as soon as the excitatory influences are imposed and as a result of the fact that concurrent activity represents an unstable system state (figure 16.3).

Noise or *randomness* in the membrane potential is found to evoke spontaneous transitions in the two-neuron model. Simulations of all the types of network architectures tested showed that increasing noise fluctuations causes a transition from a totally stable arrangement of ON/OFF states at low noise to a wide range of conditions at higher noise levels in which the arrangement of states is stable. Individual neurons exhibit steady activity for a time, then terminate for an epoch as neighboring neurons become active. Switching occurs at a time scale longer than the rate of sustained random spiking. The two-neuron model we investigated was driven by a continuous Poisson-derived noise function, and switching was approximated as a *Markov chain,* in which the probability of a switch from ON to OFF was the same after each spike of the ON neuron (Kirillov et al., 1993) (figure 16.4).

The noise-driven switching is important since it provides the physiological driving mechanism by which the recurrent feedback system evolves over time to reach a stable state. This is most readily observed in the locally interconnected cubical 1000-neuron array where ON/OFF states are set randomly at the initiation of a simulation but can be seen to evolve to an orderly arrangement where ON states become evenly spaced. In this sample system, 50 to 300 ms are required for a "readout"

Three-Dimensional Distribution of On- and Off-States: Switches Mostly Occur on a Surface Separating Two Patterns

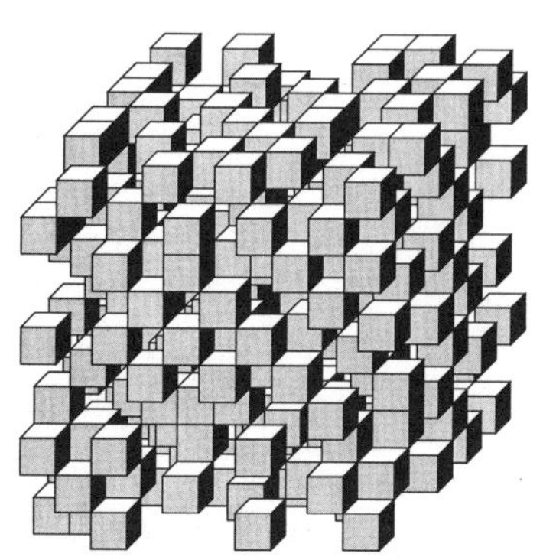

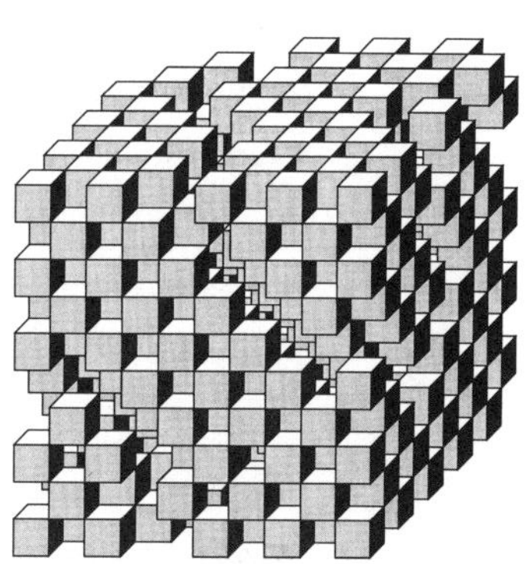

Figure 16.3 Simulation of reciprocal activity within an array of 1000 (10 × 10 × 10) neurons where each makes an inhibitory contact on its six nearest neighbors. The solid cubes represent neurons showing sustained spiking (ON state) whereas the holes represent the inhibited (OFF) neurons. When a simulation is started (*left*), all neurons are activated by excitatory synaptic noise. The noise causes spontaneous switching in some neurons in the array and eventually leads to an ordered array (*right*) of ON neurons. A part of the network may settle in one ordered checkerboard pattern, whereas another part may settle in the alternative pattern. This is evident in the *right panel,* in which the facing side of the cube has a diagonal of inactive neurons (in the *upper right sector*) that separates two patterns. This diagonal is a region of high instability where switching is frequent.

of the "memory" resident in the interconnections of equal strength. The orderly arrangement of ON states corresponds to an energy minimum to which the system converges.

Long-term memory in the neostriatal circuitry could conceivably be coded as either strengths of local inhibitory synapses or the corticostriatal excitatory connections. In the former, expression of memory as patterns of spike activity would require sustained excitation of the population to allow the stable feedback states to evolve in time. In the latter case memory could be expressed during the propagation of phasic signals from cortex through neostriatum. *Short-term* or *working memory* can be evoked if a specific pattern of ON/OFF states can be created by selective stimulation of neurons to set them to an ON state, much like the set or reset function of an electronic flip-flop. Setting of ON states by a burst of excitation in large arrays (1000 neurons) can be arranged to specify a pattern at the expense of some neighboring neurons, which must be OFF. Noise-induced switching causes evolution to the ground state and a specified fraction of neurons to the ON state (figures 16.5 and 16.6).

The reciprocal nature of inhibitory connections imposes constraints on what can be stored. For example, for a group of five totally inter-

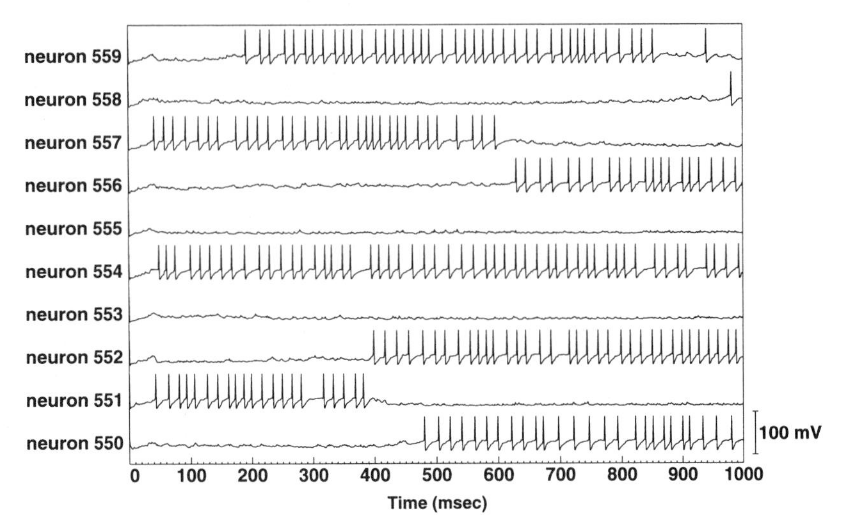

Figure 16.4 Simulated spiking activity within a subset of neurons in the interconnected inhibitory array (from same experiment as in figure 16.3). Ten neurons in a plane exhibit ON states of synaptic noise-driven activity with neighboring neurons held in an inhibited OFF state. Noise causes spontaneous reciprocal switching between states.

connected neurons, the strengths of sustained excitation and local inhibitory synaptic action may be set (in the absence of noise) so that any two, three, or four neurons are continuously firing (ON) while the other neurons are OFF. Selective stimulation provides an advantage in enabling a specified combination of three or four ON-state neurons to be set (Kirillov et al., 1992). Other constraints arise from the particular architecture of the local connectivity and the excitatory inhibitory strengths. Adaptive changes in responses to learning signals are clearly possible, but we have not yet explored these phenomena in our stimulations.

Intrinsic membrane properties have been found to reinforce synaptically mediated switching states. A striking property of medium spiny neurons is a nonlinearity of the current-voltage relation at membrane potentials progressively more hyperpolarized than the spike threshold level. The conductance is low near spike threshold, and a fast K conductance increases in these neurons upon hyperpolarization. Kawaguchi et al. (1989) and Wilson (1992) have noted that such nonlinearities are accentuated by the distribution of this fast g_K along dendritic lengths. Neurons without the fast g_K, or other nonlinear conductances, operate as linear RC circuits. Neurons with substantial g_K can exhibit two stable membrane potentials, one near E_K and the other near spike threshold. Neostriatal neurons tend to have one flat unstable zone in which synaptic influences may readily drive the membrane potential to high or low levels.

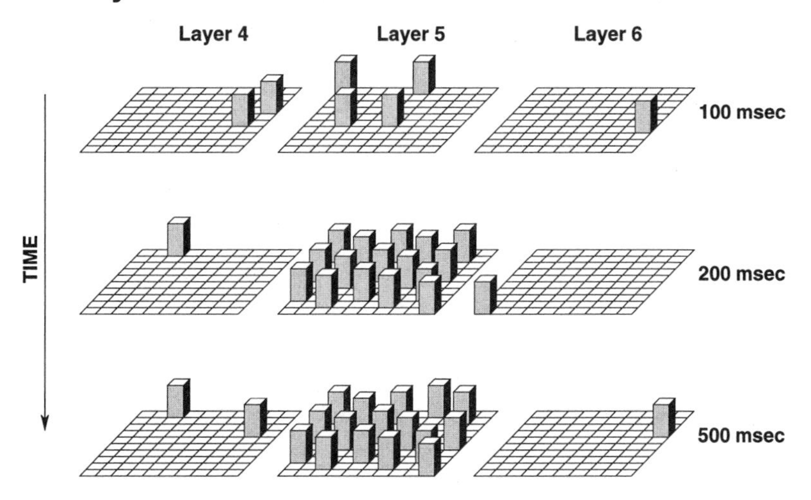

Figure 16.5 When a domain of local inhibition for each neuron is increased, a great variety of stable patterns of ON and OFF neurons is possible. Local stimulation of layer 5 in the 10 × 10 × 10 array (as in figures 16.3 and 16.4, but with every neuron inhibiting its 32 neighbors) can be adjusted to set a fraction of neurons to the ON state (here in layer 5) that remain on as a stable pattern. *Upper panels* show the distribution of ON and OFF states before the stimulation (*upper panels*), during stimulation (*middle panels*), and 300 ms after the stimulation (*lower panels*).

We have examined the influence of increasing this gK on the ON/OFF states in the two-neuron inhibitory model. Other parameters such as threshold and membrane potential were kept relatively constant. The results indicate that the intrinsically mediated switching appears to reinforce the switching mediated by the synaptic activity (figure 16.7, *lower right panel*). These intrinsic nonlinearities may provide a functional advantage in shaping the excitability function of neostriatal neurons as information propagates from cortex through the neostriatum. An alternative rationale for having such a property may reside in the capacity for facilitating a working memory property within the network. The simulation reveals, in addition, that the nonlinearities allow even very weak synaptic action of the collaterals to induce an effective switching effect. One could imagine a situation where modulation of the collateral GABAergic system could dynamically program a wide variety of scalar memory modules by generating subtle changes in synaptic efficacy.

NEOSTRIATAL SWITCHES AS A SCALAR MEMORY

The system of neostriatal circuits has the potential to function as an array of neural digital-to-analog converters. If reciprocal ON or OFF

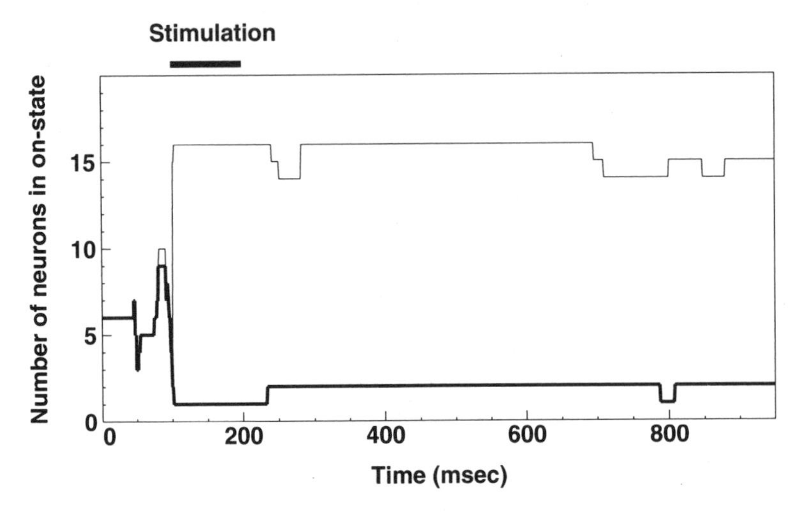

Figure 16.6 Simulation of stimulated ON states shows a capacity for the network to "store" scalar values in terms of the number of neurons that are set to ON states. This panel shows the evolution of the number of ON neurons in layer 5 of the $10 \times 10 \times 10$ array. Two stimulation patterns are applied. One pattern stimulates neurons in layer 5 (thin line); another stimulates layers 4 and 6 (thick line) and thus inhibits layer 5.

conditions (hypothesis 2) or stable states (hypothesis 3) exist, then a reason may be provided for the existence of a direct and indirect neostriatal output. There may be little value in expressing a pattern of ON/OFF states unless there were some mechanism for detecting and using the patterns. The computer simulation suggests that the direct/indirect division in outputs provides such a mechanism. The pallidum and SNr are known to have fewer but continuously active neurons than neostriatum, which has numerous but slower firing medium spiny neurons. Since a large number of neostriatal neurons converge on individual targets in pallidum and SNr, the output would summate the number of ON neurons on an incremental scale as a scalar value. Thus, the EPN (internal segment of globus pallidus, or GPi) and SNr neurons would, by this scheme, code a continuously variable degree of inhibition, i.e., "a scalar value," derived from the number of neostriatal ON-neurons projecting to the direct pathway. Information in neighboring medium spiny neurons projecting to the indirect target (globus pallidus, GP) would reside in the pattern of spike activity that is reciprocal to that of the direct pathway. Signals emitted by the GP (external segment, GPe) are inverted by the excitatory relay of the subthalamic nucleus before converging with the direct input in EPN and SNr (see figure 16.1). In this way the indirect signal can reinforce the direct signal. In this concept, the number of ON states projecting directly transmits the same scalar value as the number of OFF states projecting indirectly.

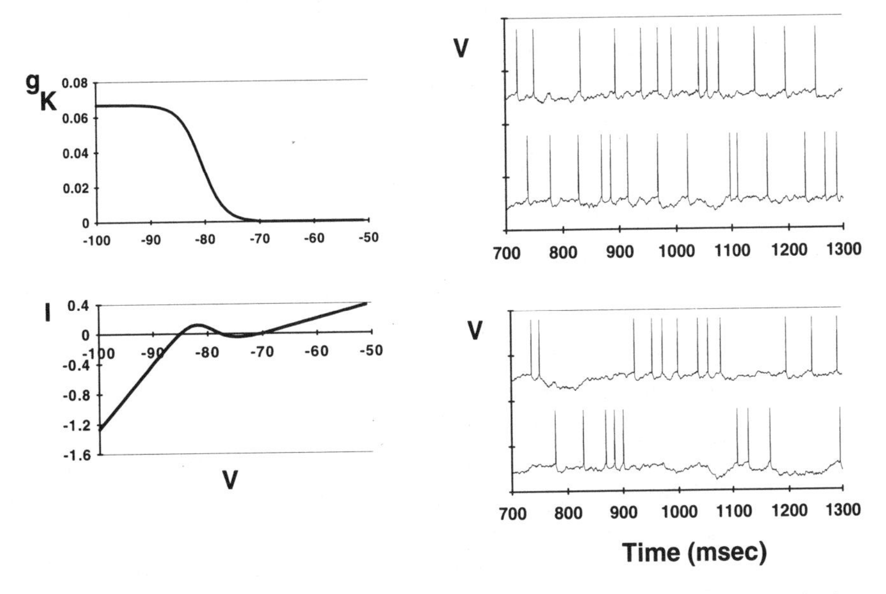

Figure 16.7 Influence of fast K+ conductance on switching. The introduction of a fast hyperpolarization-induced K+ conductance produced two stable membrane potentials, analogous to the two physiological states of the membrane of neostriatal medium spiny neurons. A two-neuron model with weak inhibitory synapses in this case does not exhibit switching (*upper right panel*), but is facilitated to do so by the presence of fast g_k (*lower right panel*).

Overall, when a distributed excitation appears across the population of medium spiny neurons, a pattern of reciprocal ON/OFF states would be expected to be generated through the action of the recurrent axon collateral network. In conditions in which both the direct and indirect pathways become active, the signals would tend to cancel. Thus, the output information of the neostriatum would be coded as *differences* in activity between the two pathways rather than in terms of the absolute output of the pathways.

This arrangement provides a way for the same envelope of global excitation of medium spiny neurons to convey different patterns of scalar values of inhibition onto the array of thalamic targets. Thus, in the case in which background excitation drives the neostriatal circuit from quiescence to sustained activity, a pattern of reciprocal states tends to emerge. To the extent that medium spiny neurons are driven to create a sustained state, this property would allow analog values to be retained after transient and asynchronous input arrives from cortex. There would be a substantial functional advantage if the neostriatal circuits were to mediate a temporary information storage function. A far more complex mode of spatiotemporal patterning and computation would be allowed by such a local circuit than if the circuitry simply relayed the convergent cortical input.

ENSEMBLE NEURON RECORDING IN NEOSTRIATUM

The possible functions of local neostriatal circuits raise serious challenges for an experimental analysis based on recording of neural spike train activity, which provides the most direct means for testing concepts of network computation. Single neuron studies of neostriatum have revealed a great deal about the specific cognitive or motor correlates, and the observation of widespread expectation signals indicates a significant role for the neostriatum in generating preparatory or contingent processing in relation to anticipated sensory and motor events (Hikosaka et al., 1989a,b; Alexander and Crutcher, 1990a,b; Apicella et al., 1992). The traditional assumption of single-neuron spike train recordings is that the behavior of the population can be reconstructed when the experimental control allows an identical sequence of brain states to be generated. However, many regions of neostriatum may exhibit highly variable activity during similar behavior patterns, and we cannot presume that identical spatiotemporal patterns are generated across repeated trials of identical behavior. In addition, little information is provided by single-neuron recording about the operations performed by local neostriatal circuitry, or about the spatiotemporal patterns of activity locally or globally within neostriatum.

Our view is that observation of a large number of concurrent spike trains from neurons whose location in neostriatum is determined with precision will be needed to define the spatial activity patterns and the time scale of reciprocal local neuron activity. Ensemble recording will allow an initial examination of the aforementioned hypotheses of neostriatal circuit function. Our initial predictions are that within ensemble activity one might observe signs of coordinated activation, reciprocal inhibition, and sequential changes between ON and OFF states of neural activity as observed in the computer simulations.

A considerable effort has been made in the past 5 years to develop instrumentation to allow routine large-scale ensemble recording. Parallel amplifiers with an array of digital signal processors have been linked to a real-time processor system for experimental control of operant behavior and acquisition of spike train data. The technique of fabrication of Teflon-coated microwire chronic recording arrays has been greatly refined. Precise reconstruction of the location of recorded neurons is accomplished with a spatial resolution of about 200 μm. Long-term recording of populations of 20 to 50 neurons in neostriatum is now routinely achieved. The availability of these data sets has necessitated development of new graphical and statistical procedures for analysis. The early preliminary results from these new recording procedures have begun to reveal information on the organization of sequential and reciprocal activity in neostriatum. A portion of these data as discussed here will delineate the advantages and difficulties of this procedure for study of local circuit function. Temporary storage of patterns would require reciprocal activity to appear across groups of neurons and in principle should be observed by these procedures. The primary difficulty, of course, arises from the fact that similar patterns may appear due to interactions within cortical circuits providing afferents to neostriatum.

Parallel spike trains recorded from groups of neurons allow large numbers of spike-triggered histograms, termed *cross-correlograms,* to be computed as an initial screen for temporal dependence between spike trains. One finding in neostriatum, given the sizes of spike counts and the locations of recorded neurons, is that few instances are found of dependencies over the short time intervals (activity windows of tens of milliseconds) characteristic of single-synapse interactions. This could be expected if single unitary inhibition synapses are weak but numerous, or if, as suggested by simulation, medium spiny neurons fire in reciprocal ON/OFF states, since this would minimize interactions observed in spike train data over short times. In contrast, substantial interactions are found in correlograms generated over longer time intervals (e.g., 4-second spike activity windows).

Four classes of two-neuron cross-correlograms can be described when the rat is engaged in simple behaviors such as alternating rest and treadmill-enslaved locomotion at 30-second intervals. A first type of common correlogram is flat, indicating a nearly complete temporal independence of activity in the 2 seconds before and after spiking. Concurrent global phasic excitation or inhibition broadcast to all neurons, as would be revealed in correlograms, does not appear to exist very frequently. A second and third type of temporal interaction consists of coincident excitation or inhibition with peaks or valleys 200 to 400 ms wide. Often a group of 2 to 5 neurons out of approximately 30 neurons may exhibit coincident phasic excitation as evidenced by a broad 200 to 300-ms central peak in the correlograms (figure 16.8). Examination of all the sets of cross-correlograms reveals that many of the excitatory correlograms are shared by the same members of a group.

The question arises as to whether these phasic excitations between neostriatal neurons, such as neurons 1, 3, 13, 14, and 15 in figure 16.8, occurs pairwise or whether groups of neurons are activated together. We were, therefore, motivated to create a form of an *ensemble event detector* to ascertain whether phasic excitatory or inhibitory events appeared synchronously within neostriatum. To detect *synchronous events,* a moving window 100 ms wide (about half the width of the of the peak in the cross-correlogram) was employed to scan across the spike events for the group of neurons that were coactivated. The number of instances were counted in which combinations of 3, 4, or 5 of the 5 neurons were

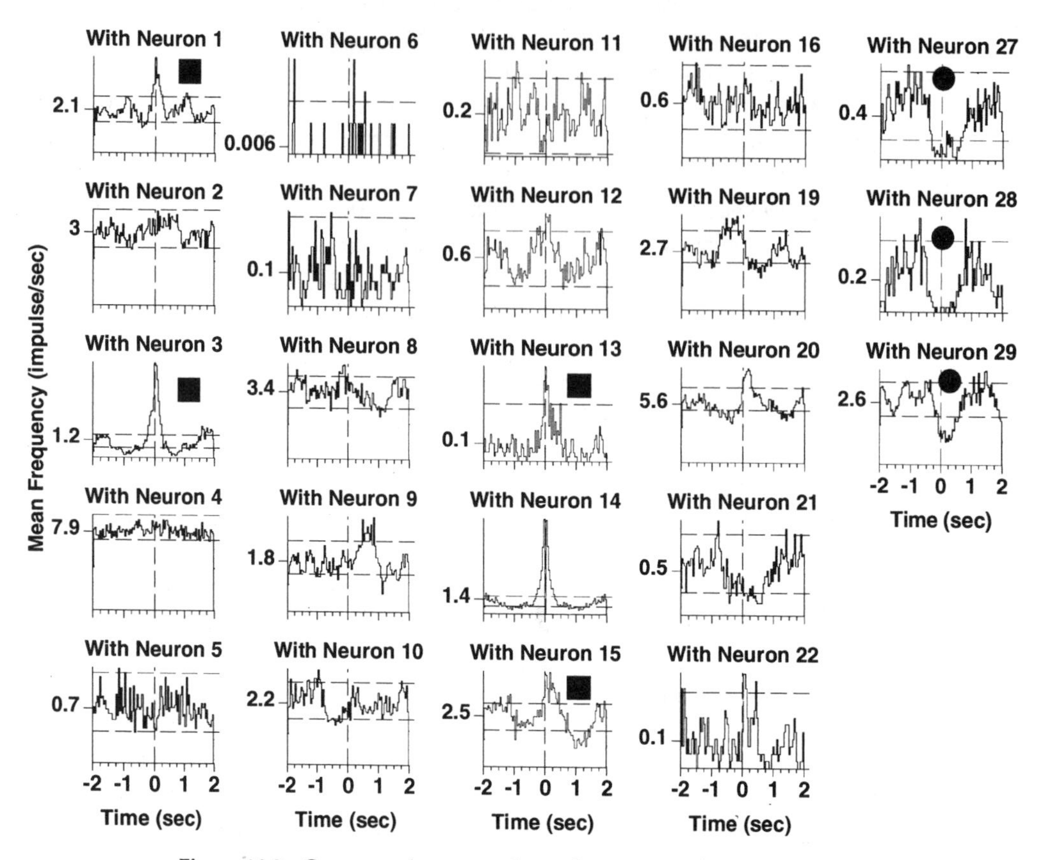

Figure 16.8 Cross-correlogram analysis of an array of 23 recorded neurons during a task consisting of alternating rest vs. treadmill-enslaved locomotion, using 20-second intervals. All pairwise cross-correlograms were computed for neuronal activity during treadmill locomotion and during rest. Cross-correlograms of 23 neurons with neuron 14 (and autocorrelogram neuron 14 vs. itself) to reveal coactivation with neurons 1, 3, 13, and 15 (squares) and coinhibition with neurons 27, 28, and 29 (circles). Total time of analysis during treadmill locomotion was 744 seconds (31 intervals of 24 seconds each, omitting the initial 5 seconds immediately after onset of treadmill activation as well as the last 1 second of the treadmill run). Some neurons increased their firing rates considerably for several seconds after the onset and offset of the treadmill locomotion. These relatively short periods of high activity would be heavily weighted in the cross-correlograms, suppressing the contribution of the long periods of low-level activity. Therefore, we excluded the initial 5 seconds after treadmill onset from the cross-correlation analysis.

found to fire together at least once during the moving window of time. Monte Carlo simulations were performed based on random distributions (Poisson) of events to study variation in the mean rates expected for each neuron. The number of occurrences of coincident firing according to each pattern was compared with that calculated from the random distributions of events. In this way it was shown that coincident events corresponding to the various patterns (as computed by z scores) appeared far more frequently than predicted by random activity. The patterns of "at least 4 of 5" appeared 3.5 times as frequently as expected (22 observed vs. 6 expected). Thus, the two-neuron cross-correlogram array has led to the discovery of multiple-neuron coincident events.

A limitation of the simple cross-correlogram computation is that no information is provided as to precisely when in time the neuron-neuron spike influence actually occurs within the spike train. In the case of the coincident event detection, the times that a pattern is detected by the moving window can be recorded in the data file as newly synthesized "events" and the distribution displayed as a histogram similar to regular spike events. A difference is that a fraction of these "events" are only random occurrences (figure 16.9) and another portion are true coincident events. For "at least 3 of 5" coincident firing, 176 events were observed but 145 were "expected" (statistically significant difference z score). A perievent histogram centered around a treadmill turning on (to enslave locomotion) revealed that the coactivated group of neurons behaved quite differently from one another during locomotion, but nevertheless exhibited synchronized coactivated events. The times of these coincident events, or pattern matches, were used to calculate perievent histograms for the entire population of recorded neurons. This procedure revealed that a group of three neurons (neurons 27, 28, and 29, figure 16.9) were coinhibited at the times when the initial group was coexcited, and neuron 20 was excited after the events. Thus, the 100-ms wide "events" (176 counts) in fact can be shown to include in this case 25% of the population of recorded neurons. A second smaller group of coactivated neurons in the right neostriatum were also detected with similar methods and these proved to be the same neurons found to be coinhibited by the examination of the initial coexcited group, an unexpected result. The marked inhibition in neurons 27, 28, and 29, and excitation in neuron 20, suggest that substantially less than 145 of the 176 events were truly random, and that a majority of spikes within the population were components of coordinated neural events. Additional analysis of statistics is in progress. A schematic diagram of the structure of the interactions between coactivated or coinhibited neurons is shown in figure 16.10.

Histological reconstruction of the recording sites of the microwire array revealed that the coactivation groups involved activity distributed widely throughout the neostriatum, including neurons on both sides,

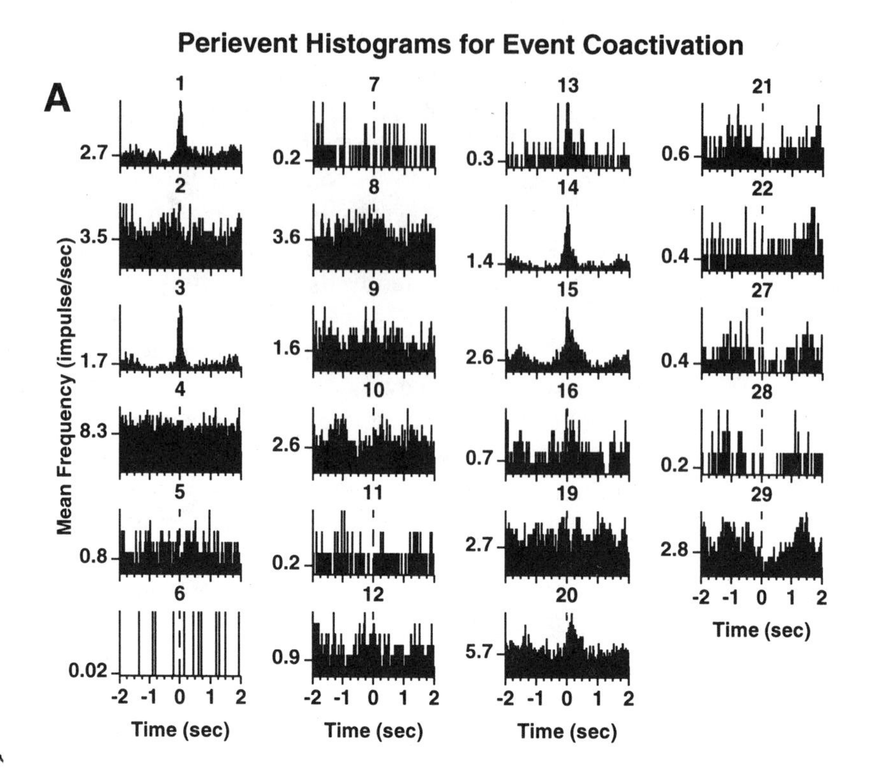

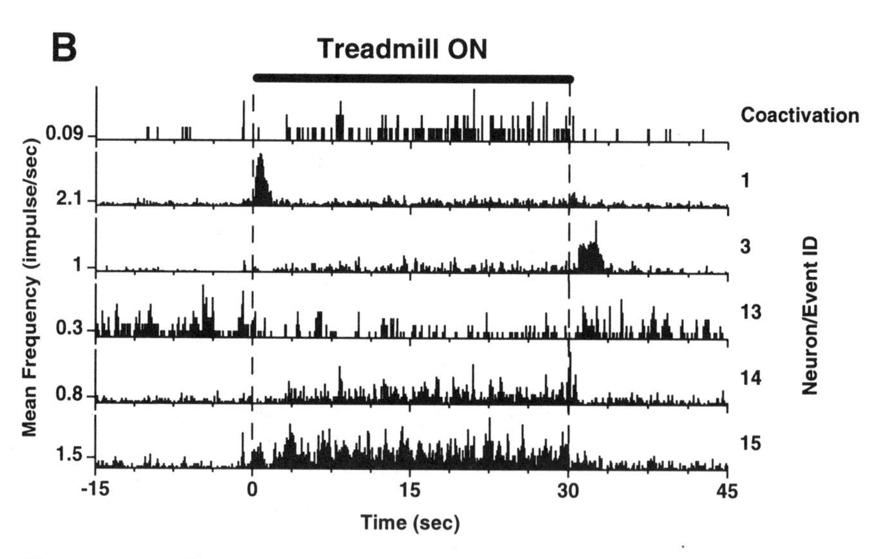

Figure 16.9 *A*, Perievent histograms around coactivation event times revealed by a pattern filter. Neurons 1, 3, 13, 14, and 15 reveal coincident activation. Neurons 27, 28, and 29 reveal coinhibition around these same event times. Neurons 20, 27, 28, and 29 were not used in establishing event times but do reveal correlated activity. *B*, Peri-treadmill-on histograms for neurons 1, 3, 13, 14, and 15. Mean firing rate during epoch is shown to left.

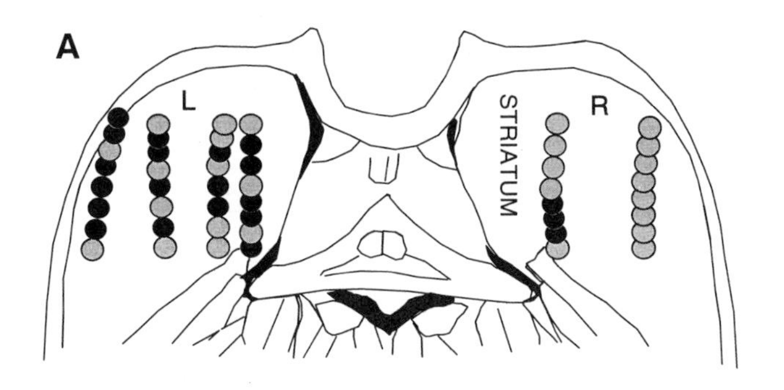

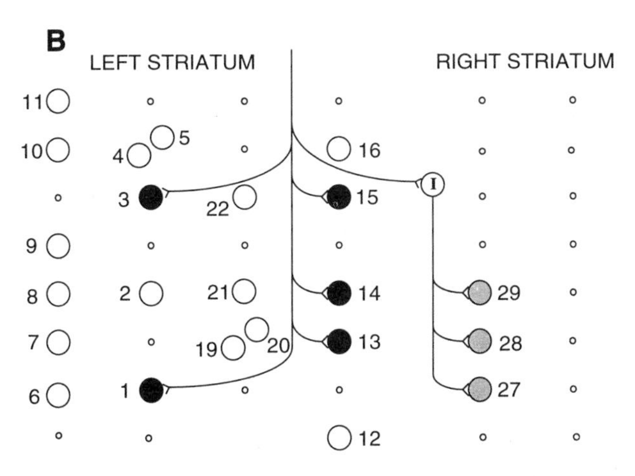

Figure 16.10 *A,* Positions of recording array mapped onto rat caudate-putamen. Thirty-two recording microwires were chronically implanted in the left neostriatum and 16 microwires were implanted in the right neostriatum. Microwires from which units were recorded are represented by black-filled circles. *B,* Relative locations of recorded neurons corresponding to those in figure 16.8. Group labeled by black-filled circles reveals a coactivation group corresponding to neurons 1, 3, 13, 14, and 15. A postulated inhibitory element, I, links coinhibition of three neurons (27, 28, 29) to the coactivation of the first group (1, 3, 13, 14, 15).

so it is unlikely that discrete highly localized corticostriatal afferent bundles could be responsible for the synchronized activity. The three coactivated and coinhibited neurons were organized along the rostrocaudal aspect of the neostriatum, suggestive of the pattern of cortical terminations in the neostriatum.

These forms of bilateral synchronized activity would appear to originate from synaptic influences coordinated by corticocoritcal or corticocallosal interactions since these connections are best situated for generating coordinated bilateral activity. However, frontal cortical regions, such as medial frontal cortex, are known to project bilaterally to the central and frontal zones of neostriatum and we have little information yet on correlated cortical phasic activity. Future work, with

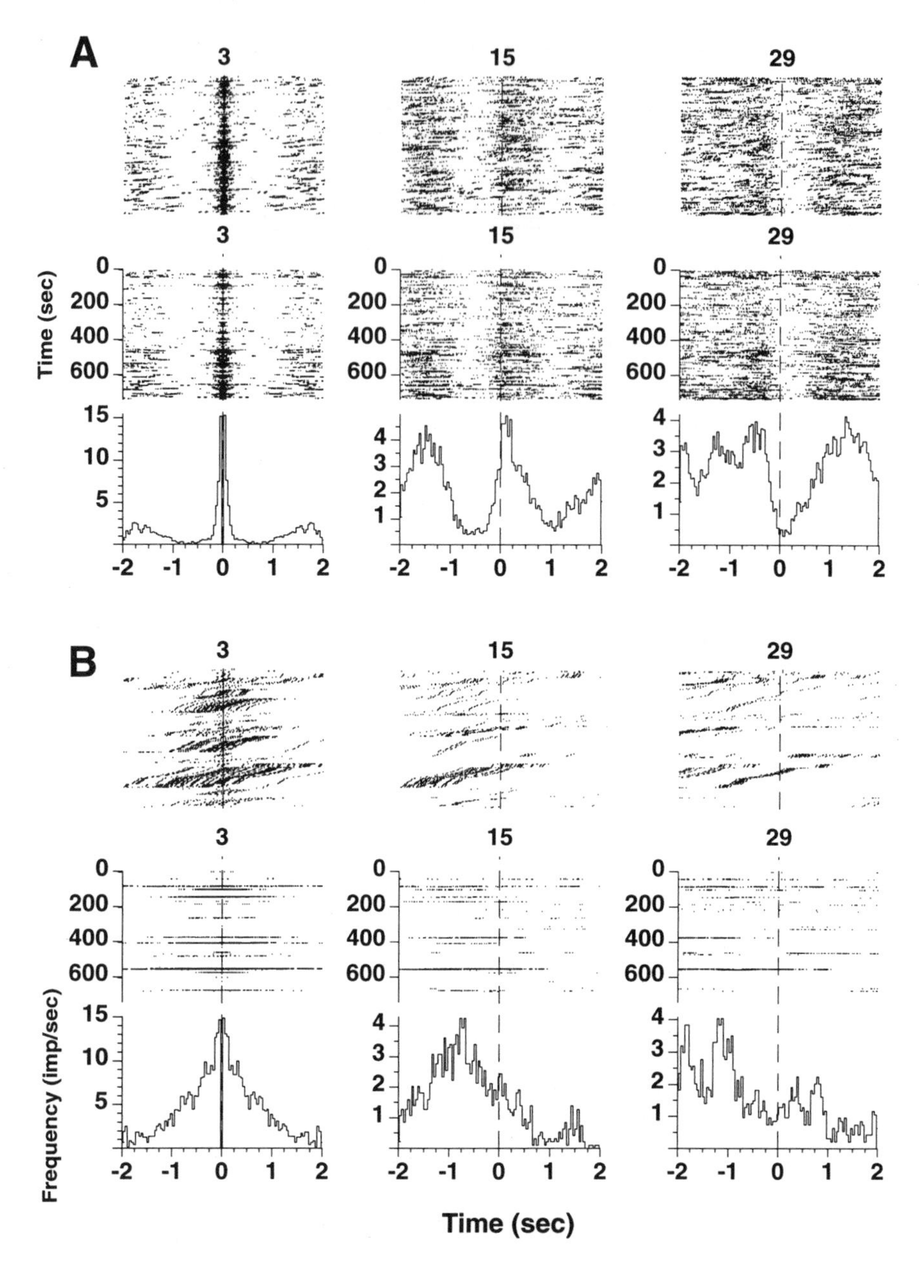

Figure 16.11 Autocorrelogram (*left column*) and cross-correlogram (*middle* and *right column*) rasters and histograms reveal temporal linkages between neurons. Spikes of neuron 3 were used as reference events for these rasters during sustained locomotion (*A*) and rest (*B*). Neuron 3 fired in bursts and produced a central peak in its autocorrelogram. Neuron 15 had an increased probability of firing following spikes by neuron 3, whereas neuron 29 exhibited inhibition after spikes or bursts of neuron 3. Sequential firing relations are well delineated during sustained locomotion but are ill defined at rest. The same experimental session is shown in figure 16.8. Time of analysis during treadmill locomotion is as described in figure 16.8. Neurons 3, 15, and 29 generated 844, 1790, and 2050 spikes, respectively, during treadmill-on cycles and 215, 310, and 753 spikes during treadmill-off cycles. Auto- and cross-correlogram rasters are plotted above the correlograms to show

additional recording arrays in cortex, will be needed to search for the origins of these cortical influences.

A fourth type of cross-correlogram is observed as an "asymmetrical" correlogram. This type of pattern is often generated when a sustained burst or group of spikes in one neuron is preceded or followed by a group of spikes in a second neuron. For example, figure 16.11 illustrates a group of three neurons during locomotion (figure 16.11A) and while at rest (figure 16.11B). Strong phasic activity with a sequential dependency is seen during steady locomotion, as excitation in one neuron is followed by an excitation or inhibition in others. Cross-correlogram rasters (see figures 16.11) show that the most spiking of neuron 3 occurred at the transition between inhibition and excitation in neuron 15 and at the onset of inhibition for neuron 29. While the study of such temporal interactions has just begun, the implication is that processing of sequential phasic information is a normal mode, sometimes even in "resting" or steady behavioral states when no clear behavioral nodes exist.

Such transitions in coordinated activity between neurons are what one might predict based on the simulations that reveal the potential for ON/OFF memory states. The central and medial neostriatal regions sampled in these data do not exhibit the marked phasic activity related to stepping observed in the dorsolateral neostriatum where sensory and motor cortex project their afferents (West et al., 1990; Carelli and West, 1991). The "events" may be interpreted as representing transitions within internal computational processes in the corticostriatal system. Sequential transitions between ON and OFF states may be regarded as changes in internal cognitive processes. These may be related to the generation of internal "intention vectors" represented by ensemble activity in primate motor cortex (Georgopoulos et al., 1993). In the future, detection of all but the simplest of such events may need to start purely from the observation of ensemble activity, with subsequent experimental work directed toward clarification of the precise role in general cognitive activity.

Overall, the ensemble recordings have started to reveal new insights into the spatiotemporal patterns of neostriatal activity. Reciprocal changes of firing rates are common with populations across transitions

spikes of neurons 3, 15, and 29, centered on spikes of neuron 3. The top rasters are plotted top to bottom around sequential occurrences of spikes by neuron 3. The bottom rasters are plotted around spikes of neuron 3 but in a linear time scale. These latter raster graphs reveal that the shape of the correlograms was more strongly influenced by more frequent spiking by neuron 3 toward the later treadmill cycles. When the animal was at rest during the treadmill-off cycles, very different correlograms are evident, resulting from infrequent episodic activity by neuron 3. These results show that correlogram analysis of sequential dependencies between interacting neuronal circuits is strongly regulated by activity specific to the behavioral state.

in behavioral states such as rest vs. locomotion. We have also found (Chang et al., 1994) reciprocal changes in spike activity by neurons in the ventral striatum during spike activity linked to self-initiated behavior of pressing a lever for cocaine self-administration. Ensemble recordings have begun to reveal spontaneous phasic coactivation events in some neurons in parallel with other groups of neurons that are coinhibited. Sequential activity patterns can be observed in functionally related groups of neurons.

SUMMARY

It should be noted that the discovery of sequential activity among neostriatal neurons is consistent with the hypothesis of significant actions of local inhibitory collaterals that act to evoke an either/or logic or to enable persistent memory effects among locally connected neurons. The time scale of at least 100 to 200 or several hundred milliseconds is sufficient to sustain short-term memory transiently, or even to allow evolution to local stable states as a mechanism for generating motor patterns. Clearly, an alternative view is that the results obtained to date suggest that neostriatal neurons simply track reciprocal, sequential activity generated in cerebral cortex. However, at this early stage, it must be said that such evidence provides only an initial survey of what will be needed to fully examine the hypotheses described. A comprehensive mapping of such temporal interactions is needed. This must be done within both dorsal and ventral neostriatum, and cross multiple behavioral tasks, to begin to clarify how local circuit interactions participate in functions mediated by the corticostriatothalamic system. The hypothesis of an array of neural digital-to-analog converters, each operating as a "scalar memory," computed by reciprocal ON/OFF states of medium spiny neurons, provides a framework for further experimental analysis.

ACKNOWLEDGMENTS

Supported by MH 44337, DA 2338, AA3901 and AFOSR F49620-92-J0301 to D. J. W.

REFERENCES

Alexander, G. E., and Crutcher, M. D. (1990a) Preparation for movement: Neural representations of intended direction in three motor areas of the monkey. *J. Neurophysiol.* 64:133–150.

Alexander, G. E., and Crutcher, M. D. (1990b) Neural representations of the target (goal) of visually guided arm movements in three motor areas of the monkey. *J. Neurophysiol.* 64:164–178.

Alexander, G. E., Crutcher, M. D., and DeLong, M. R. (1990) Basal ganglia-thalamocortical circuits: Parallel substrates for motor, oculomotor, "prefrontal" and "limbic" functions. *Prog. Brain Res.* 85:119–146.

Apicella, P., Scarnati, E., Ljungberg, T., and Schultz, W. (1992) Neuronal activity in monkey striatum related to the expectation of predictable environmental events. *J. Neurophysiol.* 68:945–959.

Carelli, R. M., and West, M. O. (1991) Representation of the body by single neurons in the dorsolateral striatum of the awake, unrestrained rat. *J. Comp. Neurol.* 309:231–249.

Chang, J-. Y., Sawyer, S. F., Lee, R. S., and Woodward, D. J. (1994) Electrophysiological and pharmacological evidence for the role of the nucleus accumbens in cocaine self-administration in freely moving rats. *J. Neurosci.* 14:1224–1244.

Felleman, D. J., and Van Essen, D. C. (1991) Distributed hierarchical processing in the primate cerebral cortex. *Cereb. Cortex* 1:1–47.

Flaherty, A. W., and Graybiel, A. M. (1993) Output architecture of the primate putamen. *J. Neurosci.* 13:3222–3237.

Gerfen, C. R. (1992) The neostriatal mosaic: Multiple levels of compartmental organization in the basal ganglia. *Annu. Rev. Neurosci.* 15:285–320.

Georgopoulos, A. P., Taira, M., and Lukashin, A. (1993) Cognitive neurophysiology of the motor cortex. *Science* 260:47–52.

Grossberg, S. (1973) Contour enhancement, short term memory and constancies in reverberating neural networks. *Stud. Appl. Math.* 52:213–257.

Groves, P. M. (1983) A theory of the functional organization of the neostriatum and the neostriatal control of voluntary movement. *Brain Res. Rev.* 5:109–132.

Hikosaka, O., Sakamoto, M., and Usui, S. (1989a) Functional properties of monkey caudate neurons. II. Visual and auditory responses. *J. Neurophysiol.* 61:799–813.

Hikosaka, O., Sakamoto, M., and Usui, S. (1989b) Functional properties of monkey caudate neurons. III. Activities related to expectation of target and reward. *J. Neurophysiol.* 61:814–832.

Hopfield, J. J. (1984) Neurons with graded responses have collective computational properties like those of two state neurons. *Proc. Natl. Acad. Sci. U. S. A.* 81:3088–3092.

Kawaguchi, Y., Wilson, C. J., and Emson, P. (1989) Intracellular recording of identified neostriatal patch and matrix spiny cells in a slice preparation preserving cortical inputs. *J. Neurophysiol.* 62:1052–1068.

Kawaguchi, Y., Wilson, C. J., and Emson, P. C. (1990) Projection subtypes of rat neostriatal matrix cells revealed by intracellular injection of biocytin. *J. Neurosci.* 10:3421–3438.

Kirillov, A. B., Myre, C. D., and Woodward, D. J. (1991a) Working memory in 3D inhibitory-feedback model inspired by neostriatum. In *Proc Int Joint Conference Neural Networks,* (IJCNN-91), Vol. 11, p. 998, Seattle Washington July 8–12, 1991, IEEE. 2:998.

Kirillov, A. B., Myre, C. D., and Woodward, D. J. (1991b) Bistable neurons and memory patterns in the inhibitory-feedback model inspired by neostriatum. *Soc. Neurosci. Abstr.* 17:124.

Kirillov, A. B., Myre, C. D., and Woodward, D. J. (1992) Working memory in small inhibitory-feedback neural networks: Bistable neurons and coding of scalar values. *Soc. Neurosci. Abstr.* 18:318.

Kirillov, A. B., Myre, C. D., and Woodward, D. J. (1993) Bistability, switches and working memory in a two-neuron inhibitory-feedback model. *Biol. Cybern.* 68:441–449.

McGeorge, A. J., Faull, R. L. M. (1989) The organization of the projection from the cerebral cortex to the striatum in the rat. *Neuroscience* 29:503–537.

Myre, C. D., and Woodward, D. J. (1990) Memory properties in the 1000-neuron inhibitory-feedback model for simulation of neostriatal circuits. *Soc. Neurosci. Abstr.* 16:1092.

Myre, C. D., Sawyer, S. F., and Woodward, D. J. (1989) Simulation of 1000-neuron inhibitory-feedback network reveals memory properties: Computer modeling inspired by neostriatum. *Soc. Neurosci. Abstr.* 15:1049.

Nisenbaum, E. S., and Berger, T. W. (1992) Functionally distinct subpopulations of striatal neurons are differentially regulated by GABAergic and dopaminergic inputs—I. *in vivo* analysis. *Neuroscience* 48:561–578.

Penny, G. R., Wilson, C. J., and Kitai, S. T. (1988) Relationship of the axonal and dendritic geometry of spiny projection neurons to the compartmental organization of the neostriatum. *J. Comp. Neurol.* 269:275–289.

West, M. O., Carelli, R. M., Pomerantz, M., Cohen, S. M., Gardner, J. P., Chapin, J. K., and Woodward, D. J. (1990) A region in the dorsolateral striatum of the rat exhibiting single-unit correlations with specific locomotor limb movements. *J. Neurophysiol.* 64:1233–1246.

Wilson, C. J. (1992) Dendritic morphology, inward rectification and the functional properties of neostriatal neurons. In T. McKenna, J. Davis, and Zoernetzer S. F. (eds), *Single Neuron Computation.* San Diego.: Academic Press, pp. 141–171.

Wilson, C. J., and Groves, P. M. (1980) Fine structure and synaptic connections of the common spiny neuron of the rat neostriatum: A study employing intracellular injection of horseradish peroxidase. *J. Comp. Neurol.* 194:599–615.

Wilson, C. J., Chang, H. T., and Kitai, S. T. (1983) Disfacilitation and long-lasting inhibition of neostriatal neurons in the rat. *Exp. Brain Res.* 51:227–235.

Wilson, C. J., Kita, H., and Kawaguchi, Y. (1989) GABAergic interneurons rather than the spiny cell axon collaterals are responsible for the IPSP responses to afferent stimulation in neostriatal spiny neurons. *Soc. Neurosci. Abstr.* 15:907.

Woodward, D. J., Moises, H. C., Waterhouse, B. D., Yeh, H. H., and Cheun, J. E. (1991) The cerebellar norepinephrine system: Inhibition, modulation and gating. In C. D. Barns, O. Pompriano (eds.), *Progress in Brain Research.* Amsterdam: Elsevier Science Publishers, pp. 331–341.

17 Sensorimotor Selection and the Basal Ganglia: A Neural Network Model

Stephen Jackson and George Houghton

Despite our growing knowledge of the anatomical and physiological organization of the basal ganglia, the function of these structures remains an enigma. The consensual view is that the basal ganglia participate in the control and programming of motor behavior (cf. Mink and Thatch, 1991). The exact nature of this participation remains to be specified, however. Several authors have recently suggested that the basal ganglia may also play a role in the control of nonmotoric, cognitive activity, and have drawn attention to the frontal association and limbic afferents of the ventral striatum within this context (Early et al., 1989; Groenewegen et al., 1991; Robbins, 1991). In this chapter we review several lines of evidence consistent with the view that the basal ganglia may perform a general computational function concerned with the allocation of attention. Specifically, we suggest that the basal ganglia facilitate the disengagement and reengagement of processing resources between different "attentional" sets, across a variety of domains, and we present a neural network model of voluntary attentional control based on the anatomical architecture of the basal ganglia.

ANATOMICAL SUBSTRATES OF VISUOSPATIAL ATTENTION

While attention has been defined as the selective aspect of perception (Kosslyn et al., 1990), traditional information-processing models have frequently considered the component processes of attention to be unitary, and largely independent of both the perceptual processes involved in the analysis of sensory information, and those processes governing action. Thus, visuospatial attention, e.g., has been likened to an attentional spotlight or a variable-focus lens, which is capable of being relocated within visual space, thereby facilitating the perceptual processing of those regions of the visual world illuminated by the spotlight or captured by the lens (Posner, 1980). However, the last decade has seen many technological advances in neuronal pathway labeling, single-cell recording, and brain imaging techniques, which have led to a number of exciting developments within the neurosciences. One consequence for the field of attention has been an increased emphasis on understanding how separate anatomical systems interact to carry out basic computational functions associated with sensorimotor selection and control. This change in emphasis has led several authors to question the unitary nature of visuospatial attention, arguing instead for several relatively independent sensorimotor circuits, each given over to the programming of different types of action, e.g., eye movements, reaching and grasping movements, and so forth, which may nevertheless need to be coordinated for the efficient fulfillment of many sensorimotor tasks (Posner and Peterson, 1990; Jackson et al., 1993; Goodale and Milner, 1992; Rizzolatti et al., 1993).

Consistent with this approach, Posner and Peterson (1990) distinguish three functionally distinct neural networks which each appear to carry out separate but complementary operations involved in selective attention. They distinguish between: an anterior attention network (anterior cingulate and supplementary motor areas), which is related to volitional control and awareness, particularly awareness of target stimuli (Posner and Rothbart, 1990); a posterior attention network (posterior parietal cortex, pulvinar, and superior colliculus), which controls spatial orienting; and a vigilance network (locus coeruleus), which functions to place the anterior and posterior systems into an alert state, thereby enhancing attentional processing in both networks.

SPATIAL ORIENTING OF ATTENTION AND THE POSTERIOR NETWORK

The computational complexity of apparently simple behaviors is often only fully illustrated when we see how the normal system is impaired following damage or disease. Attentional dysfunction is associated with a wide range of neurological and psychiatric disorders, including schizophrenia (SZ), sensorimotor neglect, and Parkinson's disease (PD). Using variants of the spatial precuing technique, Posner and his colleagues (1980, 1984, 1990, 1993) have sought to discover the functional components of visuospatial attention by examining covert orienting in several different patient groups. As a consequence they have identified three functional components of covert orienting: a "disengage" function, a "move" function, and an "engage" function. These functions are illustrated schematically in figure 17.1.

The spatial precuing technique involves presenting subjects with some form of spatially informative cue which indicates the most probable location of an impending target. Such cues may take the form of a brief change in luminance in the vicinity of the target location (figure 17.2A), or may involve the use of symbolic cues to the target's location, such as a directional arrow (figure 17.2B). The effects of precuing are typically assessed by comparing response latency to targets appearing

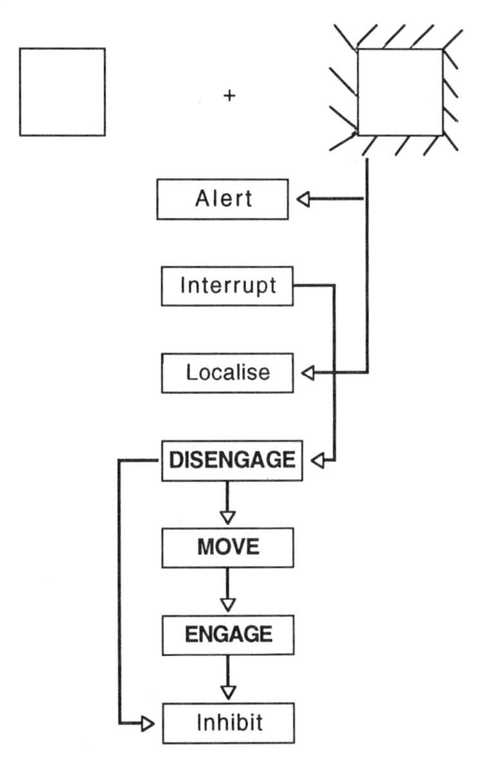

Figure 17.1 Schematic illustration of the functional components of covert orienting of attention identified by Posner and Peterson (1990).

at a cued location (valid trials) against targets appearing at unexpected, i.e., noncued, locations (invalid trials). In addition, several studies have made use of an additional condition in which a "neutral" cue is presented which informs the subject as to the onset of the next target, but is uninformative as to its spatial location (e.g., Posner, 1980). The results of many such studies have shown that latencies to detect targets appearing at validly cued locations are reduced relative to trials where a neutral cue is presented (cuing benefits), while latencies to targets appearing at unexpected locations (invalidly cued trials) are increased (cuing costs).

While spatial neglect has been reported following frontal lesions (Heilman and Valenstein, 1985; Damasio et al., 1980) and from lesions to the basal ganglia (e.g., Villardita et al., 1983), by far the most common lesion site to produce contralateral neglect for visual stimuli is the posterior parietal lobe. Although subjects may appear normal, they are frequently impaired in their ability to deal with a visual stimulus presented to their contralesional visual field, *in circumstances where they are already attending to visual information.* This impairment is revealed as a greatly magnified cost in reaction time to detect a visual target (Posner et al., 1984). Posner and Peterson (1990) have suggested that such

A. Exogenous cuing

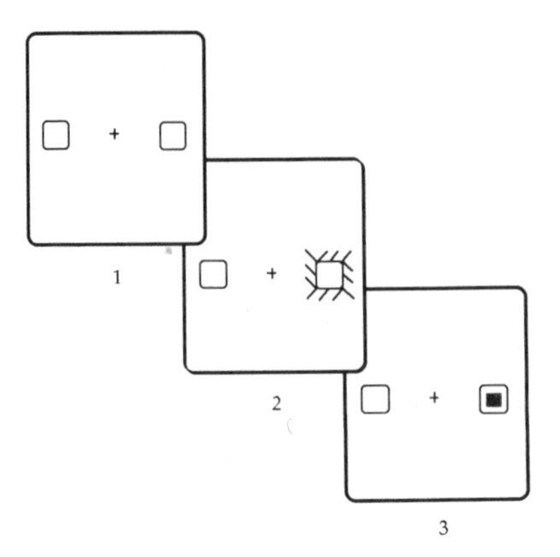

B. Endogenous cuing

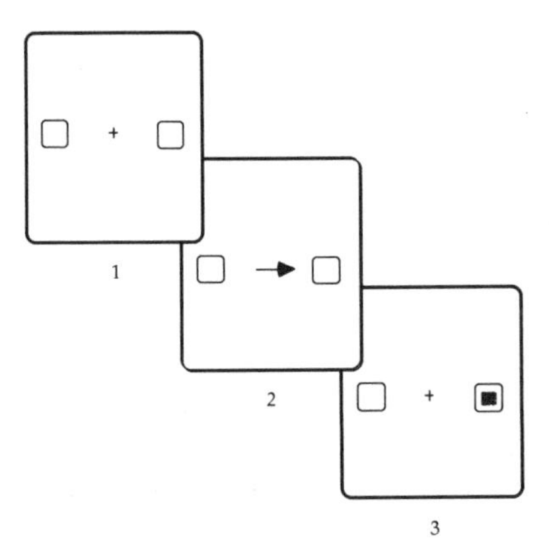

Figure 17.2 Exogenous and endogenous forms of the spatial precuing paradigm. The numbers 1–3 indicate the sequence of events occurring on each trial. In (*A*), attention is drawn to a particular space location by an abrupt, and short-lived, change in luminance of the box surrounding the target area (2), while in (*B*), attention is directed to the same target location by the appearance of an arrow cue.

patients may have a specific deficit in disengaging attention once it had been summoned to an ipsilesional location, and as a consequence, that the parietal cortex was the likely site of the disengagement operation (see figure 17.1).

While patients with damage to their posterior attentional network (e.g., parietal lesions) show greatly increased reaction time costs under circumstances such as those outlined above, damage to frontal systems—including damage to the cortical-subcortical-cortical circuits passing through the basal ganglia—appear to produce the opposite effect, i.e., a *reduction* in attentional costs. These effects are outlined below, but we shall first consider the structures of the basal ganglia within the context of Posner's attentional networks.

DO THE BASAL GANGLIA CONTROL SPATIAL ORIENTING?

While it may often be convenient to consider each of the above networks in isolation, anatomical studies indicate that these networks are part of a much larger pattern of neural pathways which span the entire neocortex and converge on key subcortical structures (Goldman-Rakic, 1987, 1988). Cortical nodes of the anterior and posterior attention networks are able to communicate with one another directly, via corticocortical pathways. Prefrontal and posterior parietal cortex are linked via topographically precise reciprocal connections, and both areas project to, and receive input from, other brain regions implicated in the control of visuospatial function (Goldman-Rakic, 1988). In addition, the cortical nodes of both the anterior and posterior attention networks receive catecholamine projections from subcortical sites implicated in the maintenance of an alert state (i.e., the locus coereuleus), and cortical nodes in both of the above networks can communicate with subcortical nodes, such as the superior colliculus, via direct excitatory projections which in all likelihood preserve spatiotopic information mapped within the cortex. Finally, in addition to the direct connections between cortical and subcortical nodes outlined above, the anterior attentional system may modulate activity within the posterior attention network via third-party structures such as the basal ganglia.[1]

The basal ganglia consist of a system of subcortical structures which lie beneath the neocortex and surround the thalamus. These structures include the caudate nucleus and putamen (jointly termed the *striatum*), the globus pallidus (lateral and medial segments), together with the substantia nigra, and subthalamic nucleus. Consistent with their anatomical location, the basal ganglia receive topographical projections from almost the entire neocortex, and project to frontal areas of the cortex via thalamocortical projections, and to certain subcortical structures, including the superior colliculus. Earlier views considered the role of the basal ganglia as that of integrating information received from diverse cortical sites. It is of interest to note here that projections originating within frontal regions (e.g., frontal eye fields and Walker's area 46) and posterior parietal areas of the cortex terminate within the striatum in a series of alternating columns, thereby maintaining the topographical integrity of spatial information originating within these two areas, and perhaps allowing for the synthesis of spatial information originating from anterior and posterior areas of the cortex.

While the basal ganglia may participate in the synthesis of some cortical information, other evidence suggests that the basal ganglia are organized into a number of largely separate corticostriatothalamocortical circuits, which appear to unite cortical and thalamic regions dedicated to a common function (Alexander and Crutcher, 1990). Alexander and co-workers have suggested that there at least five such pathways through the basal ganglia, each organized in parallel, and innervating different regions of the thalamus and frontal cortex. These include a "motor" circuit centered on the supplementary motor area and related regions of motor cortex, an "oculomotor" circuit centered on the frontal eye fields, and other circuits—limbic, orbitofrontal, and dorsolateral prefrontal—which are less obviously tied to the control of motor function. While each of these circuits, based on their cortical site of origin, appears to be dedicated to processing different kinds of information, the organization of each circuit follows a similar pattern, suggesting that the computational function of each circuit may be equivalent. When considering the computational function of the basal ganglia, it is of interest to note that unlike most other structures in the brain, the action of the basal ganglia is inhibitory. GABAergic neurons in the medial segment of the globus pallidus (GPm) and substantia nigra pars reticulata (SNr) are tonically active, and hold the thalamus and other structures (e.g., the superior colliculus), in a state of tonic inhibition. This pattern of inhibition is modulated via two distinct pathways, which appear to organized in opposition to each other, and link the striatum to output neurons in the GPm-SNr complex. One of these pathways, the *striatonigral* pathway, provides direct inhibitory modulation of activity in the SNr, while the other, the *striatopallidal* pathway, indirectly provides excitatory input to the GPm-SNr complex. Recent evidence has revealed important neurochemical differences between these two pathways, which appear under normal conditions to be sensitively balance,d and to depend critically on modulation by dopamine [DA] (Gerfen, 1992).

In summary, the basal ganglia complex can be characterized as an important component of several functionally and anatomically separate neural circuits which unite cortical and subcortical regions implicated in the processing of information in several behavioral domains. These circuits share a similar anatomical architecture, suggestive of a common computational function, and consist of two opposing pathways which respectively inhibit and facilitate the inhibitory output projections of the basal ganglia complex. Finally, these two opposing pathways are maintained in a delicate balance through modulation by dopamine.

ATTENTIONAL DYSFUNCTION IN PARKINSON'S DISEASE

Parkinson's disease is typically described as a disease of the motor system characterized by limb tremor when in a resting state, and an inability to initiate and control voluntary movement (akinesia). These physical symptoms are associated with neural degeneration of nigrostriatal DA projections within the basal ganglia, and the degree of akinesia observed is correlated with increased activity within the indirect (striatopallidal) pathway (Gerfen, 1992). PD patients show particular difficulty in the performance of complex motor behaviors. For example, they are impaired in the execution of simultaneous motor acts (Benecke et al., 1986). Moreover, such problems are compounded when patients are required to control behavior in the absence of external feedback. Thus, while PD patients can accurately track a target moving regularly and predictably, e.g., a sawtooth pattern on an oscilloscope screen, their performance declines when visual feedback is removed, requiring them to make predictive movements (Sagar and Sullivan, 1988).

These observations suggest that the motor deficit of PD includes elements of synthesis and prediction, functions that are traditionally associated more with cognition than with movement. Such findings have led several authors (e.g., Keele et al., 1992; Robbins, 1991) to suggest that the motor deficit in PD can be characterized as a dysfunction in the ability to select between, or switch, actions on the basis of interoceptive cues. Cognitive impairment in PD has often been described as similar to that found following frontal lobe damage, and involving attentional deficits such as increased interference on the Stroop task, and impaired set-shifting on the Wisconsin card-sorting task [WCST] (for a review, see Robbins, 1991). Brown and Marsden (1988) have characterized the attentional dysfunction of PD patients as an impairment in the ability to guide behavior on the basis of internal attentional control.

The distinction between internal and external modes of attentional control closely parallels that made in the context of voluntary control of action by Goldberg (1985), and by Goldman-Rakic (1987, 1988) with respect to the visuospatial function of the dorsolateral prefrontal cortex. Studies of the effects of prefrontal lesions on the performance by primates on the delayed-response task [DRT] suggest that the internal control of voluntary behavior is dependent on the "central executive" function of the prefrontal cortex, which involves three subfunctions: the ability to select appropriate information; the ability to hold that information "on-line" when the stimulus is no longer present; and, on the basis of the selected information, to initiate and execute an appropriate motor response. Primates with prefrontal lesions show specific deficits within the visuospatial domain. Using a visuospatial equivalent of the DRT, Funahashi et al. (1986) demonstrated that monkeys with unilateral prefrontal lesions were not impaired when required to move their eyes to the location of a peripheral target, but were significantly impaired when required to move their eyes to the location of a remembered target. Moreover, such deficits were mainly limited to targets which occurred in the contralesional visual field.

Similar visuospatial deficits have also been found in human subjects with frontal lobe damage, and in patients with basal ganglia abnormalities, e.g., PD and Huntington's disease [HD]. While such patients typically show normal reflex saccades to a peripheral target, they are frequently impaired on tasks which require the "voluntary" or "internal" control of orienting, e.g., on tasks which require them to make a saccadic eye movement to a remembered target (Carl and Wurtz, 1985; Crawford et al., 1989); on tasks where the location and timing of a visual event allows the subject to make a saccade in anticipation of a target appearing (predictive saccades; e.g., see Bronstein and Kennard, 1985); and on tasks where the subject is required to suppress reflexive orienting toward a peripheral stimulus (Guitton et al., 1982; Leigh et al., 1983; Lasker et al., 1987).

To date, only a limited number of studies have investigated the covert orienting of attention in patients with basal ganglia disorders. Rafal et al. (1984) showed that manipulation of medication levels in PD had little or no effect on spatial orienting of attention. However, more recent studies have reported finding deficits in covert orienting associated with PD. Wright et al. (1990) reported the slightly counterintuitive finding that while PD patients showed normal benefits for validly cued trials, they showed *decreased* costs on invalidly cued trials (figure 17.3), and Kingstone et al. (1992) reported similar findings in a study of both covert and overt orienting in a group of PD patients. Specifically, Kingstone et al. (1992) found that while endogenous orienting appeared

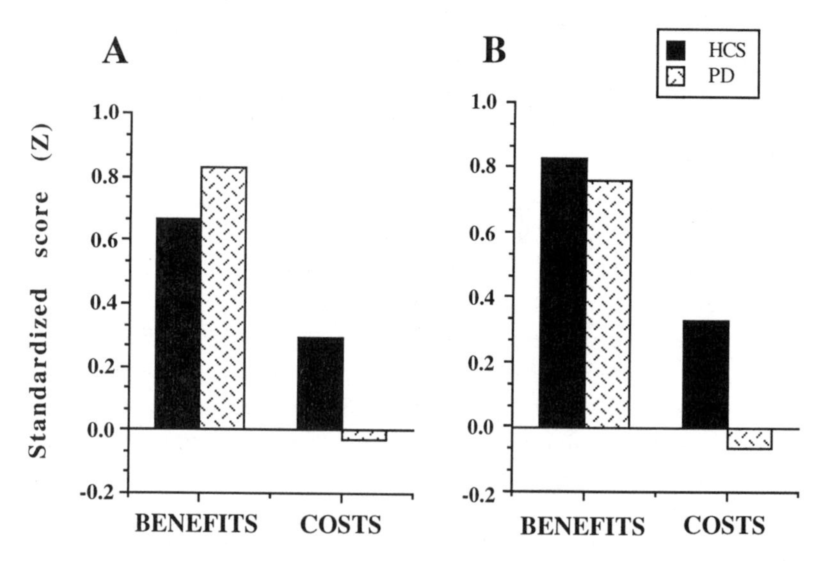

Figure 17.3 The effects of Parkinson's disease on the costs and benefits of spatial precuing. Data for age-matched healthy control subjects (HCS) and Parkinson patients (PD) were adapted from RT data obtained in two separate studies by (*A*) Wright et al., 1990; and (*B*) Sharpe, 1990. The RT data from each study have been standardized.

normal, both covert and overt orienting to an exogenously summoned target was "hyperreflexive" (i.e., paradoxically faster) in the PD group.

Henik et al. (1991) studied the effects of frontal eye field [FEF] lesions on orienting following both endogenous and exogenous cues. They showed that latencies for peripherally summoned saccades were faster for targets presented within the patients' contralesional field compared to targets appearing in the ipsilesional field. Moreover, these results were specific to both eye movements and to FEF lesions, as this pattern of results was not found on a similar task requiring a manual response, or with patients with frontal lesions which spared the FEFs. They interpreted these results as consistent with a disinhibition model whereby the FEF normally suppresses activity in the ipsilateral superior colliculus. Thus, following an FEF lesion, the ipsilateral colliculus is disinhibited, which both facilitates saccades to the contralesional field, and produces greater lateral inhibition of the contralateral colliculus, thereby resulting in increased latencies to ipsilesional targets. As we have argued elsewhere (Jackson and Lees, 1993), the most plausible means by which the FEF could suppress collicular activity in this way would be via the basal ganglia oculomotor circuit.

One important clue to the neural basis of such hyperreflexive orienting in PD patients has been obtained in a recent study by Clark et al. (1989). They manipulated catecholamine levels in normal adult subjects performing a covert orienting task through the intravenous administration of droperidol (DA antagonist) and clonidine (α_2-receptor agonist]. The results of this study are presented in figure 17.4, which clearly illustrates that the major effect of administering these two substances

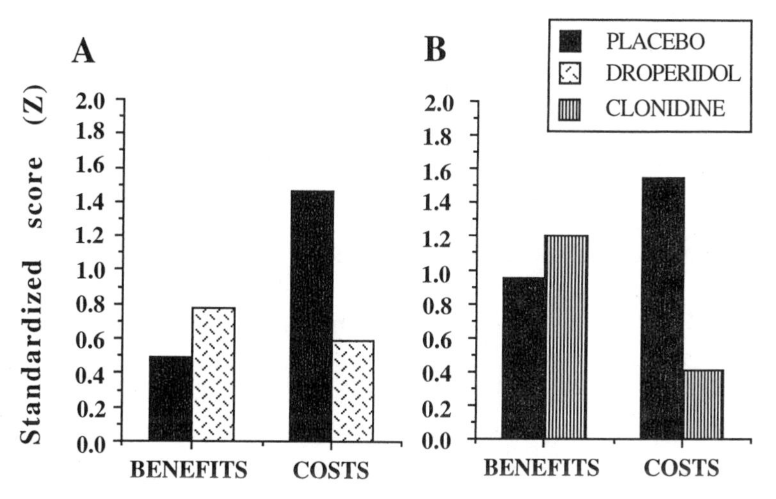

Figure 17.4 The effects of catecholamine manipulation on the costs and benefits of spatial precuing. Data were adapted from the study of Clarke et al. (1989) of the effects of droperidol and clonidine on covert orienting in healthy adults. The RT data from each study have been standardized separately.

is to drastically reduce the response time (RT) costs arising from an invalid precue.

Finally, additional support for the idea that the basal ganglia participate in the active suppression of behaviorally irrelevant perceptual dimensions comes from a study of normal subjects using the positron emission tomography (PET) procedure. Corbetta et al. (1991) had subjects make same/different discriminations to a series of pairs of colored shapes in which the members of each pair were presented successively. The members of each pair were either the same on each dimension, or differed in terms of their color, shape, or motion. Subjects performed this task under two conditions: a focused-attention condition in which the subjects were told in advance the relevant dimension on which the stimuli in each pair could differ; and a divided-attention condition in which no advance information was provided. Their results revealed that during the focused-attention trials there was increased blood flow in the caudate and pallidal regions of the basal ganglia, but this pattern of activity was not found for the divided-attention trials.

The ideas introduced above can be summarized as follows: (1) the basal ganglia appear to be anatomically organized to perform a control function, in that parallel and mutually opposed, inhibitory, and disinhibitory pathways project from the striatum to the thalamus and superior colliculus; (2) several such pathways exist, of which only two have a clear motor function; (3) the basal ganglia receive projections from all of the cortical areas implicated by Posner and Peterson (1990) in the control of visuospatial attention, and in turn project to those subcortical areas which form part of the posterior attentional system (i.e., thalamus [reticular nucleus] and superior colliculus; (4) damage to all of the components of this corticostriatothalamocortical circuit can result in some form of spatial neglect; (5) diseases of the basal ganglia, such as PD, lead to specific impairments in overt orienting, and have been reported to lead to dysfunction in the covert orienting of attention, characterized by a reduction in RT costs to targets appearing at uncued locations.

A NEURAL NETWORK MODEL

We recently proposed a functional theory of spatial attention, implemented as a neural-style computational model, based on the anatomical architecture of the basal ganglia (Jackson and Houghton, 1992). We argued that the selection of behaviorally relevant spatial locations may be carried out by two independent, but interacting, mechanisms: a general mechanism based on lateral inhibition (assumed to be a ubiquitous property of the various anatomical networks implicated in the control of attention), and a frontal mechanism, thought to selectively influence the activity of the anterior attentional network through the operation of the neural circuits passing through the basal ganglia. The latter mechanism is based on the action of a pair of antagonistic pathways, which operate to "engage" or "disengage" activity at selected loci within the anterior system, and which serve two basic computational functions: Firstly, once a behavioral sequence or event has been selected, the *direct* pathway functions as an attentional amplifier, enhancing behaviorally relevant signals. Secondly, the *indirect* path operates to maintain focal attention by suppressing the activity of irrelevant signals. A brief description of this model is given below together with the results of several simulations (a more detailed description of the model is provided in Jackson and Houghton, 1992).

The current model takes as its starting point a model of selective attention developed by Houghton (1989), a revised version of which appears as Houghton and Tipper (1992). It exploits several features of the Houghton (1989) model, such as the use of paired opponent (ON/OFF) channels, and the postulation of a match/mismatch operation, but departs from previous work in identifying components of the network model with the anatomical circuitry of the basal ganglia, and postulating that such circuitry may perform a key role in the control of attention (i.e., the engagement and disengagement operations). The model consists of a number of functionally distinct nodal fields, where each node within a given field represents the neural activity of a population of neurons coding for a particularly visuospatial location (a schematic of the different fields within the model is given in figure 17.5). Recent evidence suggests that the population coding of visuospatial parameters is a feature of several brain regions, including both motor and parietal cortex (Georgopoulos, 1986).

Architectural Details

The architecture of the model is organized to reflect the mutual competition between nodes which, within a given field, code for different regions of space. The model features two different types of field: single-layer fields, where a single node codes for a specific visuospatial location, and opponent-layer fields, where two nodes, organized into an antagonistic pair (e.g., an ON node and OFF node pair), jointly code for a single visuospatial location. For single-layer fields, each node is connected to itself by an excitatory feedback loop, and suppresses other nodes within the same field via lateral inhibitory connections (figure 17.6).

Figure 17.7 illustrates the organization of the opponent-layer fields. Within opponent-layer fields each member of a individual node pair is linked to nodes with a similar function in the same manner as described for single-layer fields, i.e., each ON or OFF node is connected to itself

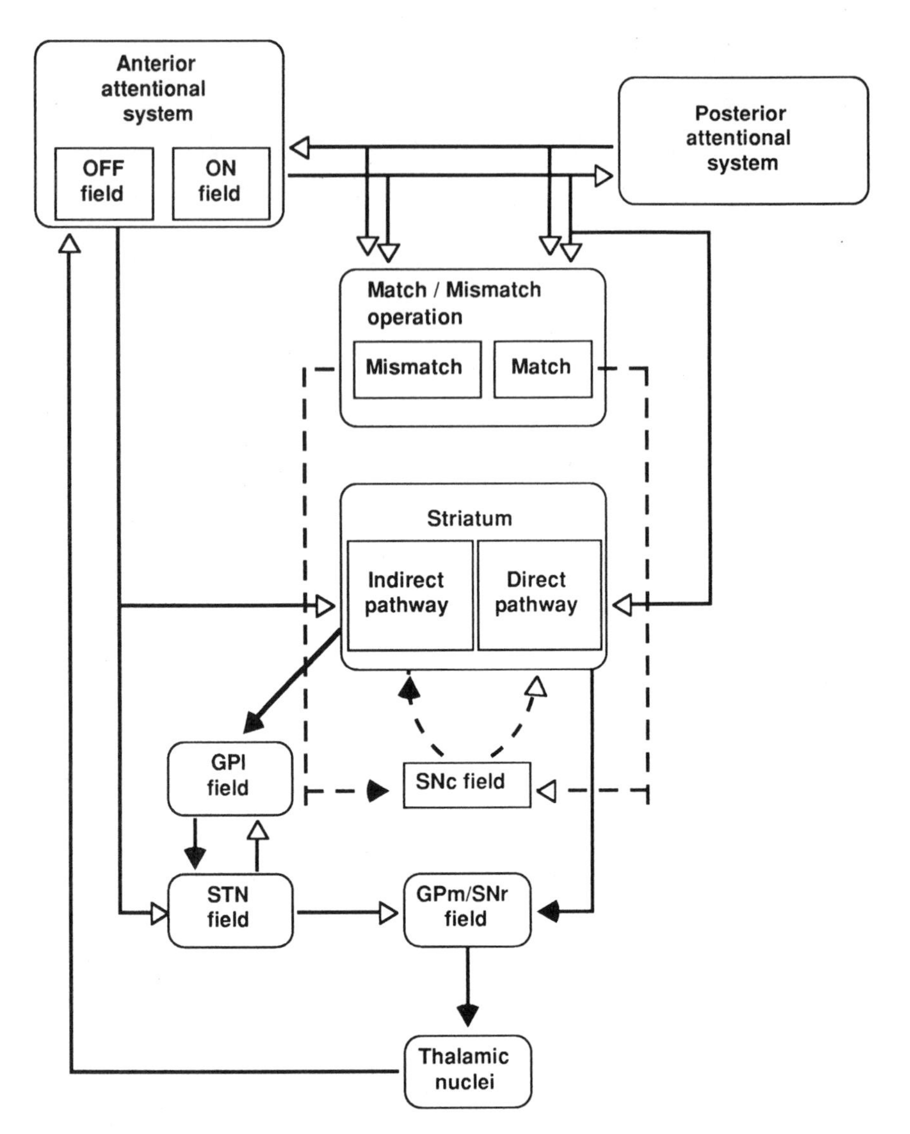

Figure 17.5 Schematic illustrating the pattern of interconnections between the different fields within the model (see text for details). GPl, lateral segment of globus pallidus; SNc, substantia nigra pars compaeta; STN, subthalamic nucleus; GPm/SNr, medial segment of globus pallidus/substantia nigra pars reticulata; MD, mediodorsal.

via an excitatory feedback loop and inhibits all other ON or OFF nodes within the same field. Opponent-layer fields also possess bidirectional inhibitory connections which link together the activity of each pair of ON andOFF nodes which together code for a specific location.

This architecture results in opponent-layer fields being particularly effective in producing an activation difference between active and suppressed regions of the field, as activity in a particular ON node not only suppresses the activity of its OFF node partner but facilitates the activity

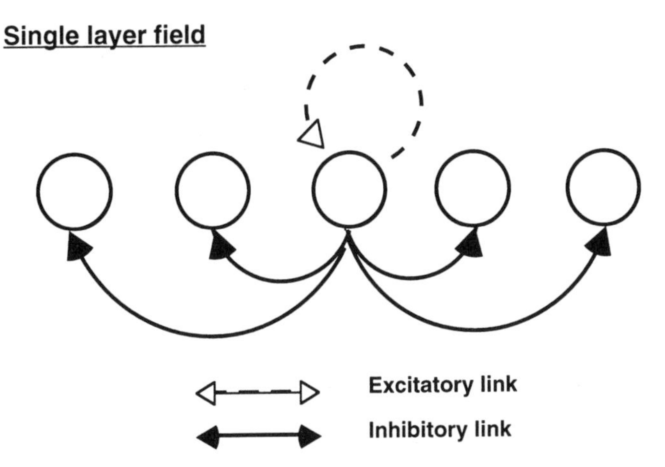

Figure 17.6 Diagram illustrating the general selection mechanism utilized within the model. All nodes within a given field suppress the activity of neighboring nodes via inhibitory connections and maintain their own activity levels by way of an excitatory link.

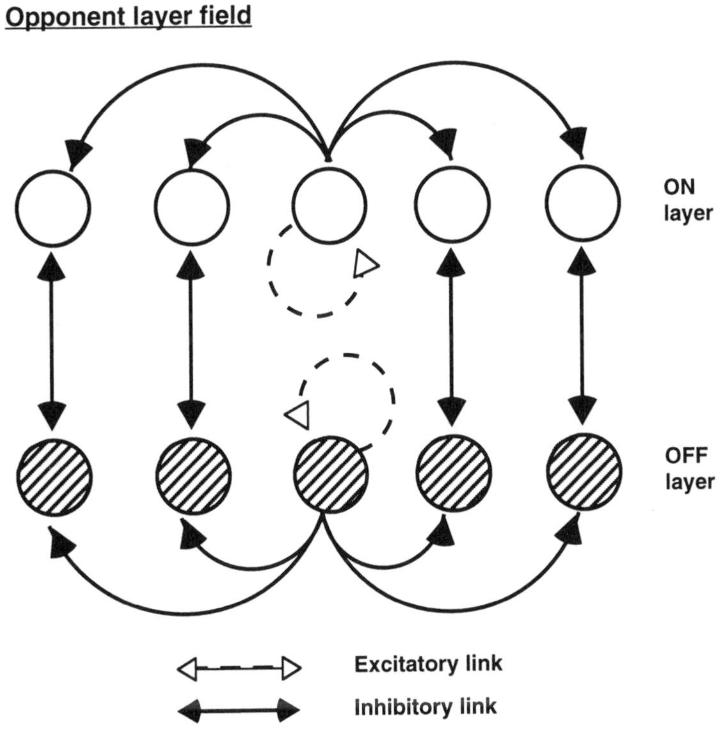

Opponent layer field

ON layer

OFF layer

Excitatory link

Inhibitory link

Figure 17.7 Diagram showing the pattern of intrafield connectivity used within opponent-layer fields. The diagram illustrates the case where the central node in each field is receiving most innervation.

of OFF nodes coding for *other* locations, via its suppression of neighboring ON nodes.

Operation of the Model

The activation of each node in the network can take on a continuous value within the range 1.0 to −1.0, where an activation of 0.0 is taken to represent baseline activity. However, negative activation values are not propagated through the network. The activation value of a node is a function both of its prior activation value and the summed input it receives from other nodes. The activation function used in the simulations described below is given by,

$$a_i(t + 1) = \delta a_i(t) + [1 - a'_i(t)]\, f(net_i(t)) \tag{1}$$

where a_i is the activation value of unit i, δ is a "decay" parameter, and net_i is the net input to unit i. It should be noted here that as activation values can become negative, two decay parameters are used within the model which represent decay from an excited state (i.e., $\delta = \delta^+$ if $a_i > 0$), and a return to baseline from an suppressed level of activity (i.e., $\delta = \delta^-$ if $a_i < 0$). The term $1 - a'_i$ is a gain term which make the change in activation proportional to the current level of activity. The term a'_i represents the *absolute* value of unit i.

In order that unit activation values remain bounded, each unit's net input is transformed by a sigmoidal function, thus insuring that all activation values lie within the range [1.0−−1.0]. This function is given by,

$$f(net_i) = \frac{2}{1 + e^{-\lambda net_i}} - 1 \tag{2}$$

The behavior of any node within a particular field will depend on the net input it receives from other nodes within the same field, and from nodes in fields to which it is connected (see figure 17.5). As all fields do not communicate with one another, the pattern of input received by any node will vary depending on its location within the network (see Jackson and Houghton, 1992, for a complete description of the pattern of connections between fields, and equations for calculating the net input to each field).

While corresponding nodes in different fields code for the same spatial location, this does not imply that such nodes are simply all coding the same thing. This point is best illustrated with reference to the nodes within the posterior and anterior attentional fields. We have assumed that the posterior attention field constructs a head-centered map of space based on *sensory input* from early visual processes, together with appropriate proprioceptive information regarding the position of the eyes and head (Andersen et al., 1993). In contrast, we have assumed that input to the anterior attentional field takes the form of an *internally generated* expectancy regarding the likely location of a impending target. Such expectancies might arise as a result of learning, or may be generated on a trial-by-trial basis as a result of priming, such as that provided by a symbolic precue, such as that illustrated in figure 2,*B*. While both types of input lead to a specific location within the appropriate field becoming activated, it is important to distinguish between the exogenous or stimulus-driven nature of posterior activity and the endogenous or internally generated activity of the anterior attentional field.

Because of the lateral inhibitory architecture, selection occurs within each field within the model, and may occur in different fields concurrently. This has important consequences for the model's dynamic characteristics. For instance, a direct input to a node within the posterior attentional field will increase that node's activity, and as a consequence of lateral inhibition, will produce a processing advantage over neighboring nodes. However, if an internally generated expectation led the anterior attentional field to receive direct input to a node different from that selected within the posterior attentional field, then this would produce a similar processing advantage for some other location, and thus lead to competition between fields. Also, as a consequence of the reciprocal connections between the anterior and posterior attentional fields, activity in either of these fields will tend, over time, to be transferred to the other field. Thus, activity at a specific node in the anterior attentional field, corresponding to an expectancy about the location of an impending target, will lead to a processing advantage at the corresponding node in the posterior attentional field, *without that node receiving sensory input.*

The Comparator Operation

As indicated above, the model essentially consists of two reverberating circuits (figure 17.8). The first of these circuits is based on reciprocal connections between nodes within the anterior and posterior attentional fields. The second circuit consists of a feedback loop which originates and terminates within the anterior attentional field, and which utilizes the various fields modeled on the anatomical architecture of the basal ganglia complex. These fields will be referred to hereafter as the *BG control circuit.* One important component of the BG control circuit is a comparator operation which functions to detect either a match or a mismatch between the spatial representations coded by the anterior and posterior attentional fields. It is important to note that the addition of this comparator operation makes the latter circuit operate quite differently from the circuit linking the anterior and posterior attentional fields. This is because the comparator mechanism can modulate activity

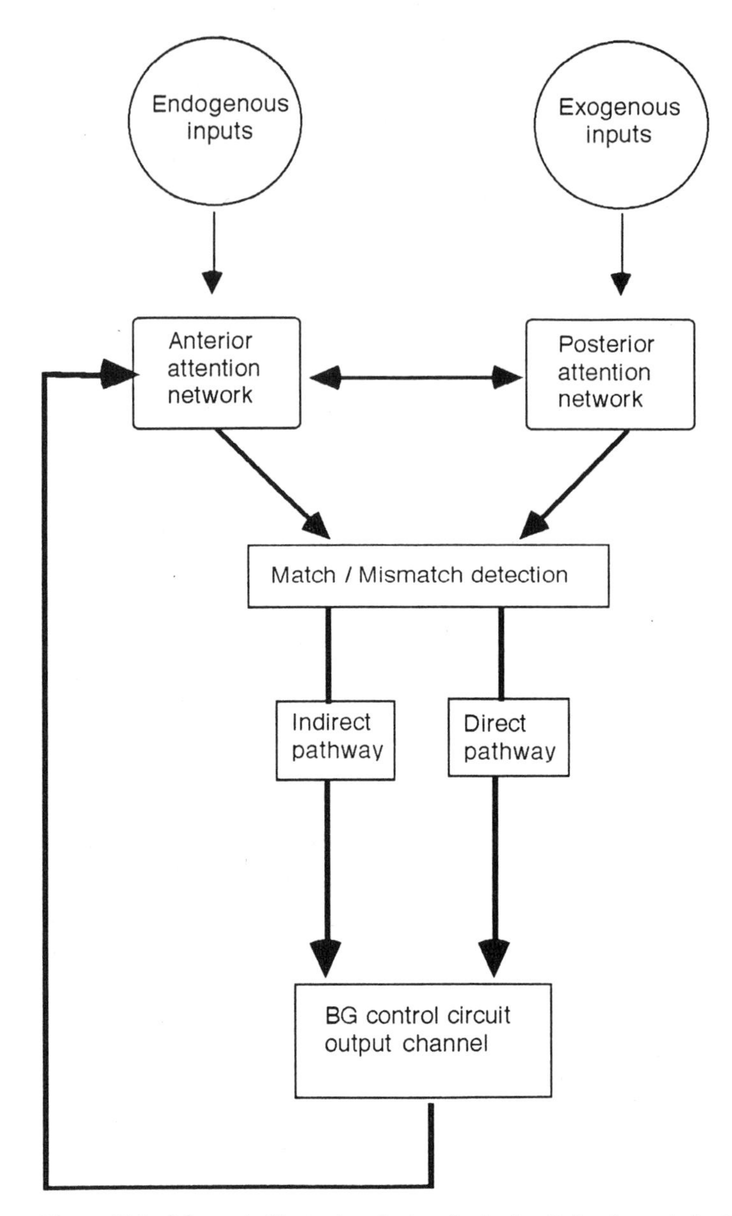

Figure 17.8 Schematic illustrating the two basic circuits implemented within the model.

within the anterior attentional field by selectively channeling activity through one or other of a pair of antagonistic pathways which operate to enhance or suppress the activity of the nodes to which they project.

These pathways serve two quite different computational functions. One pathway, the *direct pathway,* operates to facilitate or speed up selection based on winner-take-all competition within the anterior attentional field, by amplifying the activity of the current "winner." That is, once a node starts to "win" the competition with its neighbors, and *provided that a "match" between the anterior and posterior attentional fields is detected* this pathway will operate as an attentional amplifier, enhancing the winning signal within the anterior system. In contrast to the signal amplification function of the direct pathway, the second pathway, termed here the BG control circuit's *indirect* path, operates—*in the event of a mismatch*—to maintain focal activity by suppressing the activity of nonselected signals. A mismatch will be detected if there is significant (i.e., above-threshold) activity in the posterior attentional field which is not matched by similar activity at a corresponding node in the anterior attentional field. If such a mismatch is detected between winning nodes in the anterior and posterior attentional fields, then the BG control circuit's indirect pathway will function to suppress activity at any node(s) in the anterior attentional field which codes the same location(s) as nonmatching posterior attentional field nodes. This is illustrated below.

Simulation 1: Modeling the Attentional Precuing Paradigm

As previously noted, in the simplest form of the spatial precuing paradigm the subject's task is to detect the onset of a target stimulus (e.g., a luminance increment or decrement) at one of several possible spatial locations. Prior to the onset of the imperative stimulus, the subject is provided with a spatial precue indicating which of several locations is the probable site of the next target. The validity of this precue can be varied so that it is (1) a *valid* predictor of the target's location, (2) *neutral* with respect to location, or (3) an *Invalid* predictor of the target's location.

To simulate the presentation of a valid precue, we have assumed that the presentation of a symbolic precue, such as an arrow pointing to the left or right hemifield, is interpreted and subsequently coded by the anterior attentional system as an activation advantage for the cued location. In psychological terms we might be tempted to think of this as an active expectation that the target will be presented at the cued location; however, it is important to note that we are seeking to model the mechanisms responsible for selecting a particular region of space for further processing, and not those decision-making processes involved in detecting whether a target has been presented.

To simulate the presentation of a *valid precue* we presented a direct input (activation value of 1.0) to a single node within the anterior attentional field. This input was maintained for 5 processing cycles, during which time no direct input was provided to the posterior attentional field. Following presentation of the precue, the model was allowed to run on for an additional 5 processing cycles during which time no direct input was provided to either the anterior or posterior attentional fields. This is analogous to the delay between cue and target typically employed within experimental studies of this kind. At the conclusion of this delay period, a direct input (activation of 1.0) was presented to a single node in the posterior attentional field, which coded for the same location as that cued in the anterior attentional field. This input was maintained for a further 75 processing cycles. To simulate the presentation of an *invalid precue,* the same procedure was followed as for the valid trial save that the location (node) cued in the anterior attentional field was different from that receiving input in the posterior attentional field. Finally, a neutral cue condition was simulated by providing no cued input to the anterior attentional ON field.

The model's performance was evaluated by comparing the number of processing cycles needed for the activity level of the node coding for the target location in the anterior attentional field to exceed a fixed threshold (an activity level of 0.45). This procedure differs from simulations reported in Houghton (1989) where responses were determined by the activity of nodes in a dedicated "response field." However, for the simulations reported here such a field is unnecessary, and can be dispensed with given that each node in the network merely codes for the activation of a particular region of visual space. While posterior attentional field nodes code for the actual location of visual stimuli, anterior attentional field nodes code for the location of "focal" attention, and thus provide a measure of the extent to which focal attention is directed at a target location. Figure 17.9 shows the results of 25 simulated subjects performing each of the three types of trial. The figure also presents for comparison data obtained from a study of normal adult subjects performing a covert attention task following exogenous precues (Jackson, unpublished data). The model requires fewer cycles to reach threshold following a valid precue than when no precue is provided (precuing "benefits"), and an increased number of cycles to reach threshold, compared to a neutral trial, when presented with an invalid precue (cuing "costs").

To illustrate the dynamical behavior of the model, figures 17.10 and 17.11 show the activity levels of several of the model's fields during this simulation. In each case the activity levels are depicted for both valid and invalid trials. Graph*1A* of figure 17.10 shows the activity levels in the anterior attentional field (ON nodes) for a valid trial, while graph *1B* shows activity levels for an invalid trial. The latter illustrates how high levels of activity persist at the cued location long after the cue has been removed and the target presented. This occurs despite the fact that the posterior attentional field is propagating information about the target location to the anterior attentional field, and results from the joint activity of the direct and Indirect pathways of the BG control circuit.

Graph *2A* shows the activity levels in the BG control circuit's direct pathway for a valid trial. As a result of the precue, this pathway becomes active for the target location, *prior* to the target onset. Note also

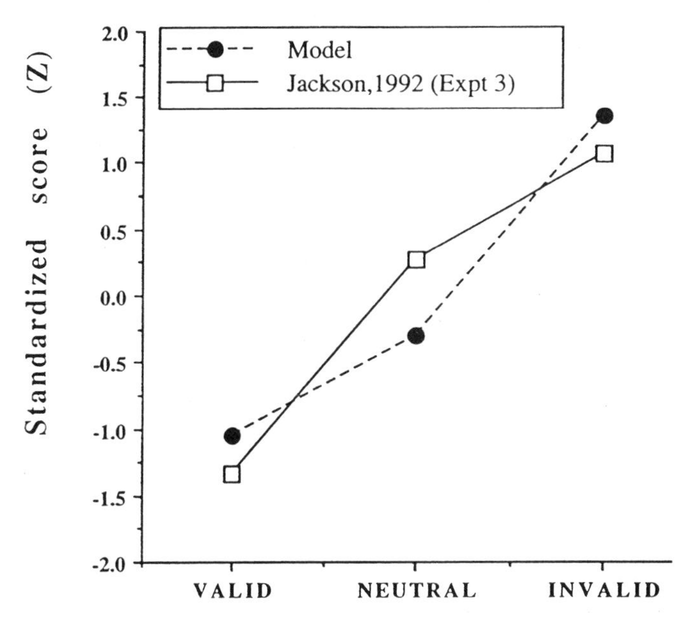

Figure 17.9 The results of the simulation of the spatial precuing paradigm (N=25 simulated subjects). The model was tested under three conditions (1) where the target location was precued (valid condition); (2) where no precue was given (neutral condition); (3) where a precue was presented at a location (invalid condition). The results of the simulation are contrasted with data obtained from adults performing a covert orienting task following peripheral precues, and have been standardized (z-transformed) for this purpose.

how activity at nontarget nodes is suppressed below baseline. In behavioral terms, we can think of the attentional channel for the target location becoming "engaged" following the precue, while attentional channels for nontarget locations are "not engaged." Graph *2B* shows the activity levels for an invalidly cued trial. It illustrates how the direct pathway node coding for the cued location initially becomes active, and remains so for a considerable period after the cue has been removed. Thus, when the target is presented, at cycle 16, the level of activity at the target location is *suppressed* below baseline.

It is important to note that at this point in the trial the target node is *not suppressed* below the level of other (nonrelevant) nodes. In contrast, between cycles 25 and 45, the activity of the target node actually becomes suppressed *below* the level of nodes which have not received input of any sort. This effect is due to the inhibition of the target node in direct pathway (ON node) by its paired node (OFF node) within the indirect pathway. This can be seen by examining graph *3B* which shows activity in the BG control circuit's indirect pathway. Note that between cycles 20 and 50, the node coding for the target location is highly active, while the node coding for the cued location is suppressed, both below baseline and below the level of nodes not receiving input.

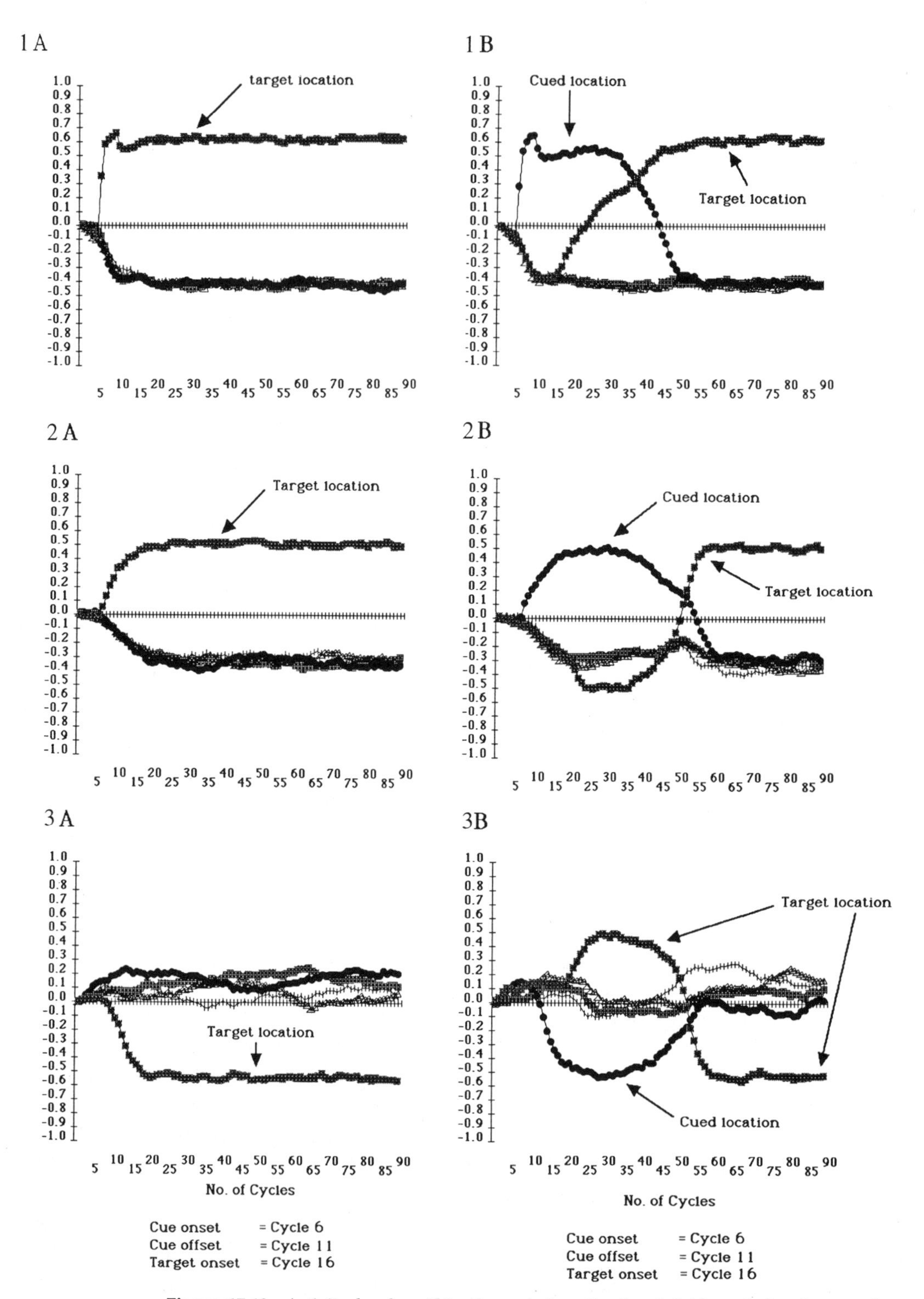

Figure 17.10 Activity levels within the anterior attentional field, and the direct and indirect pathways of the BG control circuit during the spatial precuing paradigm. Graphs marked *A* show activity levels folowing a valid precue, and those marked *B* show acitivity levels following an invalid precue (see text for further details).

Match field - Valid precue

Mismatch field - Valid precue

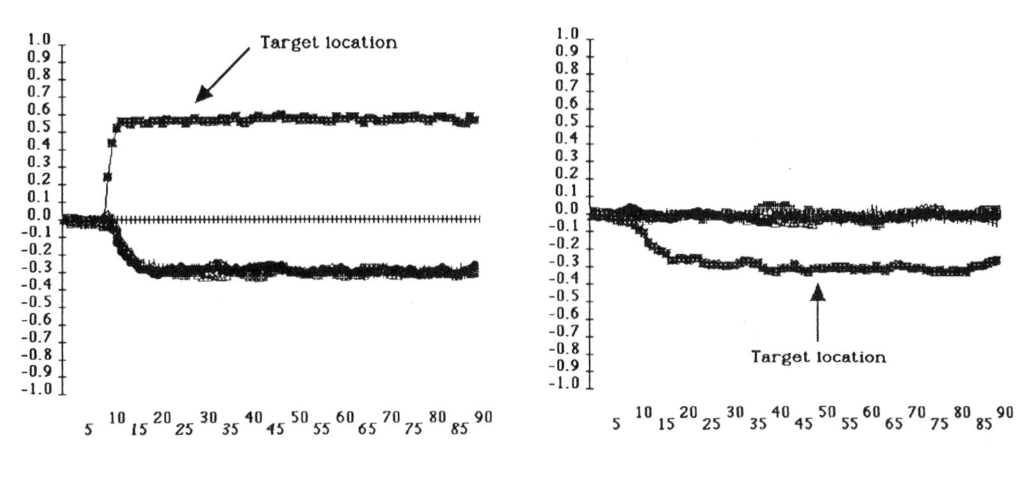

Match field - Invalid precue

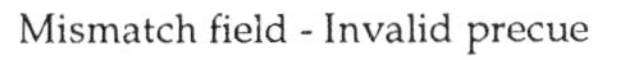

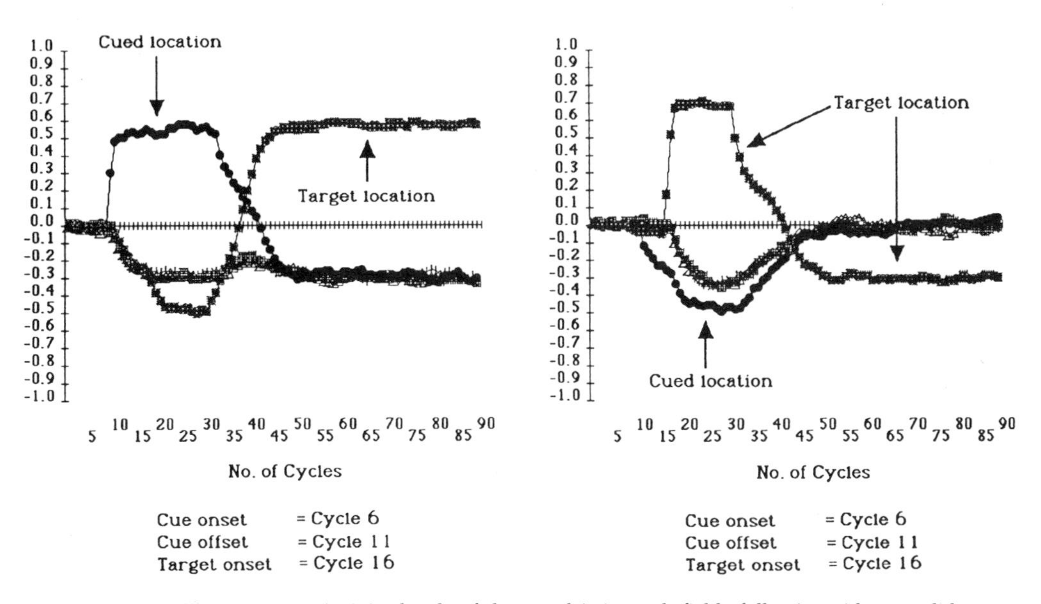

Figure 17.11 Activity levels of the match/mismatch fields following either a valid or an invalid precue.

To understand the pattern of interactions between the direct and indirect pathways within the BG control circuit, it is first necessary to recall that activity in these pathways is influenced by activity within the match and mismatch fields. Figure 17.11 shows the activity levels for these fields for both valid and invalidly cued trials. The figure illustrates how a match between anterior and posterior activity is signaled, shortly after the onset of the cue *and prior to the onset of the target,* by elevated activity within the match field at the cued (target) location, and suppressed activity for this same location within the mismatch field. Note also how activity at noncued nodes in the match field is actively suppressed.

These patterns of activity can be contrasted to those arising following an invalidly cued trial. In these circumstances an initial "match" is detected and results in elevated activity at the cued node in the match field. This remains active for some period after the cue has been removed and the target (appearing at a noncued node) presented. Following presentation of the target stimuli, target node activity (i.e., a potential match signal) in the match field is suppressed below the level of other nodes in the network (cycles 20–35). This suppression arises both as a result of inhibition from the "cued" node in the match field, and as a result of inhibition from the target node in the mismatch field.

Together, figures 17.10 and 17.11 illustrate two means by which an activation gap may develop between target and nontarget nodes in the anterior attentional field. Firstly, following direct input to a node in the posterior attentional field, a target node will, as a result of lateral inhibition, suppress other nontarget nodes within the same field. Moreover, owing to the facilitatory connection between the posterior and anterior attentional fields, this processing advantage for the target location will, over time, be transferred to the anterior field. Thus, in the paradigm simulated here, the effect of valid cuing is to initiate the selection process ahead of the target's onset, while the effect of invalid cuing is to set up increased competition both within and between fields, which will require a number of processing cycles to resolve.

A second means by which an activation gap can be generated in the anterior attentional field is through the operation of the BG control circuit, which operates to facilitate processing of a signal event through the suppression of nonsignal events, but is specific to the activity of the anterior attentional field. Thus, the combined action of these pathways can be likened to an amplification or focusing process which, consistent with the "zoom lens" analogy of spatial attention, operates to enhance a signal location within the anterior attentional field, by increasingly suppressing activity at other locations.

Within this framework, the "benefits" of precuing and the "costs" of miscuing attention arise as a joint function of two modes of selection: a general selection operation common to all fields within the model based on lateral inhibition, and a specific signal-selection/signal-amplification operation carried out by the BG control circuit.

Simulation 2: Lesioning the Posterior Attentional Field

The view of the engage-disengage operations developed in this chapter differs in several important ways from that proposed by Posner et al. (1984). For example, while Posner et al. localized the disengage operation within the posterior attentional network, we have chosen to view the acts of engaging and disengaging attention as the emergent property of activity in *both* the anterior and posterior attentional systems. Posner et al. (1984) chose to localize the attentional disengagement operation within the posterior parietal lobe based on their observation that patients with lesions to this region show increased RT costs for targets presented to their contralesional field. We now report the results of a series of trials in which the model's posterior attentional field was systematically lesioned to simulate the effects of a parietal lesion.

In this simulation the same procedure was used to construct a series of valid, neutral, and invalid trials as described above. In networks utilizing a distributed representation, lesions to the network can be induced by the selective inactivation of some subset of the unit population. In a network using a local representation, however, where the activity of a single node is assumed to represent the activity of some population of "neurons," such a scheme is not possible. Instead, one can either completely inactivate a given node, thereby assuming that all information from that node is unavailable, or else systematically limit the output of that node, thereby assuming that the output of that node is diminished rather than removed. In the simulation reported here we opted for the latter strategy, and limited the activation of the node coding for the target location in the posterior attentional field (i.e., the target node's output was reduced by some fixed percentage). We examined the model's performance at four levels of "lesion size": 0%, 25%, 50%, and 75%.

As previously noted, patients with unilateral lesions to posterior parietal cortex are not significantly impaired when detecting stimuli presented in their contralesional field, provided that the location of the target is cued in advance. However, such patients show large RT deficits to contralesional targets if attention has previously been drawn to a location in the ipsilesional field, e.g., following an invalid precue (Posner et al., 1984). Figure 17.12 shows the results of five simulated subjects. In each case target input was presented to the same lesioned node (see above), following either a valid, neutral, or invalid precue.

Examination of this figure reveals that while there are minimal effects of the lesion if the target is preceded by a valid precue, consistent

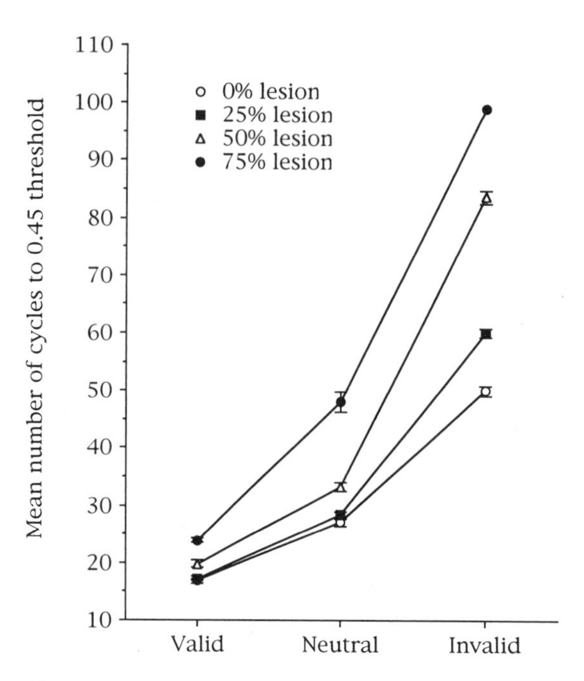

Figure 17.12 Simulation of the effects of a lesion to the posterior attentional system on the spatial precuing task. In this simulation, a node coding for a particular spatial location in the posterior attentional field was lesioned (i.e., its output was reduced by 0%, 25%, 50%, or 75%), and the model was tested, following either a valid, neutral, or invalid precue, by presenting input to the lesioned location. The results show that the effects of the lesion are limited to Invalid trials where the "attention" has been drawn to some other location by a precue.

deficits occur for invalid trials. Moreover, the size of this deficit increases with the extent of the lesion to the target location (node). Thus the model clearly demonstrates a disengaging deficit consistent with that found in human patients with parietal damage.

While we agree with Posner and his colleagues that a disengage operation may be a necessary component of an efficient attentional system, we take the view that this operation involves multiple neural maps of space and is not computed solely by the posterior parietal cortex. Within the current model, the posterior attentional system is characterized as simply coding for the location of current sensory input, which must be communicated onward to the anterior system for a shift in attention to occur. The increased costs on invalid trials following a posterior attention field lesion arise as a result of the posterior system's *decreased ability to effectively communicate information to the anterior attentional field,* and not from a deficit in a dedicated "disengage" mechanism. Although a disengage mechanism exists within our model, it does not function to disengage activity from a currently attended location. Rather, our disengage operation functions to suppress activity at a novel (target) location so as to maintain activity at a currently attended location.

Simulation 3: Disabling the Match/Mismatch Fields

We have argued thus far, that the two pathways within the BG control circuit play a specific role in modulating activity in the anterior attentional field. Specifically, we have suggested that the *direct* pathway functions as an attentional amplifier, enhancing the signal strength of behaviorally relevant spatial locations, whereas the *indirect* pathway operates to maintain focal attention by suppressing the activity of irrelevant signals. Furthermore, we have argued that the comparator function implemented by the match and mismatch fields is critical to the operation of this circuit. Therefore, disabling the operation of the comparator function would be expected to lead to inefficient focusing of attention.

The procedure used in this simulation to construct valid, neutral, and invalid trials was identical to that used previously. In this simulation both the match and mismatch fields were disabled so that they no longer propagated activity onto other fields. All other details of the simulation were identical to those reported in simulation 1. Figure 17.13 shows the results of this simulation with five simulated subjects. The figure reveals that disabling the match/mismatch fields has two effects. Firstly, there is a small but reliable increase in the number of cycles taken to reach threshold for valid trials.[2] Secondly, there is a much larger *reduction* in the number of cycles needed to reach threshold for invalid trials.

These results are consistent with data obtained in a number of quite different empirical studies of "anterior system" function. Firstly, the simulated data are consistent with the finding that occupying the anterior system with a secondary task results in a reduction of RT costs for invalidly cued targets (Posner et al., 1989). Secondly, these data are also consistent with the finding, obtained from patients with FEF lesions, of paradoxically faster saccades to targets presented in their contralesional field (Henik et al., 1991). Finally, these data are also consistent with the finding of reduced attentional costs in patients with PD (Wright et al., 1990; Kingstone et al., 1992) and following the administration of the DA antagonist droperidol (Clarke et al., 1989). Figure 17.14 contrasts the results of the model's simulation of the spatial precuing task, both with and without the match/mismatch fields, with the results obtained by Clarke et al. (1989) following administration of droperidol.

Thus far we have described the finding obtained in the studies cited above as a paradoxical decrease in the time taken to process target stimuli appearing at noncued locations. However, an alternative de-

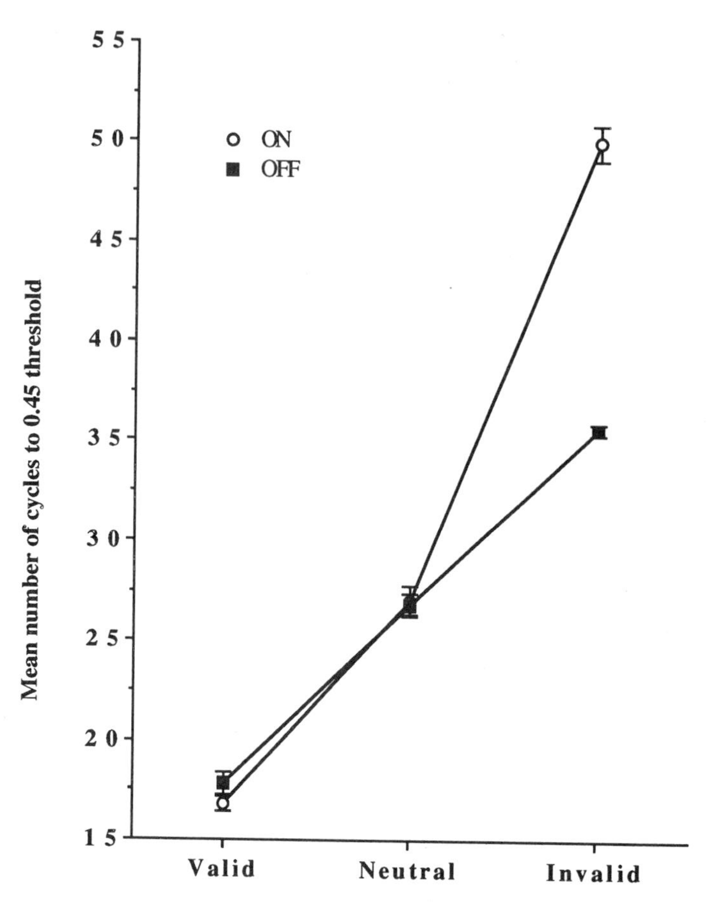

Figure 17.13 Simulation of the spatial precuing task where the match/mismatch fields have been disabled. The results show that the model is marginally slowed to reach threshold following a valid precue, but is substantially faster to reach threshold follwing an invalid precue.

scription is a *deficit* in the ability to maintain focal attention at an expected (cued) location when a target stimulus is presented at an unexpected location, i.e., increased distractibility. From the outline of the model's operation presented above, it is clear that one consequence of disabling the model's match/mismatch fields is a loss of the suppression mechanism implemented by the indirect pathway, whereby once attention has been engaged at the cued location, activity at a novel (target) location is actively inhibited so as to maintain activity at a currently attended location.

DISCUSSION

The model presented in this chapter can be interpreted at two levels. Firstly, like many such "neurally inspired" network models, it can be

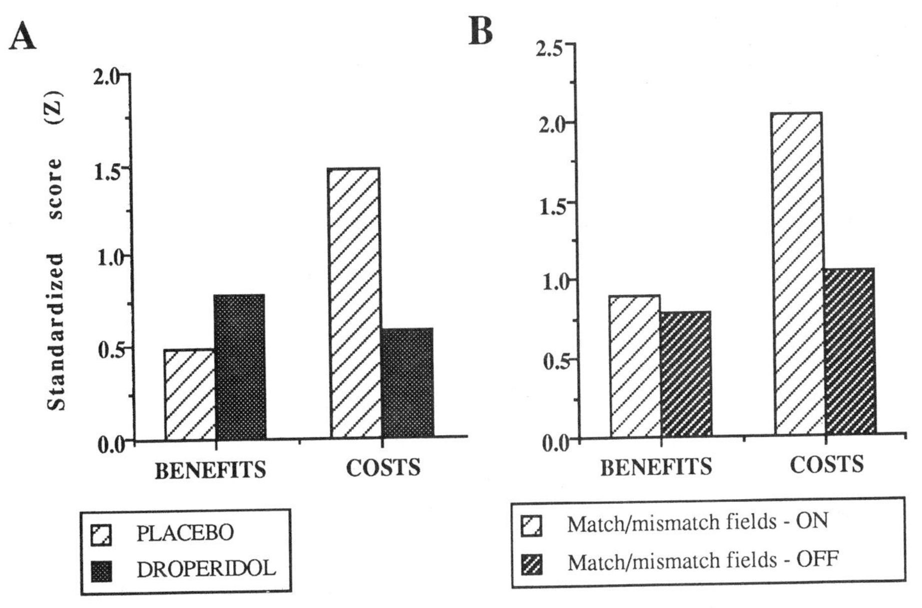

Figure 17.14 The results of the simulation of the spatial precuing task with the match/mismatch fields operating (*A*) and disabled (*B*) are contrasted here with the effects on covert orienting of placebo and droperidol injection in adult volunteers. (Data adapted from Clark et al., 1989.)

interpreted as a functional model of the computational processes under investigation, in this case the control of spatial attention. Secondly, as the architecture of the current model was based on the anatomical architecture of the basal ganglia complex and its extrinsic connections, it may also be interpreted as a more direct test of the hypothesis that one or more of the nonmotoric pathways through the basal ganglia participate in the voluntary control of spatial attention.

We suggest that as a functional model of spatial attention, the neural network model presented here succeeds in simulating many key behavioral phenomena. Thus, in addition to simulating the costs and benefits of precuing found in experimental studies of covert orienting, the model offers an integrative theoretical account of *both* behavioral phenomena associated with damage to the posterior attentional system (i.e., the disengaging deficit associated with posterior parietal damage), and phenomena associated with "frontal" syndromes (i.e., increased distractibility and the hyperreflexive orienting associated with FEF lesions and PD). The model also offers an instantiation, and in some cases an alternative account, of several key concepts featured in previous theoretical accounts of attention, i.e., the engage and disengage operations. Finally, the framework developed here also accords with existing data. For example, the operation of the general selection mechanism postulated above leads to the clear prediction that stimuli will be responded to faster when presented within an empty field compared to a cluttered field. This is supported by the finding that removing a fixation stimulus prior to the onset of a visual target leads to speeded responses (Fischer and Breitmeyer, 1987). Likewise, the operation of the specific selection mechanism suggests that attention at different loci may be actively engaged, not engaged, or actively disengaged. Tangential support for this notion can be taken from the finding that prior knowledge of the spatial location of a target appears to reduce the size or extent, or both, of the attentional "beam" (Humphreys, 1984), and that focal attention appears to prevent so-called express saccades (Fischer and Weber, 1993).

The model can also be interpreted as a more direct test of the hypothesis that one or more of the nonmotoric pathways through the basal ganglia participate in the voluntary control of spatial attention. Although we have presented several lines of argument in support of this view, we are the first to acknowledge that the question of the basal ganglia's involvement in the control of spatial attention remains largely unanswered for want of empirical evidence. However, we suggest that the model accomplishes two things at this level. First, the model succeeds in demonstrating that the overall architecture of the basal ganglia appears to be computationally *sufficient* to carry out a control function in which one behavioral event is selected, while other incompatible events are suppressed. This function has been ascribed to the basal ganglia by several authors, and would appear to be the common feature of its participation in most, if not all, of the corticostriatothalamocortical circuits described by Alexander and colleagues (Alexander and Crutcher, 1992; Alexander et al., 1986). Secondly, the model succeeds in demonstrating that this control function is also sufficient to perform the engagement/disengagement function necessary in any theory of attention.

Overall, the functional characteristics of this model suggest that the basal ganglia might serve two basic functions. First, once a behavioral sequence or event has been selected, it may function as an attentional zoom lens or amplifier, enhancing behaviorally relevant cortical signals. Secondly, it may operate to maintain focal attention by suppressing the activity of irrelevant signals.

ACKNOWLEDGMENTS

The work presented in this chapter was supported by a JCI Cognitive Science Initiative research fellowship awarded to S. J., and was carried out while S. J. was a visiting research fellow at the Institute of Cognitive and Decision Sciences, University of Oregon. S. J. is particularly grateful to Mike Posner, Steve Keele, and colleagues at the Institute of Cognitive and Decision Sciences for their many kindnesses during his time in Oregon.

NOTES

1. See Jackson and Lees (1993) for a discussion of the different signals carried by the direct and indirect connections linking cortical nodes with the superior colliculus.

2. In other simulations this effect has been shown to be dependent on the number of cycles which provide the precue, and the number of cycles which make up the delay. Reducing the number of "cuing" cycles and increasing the number of "delay" cycles both lead to an increase in the magnitude of this effect.

REFERENCES

Alexander, G. E., and Crutcher, M. D. (1990) Functional architecture of basal ganglia circuits: Neural substrates of parallel processing. *Trends Neurosci.* 13:266–271.

Andersen, R. A., Snyder, L. H., Li, C. S., and Stricanne, B. (1993) Coordinate transformations in the representation of spatial information. [Review] *Curr. Opinion Neurobiol.* 3:171–176.

Alexander, G E., DeLong, M. R., and Strick, P. L. (1986) Parallel organization of functionally segregated circuits linking basal ganglia and cortex. *Annu. Rev. Neurosci.* 9:357–381.

Benecke, R., Rothwell, J. C., Dick, J. P. R., Day, B. L., and Marsden, C. D. (1986) Performance of simultaneous movements in patients with Parkinson's disease. *Brain* 109:739–757.

Bronstein, A. M., and Kennard, C. (1985) Predictive ocular motor control in Parkinson's disease. *Brain* 105:925–940.

Brown, R. G., and Marsden, C. D. (1988) Internal versus external cues and the control of attention in Parkinson's disease. *Brain* 111:323–345.

Carl, J. R., and Wurtz, R. H. (1985) Asymmetry of saccadic control in patients with hemi-Parkinson's disease. *Assoc. Res. Vision Ophthalmol. Abstr.*258.

Clark, C. R., Geffen, G. M., and Geffen, L. B. (1989) Catecholamines and the covert orientation of attention in humans. *Neuropsychologia* 27:131–139.

Corbetta, M., Miezin, F. M., Shulman, G. L., and Peterson, S. E. (1991) Selective and divided attention during visual discrimination of shape colour and speed: Functional anatomy by positron emission tomography. *J. Neurosci.* 11:2382–2402.

Crawford, T. J., Henderson, L., and Kennard, C. (1989) Abnormalities of nonvisually guided eye movements in Parkinson's disease. Brain 112:1573–1586.

Damasio, A. R., Damasio, H., and Chui, H. C. (1980) Neglect following damage to frontal lobe or basal ganglia. *Neuropsychologia* 18:123–132.

Early, T. S., Posner, M. I., Reiman, E., and Raichle, M. E. (1989) Left stiato-pallidal hyperactivity in schizophrenia, pt II: Phenomenology and thought disorder. *Psychiatr. Dev.* 2:109–121.

Fischer, B., and Breitmeyer, B. (1987) Mechanisms of visual attention revealed by saccadic eye movements. *Neuropsychologia* 25:73–83.

Fischer, B., and Weber, H. (1993) Express saccades and visual attention. *Behav. Brain Sci.* 16:553–610.

Funahashi, S., Bruce, C. J., and Goldman-Rakic, P. S. (1986) Perimetry of spatial memory representation in primate prefrontal cortex: Evidence for a mnemonic hemianopia. *Soc. Neurosci. Abstr.* 12:554.

Georgopoulos, A. P., Schwartz, A. B., and Kettner, R. E. (1986) Neuronal population coding of movement direction. *Science* 233:1416–1419.

Gerfen, C. R. (1992) the neostriatal mosaic: Multiple levels of compartmental organization. *Trends Neurosci.* 15:133–139.

Goldberg, G. (1985) Supplementary motor area structure and function: Review and hypotheses. *Behav. Brain Sci.* 8:567–616.

Goldman-Rakic, P. S. (1987) Circuitry of primate prefrontal cortex and regulation of behavior by representational memory. In Plum, F., and Mountcastle, V. (eds.), *Handbook of Physiology—The Nervous System,* Vol 5. Bethesda, Md,: American Physiological Society.

Goldman-Rakic, P. S. (1988) Topography of cognition: Parallel distributed networks in primate association cortex. *Annu. Rev. Neurosci.* 11:137–156.

Goodale, M. A., and Milner, A. D. (1992) Separate visual pathways for perception and action. *Trends Neurosci.* 15:20–25.

Groenewegen, H. J., Berendse, H. W., Meredith, G. E., Haber, S. N., Voorn, P., Volters, J. G., and Lohman, A. H. M. (1991) Functional anatomy of the ventral, limbic system–innervated striatum. In Willner, P., and Scheel-Kruger, J. (eds.), *The Mesolimbic Dopamine System: From Motivation to Action.* New York: John Wiley & Sons.

Guitton, D., Buchtel, H. A., and Douglas, R. M. (1982) Disturbances of voluntary saccadic eye movement mechanisms following discrete unilateral lobe removals. In G. Lennerstrand and E. L. Keller (eds.), *Functional Basis of Ocular Motility Disorders.* Oxford: Pergamon Press.

Heilman, K. M., and Valenstein, E. (eds.) (1985) *Clinical Neuropsychology.* Oxford: Oxford University Press,

Henik, A., Rafal, R., and Rhodes, D. (1991) Reflexive and voluntary saccades after lesions to the human frontal eye fields. Unpublished manuscript.

Henik, A., Rafal, R., and Rhodes, D. (1992) Centrally-directed and peripherally-summoned saccades after lesions of human frontal eye fields. Unpublished manuscript.

Houghton, G. (1989) A connectionist-neuropsychological model of some aspects of selective attention. Unpublished manuscript.

Houghton, G., and Tipper, S. P. (1992) A model of the dynamics of selective attention In D. Dagenbach, and T. Carr (eds.), *Inhibitory Mechanisms in Attention, Memory, and Language.* San Diego, CA.: Academic Press.

Humphreys, G. W. (1984) Shape constancy: The effects of changing shape orientation and the effects of changing the position of focal features. *Percep. Psychophys.* 36:50–64.

Jackson, S., and Houghton, G. (1992) *Basal Ganglia Function in the Control of Visuospatial Attention: A Neural-Network Model.* Technical Report No. 92-6, Institute of Cognitive and Decision Sciences, University of Oregon, Eugene.

Jackson, S., and Lees, M. (1993) The significance of the basal ganglia in suppressing hyper-reflexive orienting, *Behav. Brain Sci.* 16.

Jackson, S., Marrocco, R., and Posner, M. I. (in press) Networks of anatomical areas controlling visuospatial attention. *Neural Networks.*

Keele, S. W., Rafal, R., and Ivry, R. (1992) An attention shifting hypothesis for the basal ganglia. Unpublished manuscript.

Kingstone, A., Klein, R., Maxner, C., and Fisk, J. (1992) Attentional systems and Parkinson's disease, Paper presented at Attention: Theoretical and Clinical Perspectives, Toronto, March 26–27, 1992.

Kosslyn, S. M., Flynn, R. A., Amsterdam, J. B., and Wang, G. (1990) Components of high-level vision: A cognitive neuroscience analysis and accounts of neurological syndromes. *Cognition* 34:203–277.

Lasker, A. G., Zee, D. S., Hain, T. C., Folstein, S. E., and Singer, H. S. (1987) Saccades in Huntington's disease: Initiation defects and distractibility. *Neurology* 37:364–370.

Leigh, R. J., Newman, S. A., Folstein, S. E., Lasker, A. G., and Jenson, B. A. (1983) Abnormal ocular motor control in Huntington's disease. *Neurology* 33:1268–1275.

Mink, J. W., and Thatch, W. T. (1991) Basal ganglia motor control. 1. Nonexclusive relation of palidal discharge to five movement modes. *J. Neurophysiol.* 65:273–300.

Posner M. I. (1980) Orienting of attention. *Q. J. Exp. Psychol.* 32:3–25.

Posner, M. I. and Peterson, S. E. (1990) The attention system of the human brain. *Annu. Rev. Neurosci.* 13:25–42.

Posner, M. I., and Rothbart, M. K. (1990) *Attentional Mechanisms and Conscious Experience.* Technical Report No. 90-17: University of Oregon,

Posner, M. I., Walker, J. A., Friedrich, F. J., and Rafal, R. (1984) Effects of parietal injury on covert orienting of visual attention. *J. Neurosci.* 4:1863–1874.

Rafal, R. D., Posner, M. I., Walker, J. A., and Friedrich, F. J. (1984) Cognition and the basal ganglia: Separating mental and motor components of performance in Parkinson's disease. *Brain* 107:1083–1094.

Rizzolatti, G., Riggio, L., and Sheliga, B. M. (1993) Space and selective attention. In *Attention and Performance* XV. Hillsdale, N.J.: Erlbaum.

Robbins, T. W. (1991) Cognitive deficits in schizophrenia and Parkinson's disease: Neural basis and the role of dopamine. In P. Willner and J. Scheel-Kruger (eds.), *The Mesolimbic Dopamine System: From Motivation to Action.* New York: John Wiley & Sons.

Sagar, H. J., and Sullivan, E. V. (1988) Patterns of cognitive impairment in dementia. In C. Kennard (ed.), *Recent Advances in Clinical Neurology,* Vol. 5. Edinburgh: Churchill Livingstone.

Sharpe, M. H. (1990) Patients with early Parkinson's disease are not impaired on spatial orientating of attention. *Cortex* 26:515–524.

Villardita, C., Smirni, P., and Zappala, G. (1993) Visual neglect in Parkinson's disease. *Arch. Neurol.* 40:737–739.

Wright, M. J., Burns, R. J., Geffen, G. M., and Geffen, L. B. (1990) Covert orientation of visual attention in Parkinson's disease: An impairment in the maintenance of attention. *Neuropsychologia* 28:151–159.

Editors' Commentary on Part IV

The behavioral tasks considered in this group of chapters emphasize cognitive functions, as contrasted with the motor emphasis in earlier chapters. However, when these more cognitive tasks are broken down into constituent information processing operations, the boundaries between cognitive and motor functions start to blur. As we will discuss momentarily, the simplest operations in the cognitive domain can be reasonably mapped onto the pattern recognition/registration model discussed in earlier commentaries and used there to explain operations in the motor behavioral domain. By considering this basic model as a modular process, one can start to build more complex cognitive and motor behaviors by stringing these modules together in various parallel and series combinations, as will be illustrated in this commentary.

SIMILARITY OF COGNITIVE AND MOTOR OPERATIONS

In chapter 14, Gabrieli compares cognitive and motor deficits in patients with Parkinson's disease. Like the deficits observed in motor tasks, the cognitive deficits of PD appear to be related to different aspects of planning such as storage, manipulation and retrieval of task information or context. In agreement with this similarity analysis, both cognitive and motor performance are improved by levodopa treatment. This drug partially compensates for the disappearance of dopamine that occurs in PD. As discussed in the commentary to part III, dopamine is thought to function as the reinforcement term in the three-factor cellular learning rule postulated for striatal spiny neurons. A reduction in dopamine would thus limit a spiny neuron's ability to recognize new contexts. As pointed out in chapter 10, the overall level of spiny neuron activity appears to be reduced in PD, due at least in part to a weakening of unreinforced pre- and postsynaptic associations. The effect of reduced spiny neuron activity on corticothalamic loops could go in either direction, because of the push-pull nature of the direct and indirect projections from the striatum to pallidal output neurons. However, observations suggest an average enhancement of pallidal discharge, leading to net inhibiton of corticothalamic loops, ultimately tipping the balance toward the overall reduction in behavioral responding that is characteristic of PD.

In the cognitive domain, it is interesting that word recognition is not impaired in PD, whereas performance is degraded on tasks that require more elaborate strategies for retrieving and using words. Perhaps this is because simple recognition tasks can be performed with a direct use of recognition/registration modules that have already consolidated their performance in this task through years of practice. In contrast, the more elaborate word retrieval tasks might be orchestrated by stringing together several arrays of modules in complex combinations that depend on an adaptive regulation driven by dopamine reinforcement. The adaptive steps might be preferentially affected in PD. One intriguing aspect of the retrieval deficits is that of decreased retrieval rate, or psychomotor speed. If we assume that PD results in a decrease in trans-striatal loop activity, this decreased retrieval rate could be explained by the cortical search mechanism hypothesized in chapter 10. According to this idea, feedforward corticostriatal loops could provide a mechanism for searching through potential cortical activation patterns. Increased activity in specific modules would deepen the basins of attraction surrounding certain states of the frontal cortex corresponding to particular solutions to the problem at hand. Alternatively, a search similar to this might take place entirely within the intrastriatal network discussed in chapter 16, the speed and accuracy of which would also be dependent on dopamine.

The deficits in motor performance discussed in chapter 14 can be analyzed in a similar fashion. The absence of a deficit in mirror tracing in PD is to be contrasted with the severe deficit in rotary pursuit. Gabrieli suggests that the main challenge of mirror tracing is the development of a new mapping between visual and motor representations, a task commonly assigned to the cerebellum. Since mirror tracing is self paced and can be executed in individual, endpoint-directed segments, it does not require elaborate predictive commands. In contrast, the rotary pursuit task is paced by the continuous rotation of a target, and some sort of predictive strategy is required for optimal performance. Normal subjects learn a predictive strategy with practice. Instead, PD patients, like unskilled control subjects, attempt to track the target using a sequence of jerky corrective movements that fail to anticipate the rotation of the target.

How might normal subjects develop the predictions necessary for skilled performance of the rotary pursuit task? Although detailed kinematic evidence was not presented, perhaps the coarse movements of an unskilled subject initially approximate the target's circular pathway with a rough polygonal trajectory. Each of the polygon's vertices could represent the goal of an individual corrective movement. In the recognition/registration model, outputs registered in the frontal cortex could designate these endpoints, which would then be transmitted to the cerebellum for execution. An adaptive adjustment of these endpoints might be regulated by nigrostriatal dopamine projections, strengthening the modules associated with appropriate movements and the inputs corresponding, for example, to visual and proprioceptive events immediately preceding the reinforced movements. As mentioned in the commentary to part III, this is a method for making first-order predictions. In subsequent trials, this method of prediction could act to sequence appropriate motor plans.

ALLOCATING ATTENTION

Attentional control refers to our ability to effectively engage desired targets, maintain attention in the face of distractions, and disengage attention when either exogenous or endogenous cues are presented. The model presented in chapter 17 promotes a role for the basal ganglia in the voluntary control of visuospatial attention. The model proposes that attention is controlled by two cortical areas, one that mediates the influence of internally generated attentional expectations or goals (anterior attentional system) and another that mediates sensory representations of potential targets (posterior attentional system). The model's connections include a reverberating corticocortical loop between the posterior attentional system of the parietal cortex and spatially correspondent nodes in the anterior attentional system of the frontal cortex. This pathway shares visuospatial representations of both targets and distractions in the posterior parietal cortex with the anterior attentional field of the frontal cortex. A second pathway consists of convergent projections from the anterior and posterior systems onto the striatum and trans-striatal loops through the frontal cortex. These connections are quite similar in form and function to the connections in the recognition/registration modules that were discussed earlier. The trans-striatal loops consist of corticostriatal projections emanating from a given region of the frontal cortex (anterior attentional system) that return, via the pallidum and thalamus, back to their zone of origin in the frontal cortex. Also, the activity in each loop is regulated by projections from a nonfrontal area of cortex, corresponding to one of the several nonfrontal corticostriatal inputs in the recognition/registration model. Activity in each transstriatal loop is gated by a match/mismatch operator which compares the representations of the two attentional systems. Chapter 17 does not assign this match/mismatch operation to a particular neural structure; however, since matching is actually a special case of pattern recognition, the spiny neurons would seem to be appropriate candidates.

As a modification to this model, we propose that two arrays of cortical-basal ganglionic modules might be involved in this precued attentional task. The first array, which we assume is connected with the anterior system, could be tuned to recognize and register the different stimuli that precue responses. The registrations of these stimuli in the corticothalamic portions of these modules would thus serve to load the anterior attentional system with a working memory of expected targets. The second array of modules would receive these expectation signals and also the sensory target inputs from the posterior attentional system, and would send its output to one of the motor cortical areas (perhaps the dorsal premotor area) so as to program a response. This second array could be tuned to detect in particular, the combined presence of the different expectations and the corresponding stimuli but might also be capable of responding, though at a slower rate, to the stimuli alone in the absence of an expectation input. This would simulate the main attentional effect. Competitive interactions between the elements of these arrays could serve a similar function to that proposed in the Jackson and Houghton model. Based on this example, it is apparent that the notion of the basal ganglia as an attentional controller fits quite naturally with the hypothesis that the basal ganglia learn to recognize salient contextual features. Indeed, attention is an important ingredient in any type of learning situation.

WORKING MEMORY STRATEGIES

In the above model, the focus of attention was controlled through a comparison of sensory-derived representations of potential target positions with working memory representations of expected target positions. Performance was enhanced by a working memory that simply provided information about expected target position. Similarly, we can imagine that the control of other attentional tasks might profit from more complex expectation models of potential targets that might include shape and color. Note that these internal models could be generated as working memories of cues presented in the environment or from long term memories. In either case, a spiny neuron-based match/mismatch operator could be used to compare the model with sensory input. However, as internal models increase in their complexity, their comparison with the outside sensory world would require increasingly more processing resources. Chapter 15 suggests that these competing influences find a balance point that favors simple internal models.

In these experiments, Ballard and colleagues analyzed both the attentional focus and cursor movements of subjects performing a pattern copying task. Colored blocks arranged in a group on a computer screen made up the model pattern. Blocks had to be moved from a resource area to another workspace where a copy of the model pattern had to be constructed one block at a time. Cerebral processing was deduced on the basis of recorded eye movements showing the sequence of foveations (attentional points). Their data suggested that the subjects sought to minimize the complexity of the internal models stored in expectation working memory. They did this by making frequent and predictable references to the model pattern. This suggests that the computational costs associated with processing, that is either developing, remembering, or comparing detailed internal models, are possibly quite high.

The spatiotemporal patterns of attention they observed suggested that subjects utilize sequences of alternating perceptual and motor actions that effectively swap simple models in and out of expectation memory, according to what is needed for the next stage in the task. Perceptual actions load models of either color or location, which are removed as soon as appropriate motor actions are taken. Environmental contexts appear to be acquired just before they are needed and discarded soon thereafter, allowing computational resources to be devoted to the most relevant features of a particular scene.

This idea of swapping memories in and out of working memory relates well to the attentional control model discussed earlier. For example, in order to locate a red colored block within the resource space, a working memory representing "red" would be established within an appropriate corticostriatal loop. Assuming that a red block is not initially foveated, a mismatch would keep the initial loop activity relatively low and attention unfocused. Foveation of a red block would increase activity in the corticostriatal loop. As mentioned in the Commentary for part II, this increased loop activity might also serve to enhance the motor command or plan for initiating actions such as "picking up." Indeed, fixation is an important component for behavioral actions as well as for strictly perceptual actions such as "identify."

SETTLING INTO WORKING MEMORIES

Several ideas have been proposed about positive feedback loops and their role in sustaining working memories. The model presented by Woodward and colleagues in chapter 16 suggests an alternate mechanism, based on a striatal network of recurrent collateral inhibition, for latching memories within the basal ganglia. The network's storage capacity rests in its tendency to settle into different stable states of activation in response to patterns of phasic cortical inputs. However, since the network is inhibitory, expression of this stable state of activation requires that each spiny neuron receive a tonic activation from some source. Based on the anatomical observation that the number of spiny neurons far exceeds the number of target pallidal neurons, the model proposes that spiny neuron activation levels are then summed and "read out" as a scalar value of pallidal activation. Perhaps each pallidal neuron is controlled by a somewhat separate recurrent network. This is an interesting departure from the idea, presented in the Commentary to part II, that pallidal neurons might be inactivated by bursts in individual spiny neurons.

How then is this scalar value utilized as a memory signal? Presumably, high activity levels of pallidal neurons exercise inhibitory control over corticothalamic loops, which are reflected as areas of low cortical activity. Rich patterns of cortical activation could result from the interaction of these local minima via corticocortical connections. This model, although motivated primarily by anatomical and behavioral evidence, simulates patterns of activation that bear general resemblance to actual striatal recordings that the authors present.

CONCLUDING REMARKS

The reader who has worked through the seventeen chapters of this book and its four commentaries will hopefully realize that there is a great potential for integration in this previously neglected area of neuroscience. As the title of this book implies, there is more than one model of information processing in the basal ganglia. The authors of this book approach the basal ganglia from a rich array of research specialities—from neuroanatomy to robotics—each with a particular set of experimental approaches, terminology, and prior assumptions. Indeed, several of the conclusions presented throughout this book appear to be somewhat incompatible at this time. Still, while these differences often make it difficult to compare results, their integration is an absolute necessity for understanding the basal ganglia. Understanding that integration often comes at the risk of oversimplification; we have nevertheless attempted to emphasize the common aspects of each chapter within our commentaries. The editors sincerely hope that this book will serve as motivation for future experimental and theoretical efforts to synthesize and eventually comprehend our growing knowledge of the basal ganglia.

Contributors

James L. Adams
Department of Physiology
Northwestern University Medical School
Chicago, Illinois

Paul Apicella
Institute of Physiology
University of Fribourg
Fribourg, Switzerland

Michael A. Arbib
Center for Neural Engineering
University of Southern California
Los Angeles, California

Dana H. Ballard
Center for Visual Science
University of Rochester
Rochester, New York

Andrew G. Barto
Department of Computer and Information Science
University of Massachusetts at Amherst
Amherst, Massachusetts

J. Brian Burns
Teleos Research, Inc.
Palo Alto, California

Christopher I. Connolly
Computer Science Department
University of Massachusetts at Amherst
Amherst, Massachusetts

Anthony Dickinson
Department of Experimental Psychology
University of Cambridge
Cambridge, England

Peter F. Dominey
Vision et Motricité
INSERM
Bron, France

Richard P. Dum
Department of Physiology
State University of New York Health Science Center at Syracuse
V. A. Medical Center
Syracuse, New York

John Gabrieli
Department of Psychology
Stanford University
Palo Alto, California

Marianela Garcia-Munoz
Department of Psychiatry and Neurosciences
University of California, San Diego, School of Medicine
San Diego, California

Patricia S. Goldman-Rakic
Section of Neurobiology
Yale University School of Medicine
New Haven, Connecticut

Ann M. Graybiel
Department of Brain and Cognitive Sciences
Massachusetts Institute of Technology
Cambridge, Massachusetts

Philip M. Groves
Department of Psychiatry and Neurosciences
University of California, San Diego, School of Medicine
San Diego, California

Mary M. Hayhoe
Department of Psychology and Center for Visual Science
University of Rochester
Rochester, New York

Jeffrey R. Hollerman
Institute of Physiology
University of Fribourg
Fribourg, Switzerland

George Houghton
Department of Psychology
University College London
London, England

James C. Houk
Department of Physiology
Northwestern University Medical School
Chicago, Illinois

Stephen Jackson
Human Movement Laboratory
Department of Psychology
University of Wales
Bangor, Wales

Minoru Kimura
Faculty of Health and Sports Sciences
Osaka University
Toyonaka, Osaka, Japan

Alexandre B. Kirillov
Biographics, Inc.
Winston-Salem, North Carolina

Rolf Kötter
Department of Anatomy and Structural Biology
Neuroscience Center
University of Otago
Dunedin, New Zealand

Jean C. Linder
Department of Psychiatry and Neurosciences
University of California, San Diego, School of Medicine
San Diego, California

Tomas Ljungberg
Institute of Physiology
University of Fribourg
Fribourg, Switzerland

Michael S. Manley
Department of Psychiatry and Neurosciences
University of California, San Diego, School of Medicine
San Diego, California

Maryann E. Martone
Department of Psychiatry and Neurosciences
University of California, San Diego, School of Medicine
San Diego, California

Jacques Mirenowicz
Institute of Physiology
University of Fribourg
Fribourg, Switzerland

Christopher D. Myre
Biographics, Inc.
Winston-Salem, North Carolina

Jeff Pelz
Department of Psychology and
Center for Visual Science
University of Rochester
Center for Imaging Science at
Rochester Institute of Technology
Rochester, New York

Nathalie Picard
Department of Physiology
State University of New York
Health Science Center at
Syracuse
V. A. Medical Center
Syracuse, New York

Ranulfo Romo
Institute of Physiology
University of Fribourg
Fribourg, Switzerland

Steven F. Sawyer
Department of Physiology and
Pharmacology
Bowman Gray School of
Medicine
Winston-Salem, North Carolina

Eugenio Scarnati
Institute of Physiology
University of Fribourg
Fribourg, Switzerland

Wolfram Schultz
Institute of Physiology
University of Fribourg
Fribourg, Switzerland

Peter L. Strick
Departments of Neurosurgery
and Physiology
State University of New York
Health Science Center at
Syracuse
V. A. Medical Center
Syracuse, New York

Jeff Wickens
Department of Anatomy and
Structural Biology
Neuroscience Center
University of Otago
Dunedin, New Zealand

Charles J. Wilson
Department of Anatomy and
Neurobiology
University of Tennessee,
Memphis
Memphis, Tennessee

Donald J. Woodward
Department of Physiology and
Pharmacology
Bowman Gray School of
Medicine
Winston-Salem, North Carolina

Stephen J. Young
Department of Psychiatry and
Neurosciences
University of California, San
Diego, School of Medicine
San Diego, California

Index